AI-Processor Electronics

Basic technology of artificial intelligence

Online at: https://doi.org/10.1088/978-0-7503-6259-7

AI-Processor Electronics

Basic technology of artificial intelligence

Vinod Kumar Khanna

Independent Researcher, Chandigarh, India

and

Retired Chief Scientist, CSIR-Central Electronics Engineering Research Institute, Pilani, India

IOP Publishing, Bristol, UK

ISBN   978-0-7503-6259-7 (ebook)
ISBN   978-0-7503-6257-3 (print)
ISBN   978-0-7503-6260-3 (myPrint)
ISBN   978-0-7503-6258-0 (mobi)

DOI   10.1088/978-0-7503-6259-7

Version: 20250101

IOP ebooks

British Library Cataloguing-in-Publication Data: A catalogue record for this book is available from the British Library.

Published by IOP Publishing, wholly owned by The Institute of Physics, London

IOP Publishing, No.2 The Distillery, Glassfields, Avon Street, Bristol, BS2 0GR, UK

US Office: IOP Publishing, Inc., 190 North Independence Mall West, Suite 601, Philadelphia, PA 19106, USA

Dedicated with great reverence to my parents, the late Shri Amarnath Khanna and Shrimati Pushpa Khanna, in cherished memory of their motivation and guidance to face the challenges of life; and most affectionately to my grandson Hansh, daughter Aloka and wife Amita, for their love, encouragement and unwavering support.

Contents

Preface

Computers have penetrated into every nook and corner of our homes, offices, industries, markets and entertainment media. They have been embraced and adored in all aspects of human life and activities from school homework assignments, online shopping and bank transactions to working from home. But computer progression and proliferation has now entered a new phase. From the epochs of mainframe to personal computing and the omnipresent laptops, tablets, smartphones and mobile devices, a new era of intelligent computers has dawned. These are the computers which can not only learn from experience, but think, reason and make decisions like intelligent human beings. Thanks to the artificial intelligence (AI) revolution, we are living in the age of intelligent machines that pass the Turing test in which an interrogator asks questions from a human and a machine but fails to ascertain which is a human and which is a machine.

Pioneered in the 1950s, the dream of AI could not be realized at that time because of the non-availability of the required computing capabilities and the supporting database. AI evolved with the booming of data like a tsunami, maturation of AI algorithms and the leap forward in the semiconductor industry for microchip fabrication. Machine learning algorithms, particularly deep neural networks, pose unique challenges regarding the resources and execution time of training/inference calculations. The processing unit or processor of a computer is its principal component which is the point of convergence of all computational activities and on which the performance of the computer crucially depends. Therefore, it is the focus of attention of all computer scientists and engineers. All-purpose processors like the central processing unit (CPU) will always remain at the helm of affairs but they are deficient in dealing with the complex computations of the AI age. Therefore, development of specialized processing units for computationally-intensive AI algorithms is the hot topic of research worldwide. Particularly, without the stupendous progress in electronic processing units for computing, the great strides made by AI, machine learning and strikingly its subset 'deep learning' in recent years would not have been possible. Several off-the-shelf and emerging processors from different manufacturers have been used. These processors differ in their performance and energy efficiency. They are used for initial development and refinement of AI algorithms known as the training process as well as to apply trained AI algorithms to real-world data inputs to solve practical problems, viz., the inference phase.

Electronics of AI is a subject whose impact on modern society needs hardly to be emphasized. It is out of the question to ignore the colossal impact of AI processor electronics on the advancement of AI because AI computing is not a bed of roses. It goes without saying that progress in AI largely depends on the innovations in AI processor electronics. Growth of AI applications will slow down with slackening of progress in AI processor technology. Being an area of research and development, books on this topic are few. Most of the information is only available on the websites of the manufacturers. So, a book on this topic is the need of the hour.

The book is organized into 15 chapters. Chapters 1 and 2 build the foundations of the subject. Chapter 1 provides a survey of preliminary concepts in AI, pointing out the shortcomings of existing processors in coping with the enormous workload of AI computations. Chapter 2 recalls the fundamentals of digital computing electronics, viz., the logic gates and arithmetic circuits. Chapter 3 on the CPU explains its architecture and workflow, clearly indicating its flexibilities and underlining its limitations which have become the driving force towards the evolution of a new family of processors that can execute AI algorithms speedily, efficiently and at low power consumption to fulfil the ever-increasing user demands. Chapters 4 and 5 lay down the groundwork on which the magnificent edifice of AI processors is built, viz., parallel computing and AI-optimized hardware concepts. The forthcoming chapters make a beeline for specialized AI processors such as the graphical/graphics processing unit (GPU) (chapter 6), the tensor processing unit (TPU) (chapter 7), the neural processing unit (NPU) (chapter 8), the convolutional neural network (CNN) processor and vision processing unit (VPU) (chapter 9), the sparse neural network (NN) processor (chapter 10), the graph processor (chapter 11), and the associative processing unit (APU) (chapter 12). Chapters 13–15 take a departure from the classical processors. They are devoted to the emerging and promising, multidisciplinary field of quantum processing units, unifying aspects of computer science with mathematics and physics.

The book is the result of a humble attempt by the author to peer into the almost boundless and amazing world of AI processors and to provide the main points of observations. Some important segments of this world may have escaped the author's attention. No effort to achieve completeness is made. Only a handful of the vast diversity of AI processors in the market are covered in this book for the sole purpose of illustrating scientific principles. A comprehensive list of processors is not made because the sky is the limit when we talk about AI. Every moment, the technology is upgrading. In this fast-moving competitive scenario, new chips are being commercialized every now and then, speaking volumes of the backbreaking efforts being made in this direction. Therefore, the approach is only to learn the underlying science and technology inside out about the advancements made.

In view of the defined goals, the book seeks to provide a comprehensive survey of the principles of operation and technology of a few representative AI processors among the diverse chips available, along with the related technologies, to highlight the key concepts and innovations underlying their development. A broad perspective of the underlying principles, technologies, the available opportunities and future prospects is sketched.

Encompassing examples from the wide range of processors based on different principles and technologies from classical to quantum mechanical theory, the book traces out an elementary picture of the processor scenario for AI. The book will be of interest to the students and teachers of electrical engineering, electronics, computer science and information technology disciplines. It is also intended to be a useful reference book for research students, professionals, and R&D scientists. Therefore, it is earnestly hoped that the book will cater to the needs of a wide audience comprising students and professional workers engaged in this field. It will serve as an overarching bridge connecting intricate topics and accessible understanding for experts as well as non-experts.

Bogged down by the din and bustle of algoritmms and software;
In the cloud and haze of programming;
Let us not forget the underlying hardware,
Of artificial intelligence and machine learning;
The soldiers on whose shoulders the programs run,
And all amazing magic is done.
With the rapid knowledge upsurge;
Daily new technologies of processors emerge;
To enable faster computations;
Bringing new hopes and aspirations;
With novel techniques and innovations;
Beyond all limits and expectations;
For advancement of civilization;
Three cheers for AI processors;
The present generation AI chips and their upcoming successors.
Hip, hip hooray!

Vinod Kumar Khanna
Chandigarh, India

Acknowledgments

First and foremost, I thank The Almighty God for giving me the strength and motivation to complete this work.

I sincerely appreciate the endeavors of the pioneering computer scientists and engineers, research institutes and companies whose wisdom and hard work have helped us to reach the present state of expertise and knowledge in AI processor electronics. I tip my hat and extend my thanks to them for their service to the progress of AI by devoting their blood, sweat and tears. Their works are cited in the end-of-chapter bibliographies throughout the book.

I am also thankful to the commissioning and production editors and staff of IoP Publishing for their kind cooperation and support throughout the project

Last but not the least, I extend my gratitude to my wife and family for providing the serene homely environment and backing me up ardently and tirelessly to pursue my work.

Vinod Kumar Khanna
Chandigarh, India

Author biography

Vinod Kumar Khanna

 **Vinod Kumar Khanna** is an independent researcher from Chandigarh, India. He is a retired Chief Scientist from the Council of Scientific and Industrial Research (CSIR)—Central Electronics Engineering Research Institute (CEERI), Pilani, India, and a retired Professor from the Academy of Scientific and Innovative Research (AcSIR), Ghaziabad, India. He is a former Emeritus Scientist, CSIR, and Professor Emeritus, AcSIR, India. His broad areas of research include the design, fabrication, and characterization of power semiconductor devices and micro- and nanosensors.

Academic qualifications

He received an MSc degree in physics with a specialization in electronics from the University of Lucknow in 1975 and a PhD degree in physics from Kurukshetra University in 1988 for the thesis titled, 'Development, Characterization and Modeling of the Porous Alumina Humidity Sensor'.

Research/teaching experience and accomplishments

His research experience spans over a period of 40 years, from 1977 to 2017. Starting his career as a Research Assistant in the Department of Physics, University of Lucknow, from 1977 to 1980, he joined CSIR-CEERI, Pilani (Rajasthan), in April 1980. There, he worked on several CSIR-funded as well as sponsored research and development projects. His major fields of research included power semiconductor devices and microelectronics/MEMS and nanotechnology-based sensors and dosimeters.

In the power semiconductor devices area, he worked on the high-voltage and high-current rectifier (600 A, 4300 V) for railway traction, high-voltage TV deflection transistor (5 A, 1600 V), power Darlington transistor for AC motor drives (100 A, 500 V), fast-switching thyristor (1300 A, 1700 V), power DMOSFET, and IGBT. He contributed toward the development of sealed tube Ga/Al diffusion for deep junctions, surface electric field control techniques using edge beveling and contouring of large-area devices, and floating field limiting ring design. He carried out an extensive characterization of minority-carrier lifetime in power semiconductor devices as a function of process steps. He also contributed toward the development of the P–I–N diode neutron dosimeter and PMOSFET-based gamma-ray dosimeter.

In the area of sensor technology, he worked on the nanoporous aluminum oxide humidity sensor, ion-sensitive field-effect transistor-based microsensors for biomedical, food, and environmental applications; microheater-embedded gas sensor for automotive electronics, MEMS acoustic sensor for satellite launch vehicles, and capacitive MEMS ultrasonic transducer for medical applications.

As an AcSIR faculty member, he was the course coordinator of MEMS/IC Technology for the advanced semiconductor electronics program (2011–13) and taught 'MEMS Technology' to students pursuing MTech degree. As an adjunct faculty, BESU, Kolkata, he taught 'MEMS Technology & Design' to MTech (Mechatronics) students. He was invited by IIT, Jodhpur, to deliver lectures on 'Semiconductor Fundamentals and Technology' to BTech students during February 2011. He guided BTech/MTech theses of students from BITS, Pilani; VIT, Vellore; and Kurukshetra University. He also guided a PhD thesis on 'MEMS acoustic sensor', MNIT, Jaipur.

Semiconductor facility creation and maintenance

He was responsible for setting up and looking after diffusion/oxidation facilities, edge beveling and contouring, reactive sputtering, and carrier lifetime measurement facilities. As the Head of the MEMS and Microsensors Group, he looked after the maintenance of a six-inch MEMS fabrication facility for R&D projects as well as the augmentation of processing equipment under this facility at CSIR-CEERI.

Scientific positions held

During his tenure of service at CSIR-CEERI from April 1980 till superannuation in November 2014, he was promoted to various positions including one merit promotion. He retired as a Chief Scientist and Professor (AcSIR) and as the Head of MEMS and Microsensors Group. Subsequently, he worked for three years as an Emeritus Scientist, CSIR, and Emeritus Professor, AcSIR, from November 2014 to November 2017. After the completion of the emeritus scientist scheme, he now lives in Chandigarh. He is a passionate author and enjoys reading and writing.

Membership of professional societies

He is a Fellow and Life Member of the Institution of Electronics and Telecommunication Engineers (IETE), India. He is a life member of the Indian Physics Association (IPA), Semiconductor Society, India (SSI), and Indo-French Technical Association (IFTA).

Foreign travel

He has traveled widely. He had participated in and presented research papers at the IEEE Industry Application Society (IEEE-IAS) Annual Meeting in Denver, Colorado, USA, in September–October 1986. His short-term research assignments include deputations to Technische Universität Darmstadt, Germany, in 1999; at Kurt-Schwabe-Institut fur Mess-und Sensortechnike e.V., Meinsberg, Germany, in 2008; and at Fondazione Bruno Kessler, Trento, Italy, in 2011, under collaborative programs. He was a member of the Indian Delegation to the Institute of Chemical Physics, Novosibirsk, Russia, in 2009.

Scholarships and awards

He was awarded a National Scholarship by the Ministry of Education and Social Welfare, Government of India, on the basis of a Higher Secondary result, 1970; CEERI Foundation Day Merit Team Award for projects on fast-switching thyristor (1986); for power Darlington transistor for transportation (1988), for P–I–N diode neutron dosimeter (1992); and for high-voltage TV deflection transistor (1994); Dr N G Patel Prize for best poster presentation in the 12th National Seminar on Physics and Technology of Sensors, 2007, BARC, Mumbai; and the CSIR-DAAD Fellowship in 2008 under Indo-German Bilateral Exchange Programme of Senior scientists, 2008. He is featured in the Stanford–Elsevier prestigious list of the world's top 2% scientists (2022, Elsevier Data Repository, V4, doi:10.17632/btchxktzyw.4). He is named as a Highly Ranked Scholar-Lifetime: # 3, Nanoelectronics by ScholarGPS (https://scholargps.com/scholars/25423546982929/vinod-kumar-khanna).

Research publications and books

He has published 194 research papers in leading peer-reviewed national/international journals and conference proceedings. He has authored 21 books and has also contributed 6 chapters to edited books. He has five granted patents to his credit, including two US patents.

About the book

High-performance AI processors deployed in data centers or at the edge can help unlock a world of new opportunities for businesses and industries. These chips have been the enablers of major AI advancements. Such leading-edge, specialized AI chips are essential for cost-effective AI implementation on a large scale. With ever-increasing sophistication of AI algorithms, the need for higher processing power, speed and efficiency has grown rapidly, raising the expectations form these devices for autonomous, smart processes in robotics, internet of things, computer vision, voice recognition and natural language understanding and many more fields.

The book apprises the reader about the type of challenges faced in AI computing. After a quick revision of the computing applications of digital electronics in a nutshell, the operation of CPU is explained and the versatility advantages vis-à-vis inadequacies of the general-purpose processors for handling AI tasks are pointed out. Then the ubiquitously prevalent parallel processing architecture in AI chips is presented followed by highlighting the approaches that must be incorporated for efficient computing keeping in view the deceleration of Moore's law. A diversity of specialized processors is successively described such as the GPU, the TPU, the NPU, the CNN processor and VPU, the sparse NN processor, the graph processor and the in-memory processor. Finally, the rudimentary principles of quantum computing are introduced. Quantum computing devices, logic gates and circuits are dealt with and quantum processing unit (QPU) operation is summarized.

All in all, the book sketches a broad all-inclusive perspective of electronics for AI processing units, extending from classical to quantum-mechanical computing. It elucidates the various optimization approaches for AI processors including the use of parallel processing architecture in concurrently performing multiple calculations across data streams, besides low-precision computing, domain-specific architecture and the neural network sparsity concept. It offers a compendious exposition of the principles and technology of processors capable of performing fast and energy-efficient large-scale data processing and complex computations.

A large number of semiconductor chip manufactures are working on AI processors and a vast spectrum of processors are available in the market. It is a highly competitive and fast-developing field progressing by leaps and bounds. As new chips are introduced, older versions go into obsolescence and oblivion. It is well-nigh impossible to attempt an exhaustive treatment of the subject.

The book will help the reader in understanding the core foundational principles of AI processors, and their enabling fabrication technologies, and getting glimpses of the scenario. The book will be of immense value to graduate/post graduate and PhD students in electronics and computer engineering and related disciplines, and equally relevant to the practicing engineers and scientists engaged in this field.

Abbreviations, acronyms, initialisms and symbols

A

A_1, A_2, ...	Access transistors
A* algorithm	A-star graph traversal and pathfinding algorithm
AAM	Active appearance model
AC	Alternating current
ACC	Accumulator
ADAS	Advancements in driver-assistance systems
ADC	Analog-to-digital conversion
AGU	Address generator unit
AI	Artificial intelligence
AIM	Associative index matching (unit)
AiMP	Associative in-memory processor
Al	Aluminum
ALU	Arithmetic logic unit
Amazon EC2	Amazon Elastic Compute Cloud2
AMD	Advanced Micro Devices, Inc.
AMP	Analog matrix processor
AMX	Advanced matrix extensions
ANE	Apple's Neural Engine
ANN	Artificial neural network
APU	Associative processing unit
ARM	Advanced RISC machine or Acorn RISC machine
ASCII	American standard code for information interchange
ASIC	Application-specific integrated circuit
AVX	Advanced vector extensions
AWS	Amazon Web Services, Inc.
AXNet	ApproXimate Net

B

BGP	Border gateway protocol
BF16	Brain floating-point computer number format occupying 16 bits in memory
BIOS	Basic input/output system
BL	Bit line
BN	Bayesian network

C

C++	C plus 'features'
C_1, C_2, ...	Comparison transistors
Ca	Calcium
CAM	Content addressable memory

CAWS	Critically aware warp scheduling
CCU	Central control unit
CDNA	Compute DNA, a compute-centered graphics processing unit microarchitecture
CI	Compute intensity
CIR	Current instruction register
CISC	Complex instructions set computer
CLB	Configurable logic block
CLK	Clock
CMOS	Complementary metal-oxide semiconductor (FET)
CNN	Convolution neural network
CNOT	Controlled NOT gate
Co	Cobalt
CONV	Convolution
CPU	Central processing unit
CQUIP	Control quasi-independent point
CSC	Compressed sparse column (format)
CSR	Compressed sparse row (format)
CSS	Coarse-grained structured sparsity
CU	Coalescing unit
CUDA	Compute unified device architecture
CXL	Compute express link
CZ gate	Controlled Z-gate

D

D	Data input
1D	One-dimensional
2D	Two-dimensional
6D	Six-dimensional
DAC	Digital-to-analog conversion
DAG	Directed acyclic graph
DARPA	Defense Agencies Research Program Agency
DBSCAN	Density-based spatial clustering of applications with noise
DC	Direct current
DCe	Data cache
DCNN	Deep convolution neural network (accelerator)
DDR	Double data rate
DDR3	Double data rate third generation memory
DDR-SDRAM	Double data rate-synchronous dynamic random-access memory
DFT	Discrete Fourier transform
DIMM	Dual in-line memory module
DL	Deep learning
DMA	Direct memory access
DNN	Deep neural network
DNPU	Deep neural network processing unit
DQD	Double quantum dot

DRAM	Dynamic random-access memory
DSA	Domain-specific architecture
DSL	Domain-specific language
DVPP	Digital vision pre-processing (module)

E

ECC	Error correcting code
Eclat	Equivalence class clustering and bottom-up lattice traversal
E-core	Efficiency core
EEPROM	Electrically erasable programmable read-only memory
EIE	Energy-efficient inference engine
ELSI	Extra large-scale integration
EMI	Electromagnetic interference
EPROM	Erasable programmable read-only memory
EROM	Erasable read-only memory

F

FA	Full adder
FC	Fully connected (layer)
FCNN	Fully-connected neural network
FET	Field-effect transistor
FF	Flip-flop
FFNN or FNN	Feed-forward neural network
FIFO	First in, first out
FINFET	Fin field-effect transistor
Float 32	FP32, Single-precision floating-point data format
FLOP byte^{-1}	Floating point operations per byte read
FLOPS	Floating point operations per second
FP	Frequent pattern (growth)
FP8	8-Bit floating point
FP16	Half-precision floating-point format that occupies 16 bits in memory
FP32	Single-precision floating-point format using 32 bits to represent real numbers
FPGA	Field-programmable gate array
FPU	Floating point unit
FS	Full subtractor
FSS	Fine-grained structured sparsity

G

GA	Global accelerator
GALS	Globally asynchronous locally synchronous
GAN	Generative adversarial network
GB or Gbyte	Gigabyte
GCC	GNU compiler collection
GEMM	General matrix multiplication
GenAI	Generative artificial intelligence

GH	Grace Hopper
GHz	Gigahertz
GMAC	Giga multiply accumulate operations per second
GNU	GNU's not Unix (a recursive acronym)
Google	A word derived from a mathematical term 'Googol', representing a number '1 followed by 100 zeros or 10^{100}.
GOPS	Giga (billion) operations per second
GPC	Graphics processor cluster
GPU	Graphical/Graphics processing unit
GSI	Giant scale integration
GTX	Giga texel shader eXtreme

H

HA	Half adder
HBM	High-bandwidth memory
HBM2e	2nd generation high-bandwidth memory
HDD	Hard disk drive
HDL	Hardware description language
HIVE	Hierarchical identify verify exploit
HL	Hidden layer
HPC	High-performance computing
HS	Half subtractor

I

IARAM	Input activation RAM
IBM	International Business Machines Corporation
IC	Integrated circuit
ICe	Instruction cache
ID	Instruction decode (in control unit)
IE	Instruction execution (in ALU)
IF	Instruction fetch (from memory), Infrared
ILP	Instruction-level parallelism
IMT	Implicitly-multithreaded (control flow)
INT8	8-bit integer precision
I/O	Input/output
IoT	Internet of things
IP	Intellectual property
IPU	Intelligence processing unit
IQFT	Inverse quantum Fourier transform
ISA	Instruction set architecture
ISP	Image signal processing

J

JIT	Just-in-time (compiler)
JPG (or JPEG)	Joint Photographic Experts Group

K

K	Kelvin scale, an absolute temperature scale

kB	Kilobyte
keV	Kilo electron volt
KNN	K-Nearest neighbors

L

L1 cache	Level 1 cache
L1 regularization	Lasso regression
L2 cache	Level 2 cache
L3 cache	Level 3 cache
LASSO	Least absolute shrinkage and selection operator
LD/ST	Load/store (unit)
LLM	Large language model
LNZD	Leading non-zero detection (node)
LPU	Language processing unit
LReLU	Leaky rectified linear unit
LSB	Least significant bit
LSI	Large-scale integration
LSM	Liquid state machine
LSTM	Long short-term memory (network)
LUT	Lookup table
LVCSR	Large vocabulary continuous speech recognition

M

M_1, M_2, ...	Memory transistors
MAC	Multiply-and-accumulate (unit), multiplication and accumulation
MAR	Memory address register
MB	Megabyte
MBC	Memristor-based crossbar
MB s^{-1}	Megabyte per second
MCR	Multiplexer combined rank
MDR	Memory data register
Memristor	Memory resistor
MF	Matrix factorization
MHz	Megahertz
MIMD	Multiple instructions, multiple data
MISD	Multiple instructions, single data
ML	Machine learning, match line
MLP	Multilayer perceptron
MM	Memory module
mm	Millimeter
MME	Matrix multiplication engine
M-net	Mixed-signal interconnection network
MNN	Modular neural network
MOS	Metal-oxide-semiconductor
MOSFET	Metal-oxide-semiconductor field-effect transistor
MRF	Markov random field

MRI	Magnetic resonance imaging
ms	Millisecond
MSB	Most significant bit
MSI	Medium scale integration
MTIA	Meta training and inference accelerator
MTC	Multi-threaded core
MT s^{-1}	Mega transfers per second
mW	Milliwatt
MXU	Matrix multiplier unit

N

NCE	Neural compute engine
NFU	Neural functional unit
Ni	Nickel
NIC	Network interface card
NLP	Natural language processing
nm	Nano meter
NMOS	N-channel metal-oxide semiconductor (FET)
NN	Neural network
NoC	Network on chip
NPU	Neural processing unit
NRE	Non-recurring engineering (cost)
NUMA	Non-uniform memory access
NVCC	NVIDIA CUDA compiler
NVDLA	NVIDIA deep learning accelerator
NVIDIA	'invidia' (latin word for envy) + NV (next vision)

O

OARAM	Output activation RAM
OS	Operating system

P

PC	Program counter, Personal computer
PCA	Principal component analysis
PCI	Peripheral component interconnect (bus)
PCI-e	Peripheral component interconnect express
P-core	Performance core
PCU	Pattern compute unit
PE	Processing element, processing engine
PEB	Processing element block
PFLOP s^{-1}	Petaflops (10^{15} floating-point operations per second)
PIM	Processing in-memory
PIUMA	Programmable integrated unified memory architecture
PLD	Programmable logic device
PM	Programmable multiplexer
PMMA	Poly (methyl methacrylate)
PMOS	P-channel metal-oxide semiconductor (FET)

PMU	Pattern memory unit
PPU	Post processing unit
PReLU	Parametric rectified linear unit
ps	Picosecond
PSM	Programmable multiplexer
psum	Partial sum
PT-IS-CP	Planar Tiled-Input stationary-Cartesian product
PTX	Parallel thread execution
PU	Processing unit
PyTorch	Python and Torch-based optimized deep learning tensor library

Q

QFT	Quantum Fourier transform
QPU	Quantum processing unit
QRAM	Quantum random-access memory

R

RAM	Random-access memory
RBF	Radial basis function
RBFNN	Radial basis function neural network
RDMA-NIC	Remote direct memory access-network interface card
RDU	Reconfigurable dataflow unit
ReLU	Rectified linear unit
RENO	Reconfigurable network on chip (NoC), a neuromorphic computing accelerator
ReRAM	Resistive random-access memory
ResNet	Residual neural network
RF	Register file, Radio frequency
RGB-RAW	Red, green, blue unprocessed data (RAW is not an acronym, it means 'natural or unaltered')
RISC	Reduced instruction set computer
RL	Reinforcement learning
RNN	Recurrent neural network
ROM	Read-only memory
RR	Round-Robin (warp scheduling)
RS	Row stationary
RTX	Ray tracing texel eXtreme
RW	Result writing (in register or memory)

S

s, (or sec)	Second
|S>	Singlet state of electron spin
SARSA	State-action-reward-state-action
SASS	Scalable assembly
SCNN	Sparse convolution neural network (accelerator)

SDK	Software development kit
SDRAM, DDR	Synchronous dynamic random-access memory, Double data rate
SFU	Special function unit
SHAVE	Streaming hybrid architecture vector engine
Si	Silicon
SiGe	Silicon-germanium
SIMD	Single-instruction, multiple-data
SIMT	Single-instruction, multiple-threads
SiO_2	Silicon dioxide
SISD	Single-instruction, single data
SL	Supervised learning, search line
$\overline{SL}$	Search line bar (complementary search line)
SLSI	Super large-scale integration
SM	Streaming multiprocessor
SNAP	Sparse neural accelerator processor
SoC	System on chip
SOTA	State of the art
SP	Streaming processor, Spawn point
SPAD	Scratch pad
SpMT	Speculative multithreading
SpMV	Sparse matrix dense vector multiplication
SQUID	Superconducting quantum interference device
SR	Stochastic rounding
S–R	Set–Reset
SRAM	Static random-access memory
SSI	Small-scale integration
SSL	Semi-supervised learning
STC	Single-threaded core
SW	Software

T

TB	Terabyte
Tb	Terabit
TBS	Thread-block scheduler
10T CAM	10-transistor content-addressable memory
TD	Temporal difference (learning)
TDP	Thermal design power
TFLOPS (also: TeraFLOPS, TFLOPs, TeraFLOPs, TFLOP s^{-1}, or TeraFLOP s^{-1})	Tera (or trillion) floating-point operations per second; trillion = 10^{12}
THz	Terahertz (10^{12} hertz)
Ti	Titanium
TLS	Thread-level speculation
TMUL	Tile matrix multiply (accelerator)
TOPS	Tera operations per second
TOPS W^{-1}	Trillion operations per second per watt
TPC	Tensor processor core
TPU	Tensor processing unit

| TSMC | Taiwan Semiconductor Manufacturing Company (Limited) |
| 6T SRAM | 6-transistor static random-access memory |

U

ULSI	Ultra large-scale integration
US	Unstructured sparsity
UVM	Unified virtual memory

V

v1, v2, v3, ...	Versions 1,2 3, ...
VDU	Visual display unit
VHDL	VHIC hardware description language or Very high-speed integrated circuit hardware description language
VLIW	Very long instruction word
VLSI	Very large-scale integration
VNNI	Vector neural network instructions
VPU	Vision processing unit

W

W	Watt
WL	Word line
WSE-3	Wafer scale engine, third generation

X

| XOR | Exclusive OR gate |
| XNOR | Exclusive NOR gate |

Y

| Yb^+ | Ytterbium ion |
| μCPU | Micro CPU |

Mathematical symbols

A

$\overline{A}$	Bar over A means NOT function on A
$a^{(i)}$	ith activation vector
$a^{(i+1)}$	$(i+1)$th activation vector
A_{ij}	The matrix element representing the edge joining vertex i to vertex j

B

b	Bias
$b^{(i)}$	ith bias vector
$B_{External}$	External magnetic field
B_{IN}	Borrow in
B_{OUT}	Borrow out
B_z	Driving magnetic field

C

C	Carry, loss function of CNN, capacitance
$CCNOT$	Toffoli gate matrix
C_G	Gate capacitance
$c^{(i)}$	ith control vector
c in $CZ_{c,t}$	Index of the control state c in Controlled-Z gate
C_{IN} or C-in	Carry in
$CNOT$	Controlled NOT gate or CNOT gate matrix
C_{OUT} or C-out	Carry out
$CSWAP$	Controlled swap gate or Fredkin gate matrix

D

D	Difference
D, d'	Dimensions

E

E	Energy of an electron
E_{01}	Qubit splitting
E_C	Charging energy of the superconducting island
E_J	Josephson energy of the superconducting island
E_J/E_C	Ratio between the charging energy and Josephson energy of the superconducting island

F

f	Activation function of the hidden layer i,
$[f \otimes g](t)$	Convolution of functions f and g of time t
$F(x)$	Residual function, Identity function

G

$G(V,E)$	Graph (Vertices, Edges)

H

H	Hadamard gate matrix
$\hbar$	Reduced Planck constant

I

I	Identity matrix
I_{Bias}	Biasing current

L

L	Width of the potential well, Josephson inductance

M

m	Mass of an electron
$m \times n$ matrix	A matrix with m rows and n columns

N

n	A number
N	Number of memory cycles

O

O_d	Output data
O-output	The output layer
O_{t-1}, O_t, O_{t+1}	The outputs at time instants $t-1$, t and $t+1$, respectively

P

p_j, p_{j+1}	The beginning and end pointers for the arrays
$P(x)$	Prediction result

Q

Q (and $\overline{Q}$)	Stored bit (and complementary stored bit)
q	Electronic charge
q_0	Qubit register for input A
q_1	Qubit register for input B
q_2	Qubit register for C-in
q_3	Qubit register for Sum
q_4	Qubit register for C-out

R

r	Rank of a matrix

S

S	Sum, S gate or Z90 gate matrix
$S^\dagger$	Conjugate transpose of S gate matrix
S_P	Partial sum

T

T	T gate matrix
t_1, t_2, t_3, ...	Time instants
$T^\dagger$	Conjugate transpose of T gate matrix
$|T_0\rangle$, $|T_-\rangle$ and $|T_+\rangle$	Triplet states of electron spin
T_{CAl}	Critical temperature for aluminum
t in $CZ_{c,t}$	Index of the target state t in Controlled-Z gate

U

U	Unitary matrix
U^{-1}	Inverse matrix of unitary matrix U
U^H	Complex conjugate transpose of unitary matrix U
U, V, W	The weight matrices of the input-to-hidden layer connection, the hidden-to-output layer connection and hidden-to-hidden layer connection in LSTM network

V

V	Volt
V_1 and V_2	End cap potentials
V_{DD}	Drain voltage supply
V_G	Gate voltage
V_{Left}, V_{Middle} and V_{Right}	Potentials of the left, middle and right gates
$V(x)$	Potential due to an applied electric field

W

W_1, W_2, W_3 or W_{t1}, W_{t2}, W_{t3}	Weights
$W_{\text{CORRECTED}}$	Corrected weight
W_i or w_i	Weight
$W^{(i)}$	ith vector of weight matrix
W_{PREVIOUS}	Previous weight
$[W_t]$	Weight matrix
W_{ti} or w_{ti}	Weight

X

X	Pauli X-gate matrix
x_1, x_2, ..., x_N	N integers
$[X_d]$	Data matrix
X_{di} or x_{di}; X_i or x_i	Input data
X-Input	The input layer
$\{x_j\}$, $\{X_k\}$	n-number sequences
X_{t-1}, X_t, X_{t+1}	The inputs at time instants $t-1$, t and $t+1$

Y

$[Y_d]$	Data matrix

Greek letters

$\boldsymbol{\alpha}$

α	Complex probability amplitude

$\boldsymbol{\beta}$

β	Complex probability amplitude

$\boldsymbol{\eta}$

η	Learning rate

$\boldsymbol{\psi}$

$|\psi_{\text{L}}\rangle$ and $|\psi_{\text{R}}\rangle$	Quantum states of qubit $|\psi\rangle$
$\psi(x)$	Wave function of an electron

$\boldsymbol{\theta}$

θ	Angle subtended by the wave function of the qubit with the Z-axis
$|\theta_j(\alpha)\rangle$	jth qubit of the phase state $|\theta(\alpha)\rangle$ encoding α

$\boldsymbol{\phi}$

Φ_{External}	External flux
ϕ	Angle made by the wave function of the qubit with the X-axis

Special symbols

$\odot$	Hadamard product, XNOR gate
$\otimes$	Convolution
$\wedge$	Conjunction (AND function)
$\vee$	Disjunction (OR function)
$\leftarrow$	Negation (NOT function)
$|0\rangle$ and $|1\rangle$	Quantum states of a qubit
$|\cdot\rangle$, $|0\rangle$, and $|1\rangle$	States of a three-level qutrit system

IOP Publishing

AI-Processor Electronics
Basic technology of artificial intelligence
Vinod Kumar Khanna

Chapter 1

Artificial intelligence, machine learning, deep learning and generative artificial intelligence

The terms 'artificial intelligence (AI)', 'machine learning (ML)', 'deep learning (DL)' and 'generative AI (Gen AI)' are introduced. Their specific domains and usage areas are mentioned. Different types of ML, namely, supervised (SL), unsupervised, semi-supervised (SSL) and reinforcement learning (RL) are defined. Amongst artificial neural networks (ANNs), the perceptron, multilayer perceptron, liquid state machine, feed-forward, recurrent, long short-term memory, convolutional, modular, radial basis function, residual and generative adversarial neural networks are briefly touched upon. Procedures and applications of AI computing are examined. AI-focused processor design considerations are pointed out. The state-of-the-art scenario of AI processors is sketched by discussion of salient features of a few of the vast variety of AI chips in the market or new chip announcements. Aims, scope and organization of the book are laid out. The contents of the different chapters are summarized.

1.1 Introduction

The field of AI has made great strides through countless breakthroughs and innovations since its humble beginnings in the 1950 with a seminal paper (Turing 1950). It evolved and turned over a new leaf to become an indispensable technology. Today it is hugely impacting human lives and modern industries worldwide. The advancements in AI appear as if we were watching a science fiction movie or have moved into a state of hallucination. We look amazed and flabbergasted as we watch AI achievements with bated breath. We are delighted when we see what AI has done for us. This chapter introduces the preliminary concepts and terminology of AI. It sketches the outline plan of the book, the organization of chapters, and their contents.

1.2 AI versus ML vs DL vs Gen AI

AI, ML, DL and Gen AI are four closely related concepts. They have different overlapping meanings that are widely mistakenly used, often interchangeably. AI is the umbrella term under which comes ML. Under ML falls DL and within DL comes Gen AI. It must be accentuated that ML is a subset of AI, DL is a subset of ML and Gen AI is a subset of DL (figure 1.1(a)).

AI deals with machines mimicking human intelligence and cognitive functions. The interesting fields of study under AI are reasoning, learning and solving problems with minimal human intervention (Rothman 2020, Oliver 2023).

ML uses algorithms for learning patterns from data, often labeled data, and making predictions based on the training received. To clarify with an example, ML learns to identify particular fruits in a collection of assorted items (Theobald 2021).

DL works on deep neural networks. Huge sets of complex data are used for training a model for recognizing intricate patterns to execute complicated tasks. Continuing with the above example given for ML, DL not only performs fruit identification but also assesses its quality, fresh or rotten, and the degree of ripeness (Goodfellow *et al* 2016).

Gen AI is a part of DL that can create new content by extrapolation from its training data (Emerson 2023). Gen AI's focus on new content creation differentiates it from its predecessors owing to the fact that traditional AI is largely devoted to data analysis and interpretation. Some examples of uses of Gen AI are:

   (i) Producing human-like writing.
   (ii) Developing songs and composing a symphony in the style of a famous composer.
  (iii) Making snippets of audio and video clips from a simple textual description, which are indistinguishable from similar content created by humans.
  (iv) Creating graphs showing new chemical compounds and molecules and developing new protein sequences, thus aiding in drug discovery and medical research.
   (v) Drawing realistic images for virtual or augmented reality.
  (vi) Performing simulations of the planet for accurate weather forecasting and natural disaster prediction.
 (vii) Producing synthetic data to train AI models in situations where data is unavailable.

Gen AI starts by preparing a large dataset containing examples of the type of content to be produced. The Gen AI model is trained on the collected dataset by applying techniques such as DL. In particular, models like generative adversarial networks (GANs) are used, section 1.5.11. During training, the model learns and understands the key underlying characteristics of the dataset. The learning process hinges on analyzing patterns, structures, and features in the dataset. From the core and crux essentials of the dataset, the AI model is able to form a compressed, abstract, mathematical representation of the patterns and features learned from the training data. This representation is known as the latent space. The Gen AI model

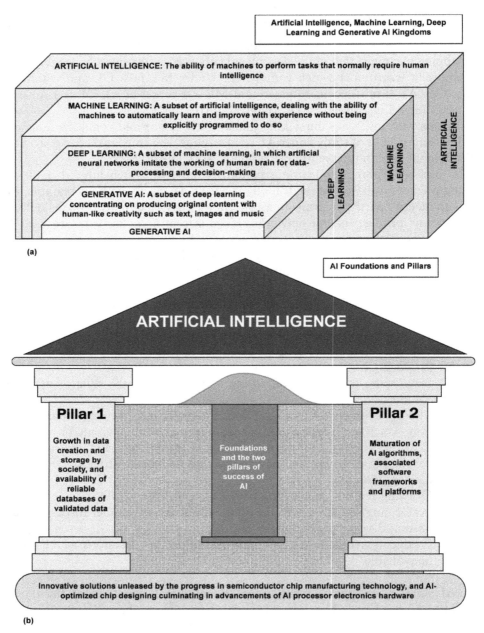

Figure 1.1. Domains of AI, ML, DL and Gen AI, their foundational infrastructure and supports: (a) AI, a vast field covering machines imitating human thinking behavior is represented by a large cuboid. ML, a part of AI concentrating on machines leaning through experience is represented by a smaller cuboid inscribed inside the AI cuboid. DL, a part of ML focused on human brain functionality is embodied in a still smaller cuboid enclosed within ML. Gen AI, a fragment of DL is shown as a cuboid inside DL. (b) The foundations and the two pillars supporting the AI boom.

creates a new content with the help of the learned latent space representation. For this purpose, points are sampled in latent space and decoded back into the original content format. The content thus generated is iteratively refined by adjusting the parameters of the model to improve its quality and realism until found satisfactory.

In Gen AI, probabilistic modeling is applied to capture the distribution of the training data. Autoregressive models such as recurrent neural networks produce sequences of data points for text, images or a time series. They work by modeling the conditional probability of each data point in a sequence from the given previously generated data points, i.e., the current values of a series as a function of the past values. Models like variational autoencoders are engaged in the reconstruction of original input data from a lower-dimensional latent space.

To summarize the main takeaways from this discussion, AI is a big idea influencing our quality of lives, helping in industrial progress and revolutionizing human-technology interaction. ML is a technique within AI where machines learn from structured, labeled data or unstructured data after its pre-processing for organization into a structured format. ML algorithms use computational methods for learning without dependence on a predetermined equation as a model. The ML algorithms incessantly improve their performance upon availability of a larger number of samples for learning.

DL is a specialized type of ML using deep neural networks for implementing more intricate tasks. It is distinguished from ML by the type of data it can use and the learning methods applied. The methods applied eliminate some data pre-processing necessary in ML. The algorithms of DL can process unstructured data like text and images to extract useful features of information from them. Thereby the dependence on human experts is reduced.

Gen AI is AI capable of humanoid creative activities for producing text, images, audio and video from a given dataset containing precedent examples.

Thus, AI is the comprehensive book of procedures, ML is the procedure for one particular task, DL is a specialized procedure for carrying out a difficult task and Gen AI is the procedure for emulating human creativity. The magnificent palace of AI is erected on the groundwork of AI processor electronics and accompanying hardware and two strong supporting pillars of a huge database of knowledge (pillar 1), and ML algorithms and related software (pillar 2) (figure 1.1(b)). This book is dedicated to the electronic processors underpinning AI advancements.

1.3 Ethics of AI

Owing to the widespread adoption of AI techniques in healthcare, financial trans-actions and justice delivery, research institutions, government and private bodies have issued ethical guidelines on transparency, justice and fairness, responsibility and accountability, privacy, and non-maleficence for proper utilization of AI for human welfare and development. Development and use of AI should be done responsibly, adhering to ethical principles in letter and spirit (Chahal 2022).

1.4 Types of machine learning

ML is divided into four classes (figure 1.2).

1.4.1 Supervised learning

In supervised learning (SL), the ML program called the model is trained on labeled datasets. Its subclasses are:
  (i) Classification: It works on the prediction of categorical target variables. These represent discrete classes or characteristics to forecast distinct values, e.g., male/female, spam email/not spam.
  (ii) Regression: It works on the prediction of continuous target variables. These variables represent numerical values to forecast continuous values of price, income, age, etc.

1.4.2 Unsupervised learning

Here the ML program discovers patterns and relationships in unlabeled data. Its two branches are:

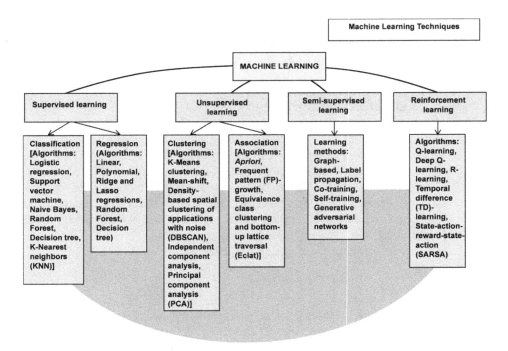

Figure 1.2. Classification of ML into supervised, unsupervised, semi-supervised and reinforcement types. Subdivisions of supervised learning are classification and regression. The two subclasses of unsupervised learning are: cluster analysis and association. Semi-supervised learning bridges supervised and unsupervised learning methods. RL is like the trial-and-error learning process of humans. Important algorithms for each type of learning are mentioned. An algorithm is a step-by-step instruction sequence followed in calculations for solving problems.

(i) Clustering: The cluster analysis involves grouping a set of objects in such a manner that objects in the group thus formed bear more similarity to each other than those in another group. Hence, it groups unlabeled objects in a dataset according to their common features or commonalities.

(ii) Association: This is a rule-based learning method which aims to discover interesting relationships or associations (one-to-one, one-to-many, many-to-many, etc) among different items in a large dataset.

1.4.3 Semi-supervised learning

In semi-supervised learning (SSL), labeled as well as unlabeled data are used for learning.

1.4.4 Reinforcement learning

In RL, the learning is done through interaction with the environment. The learning method entails producing actions, in tandem with discovering errors, and encompasses rewarding and feedback processes. Rewards are the numerical values, positive, negative or neutral given to an agent after it performs an action in an environment to help the agent in learning which actions are favorable and which are unfavorable.

1.5 Artificial neural networks

An ANN or simply NN is an ML program or model. It works in a way resembling the human brain. It mimics the cooperative approach of neurons in the human brain for identification of phenomena. It arrives at decisions after pondering over the relative significance of various possibilities and options. These operational mechanisms resemble human problem-solving methodology.

An ANN is organized into multiple layers with the input layer as the first layer and the output layer as the final layer. One or more hidden layers constitute the intermediate layers (figure 1.3). There are two classes of neural networks named the shallow and deep types. The shallow network has one hidden layer. A deep network has several hidden layers.

Each layer in an NN consists of several processing nodes where mathematical operations are performed. These nodes designated as artificial neurons are interconnected by edges which are assigned weights. During network operation, a node receives input data over its connections. The data values are multiplied with the associated weights, and the resulting products are added together. Whensoever the output of a node exceeds a threshold value, the particular node is activated. Node activation results in transmission of data to the next layer in analogy to the exchange of electrical impulses between biological neurons in the brain.

1.5.1 Feed-forward neural network

The feed-forward neural network (FNN) is a neural network in which information flows in a single direction from the input to output. The NN shown in figure 1.3 belongs to the FNN class.

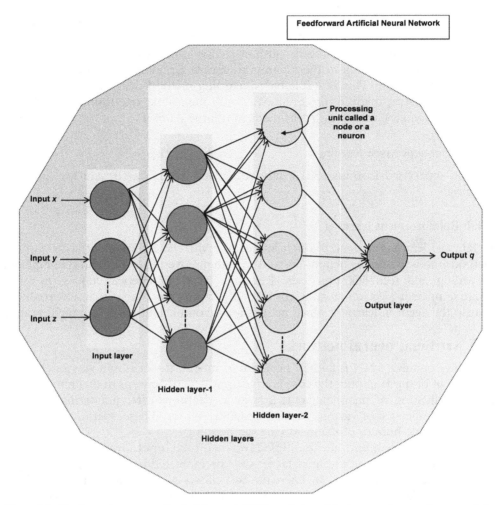

Figure 1.3. The layered structure of a feed-forward ANN consisting of the input and output layers, and the hidden layers. Inputs x, y and z are fed to the input layer. The output produced is supplied to the output layer, and delivered as output q of the neural network. Two hidden layers are shown but there can be more than two layers for a deep network.

1.5.2 Perceptron

The perceptron (perceiving and recognizing automaton) is the simplest NN (figure 1.4). It is a single-layer FNN in which the primary components are:

   (i) The input nodes: They contain real numerical values representing the characteristics or attributes of input data.
   (ii) The weights and bias: The weights determine the relative influence of each input feature on the output proclaiming the comparative strengths of the connections in an NN. The bias is a parameter used for making adjustments independent of the input.

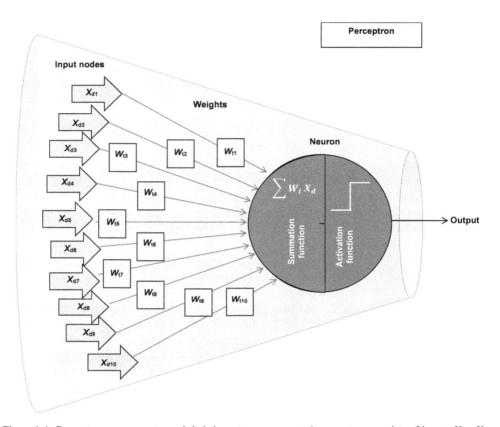

Figure 1.4. Perceptron components, and their layout arrangement. A perceptron consists of inputs X_{d1}, X_{d2}, X_{d3}, ...; the weights W_{t1}, W_{t2}, W_{t3}, ... corresponding to the inputs X_{d1}, X_{d2}, X_{d3}, ... The summation function performs addition of the weighted products $W_{t1}X_{d1}$, $W_{t2}X_{d2}$, $W_{t3} X_{d3}$, ... for each input. The weighted sum $\sum W_t X_d$, thus generated is supplied to the activation function. The activation function produces the output by comparison of the weighted sum with a threshold value. The summation and activation function tasks are done by the neuron on receipt of the input signals.

(iii) The net input function: It is obtained by combining each input feature with its respective weight to produce a weighted sum of inputs.

(iv) An activation function: It receives the weighted sum as the input. Then it compares it with a preselected threshold value, and determines whether the neuron will fire or not.

(v) The output: It represents a predicted class, 0 or 1.

(vi) The learning algorithm: It is the weight update rule through which the perceptron learns by adjusting its weights and biases.

1.5.3 Multilayer perceptron

A multilayer perceptron (MLP) is a perceptron containing a hidden layer to handle more complex tasks.

1.5.4 Recurrent neural network

The recurrent NN (RNN) is an NN which uses sequential data or time series data for solving temporal problems. Information from prior inputs is used to influence the current input and output. This characteristic differentiates the RNN from a traditional network. In the traditional network, the inputs and outputs are independent of each other.

1.5.4.1 The layers and the weight matrices
Figure 1.5 shows an RNN. The RNN is made of three layers: the input layer X-input, the hidden layer HL and output layer O-output. The symbols U, V, W represent weight matrices. The symbol U denotes the weight matrix of the input-to-hidden layer connection, V that for the hidden-to-output layer connection and W for the hidden-to-hidden layer connection. The same set of weight matrices U, V, W are used consistently all throughout.

1.5.4.2 Inputs and outputs at different instants of time
At time instant $t - 1$, the network takes the input X_{t-1} and updates the hidden layer to produce the output O_{t-1}. At time instant t, the network takes the input X_t and updates the hidden layer to produce the output O_t. At time instant $t + 1$, the network takes the input X_{t+1} and updates the hidden layer to produce the output O_{t+1}.

Thus, X_{t-1} is the input at time $t - 1$ and O_{t-1} is the resulting output at this instant. Similarly, X_t, O_t are the input and output, respectively, at time t. Likewise, X_{t+1}, O_{t+1} are the input and output at time $t + 1$.

Therefore, the RNN takes an input vector X and generates an output vector O by taking the input data sequentially as we move from left to right steps in time. In each time step, the hidden state is updated and the output is produced.

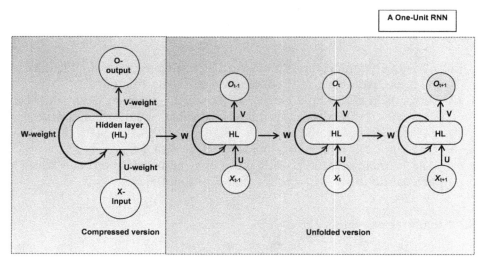

Figure 1.5. The compressed and unfolded versions of an RNN.

1.5.5 Long short-term memory network

The long short-term memory (LSTM) network is a special case of RNN. It has the same chain-like structure as an RNN but with a different repeating module structure. This structure allows the network to preserve a much larger proportion of values of the preceding stage.

1.5.6 Convolutional neural network

This is also called a ConvNet. The CNN is an NN specialized in analyzing data arranged in a grid topology. The data about the pixels of an image fall under this topology (figure 1.6).

1.5.7 Modular neural network

The modular neural network (MNN) is a neural network that uses two or more NNs to produce the output (figure 1.7). The network comprises modules 1, 2, ..., N fed by inputs p, q, r, ..., z. A gating network supplies gate signals G_1, G_2, ..., G_N to the modules. The outputs of the modules are summed together to give the output y.

1.5.8 Radial basis function neural network

The radial basis function NN (RBFNN) is an NN using a mathematical function known as the radial basis function. The radial basis function (RBF) is a real-valued

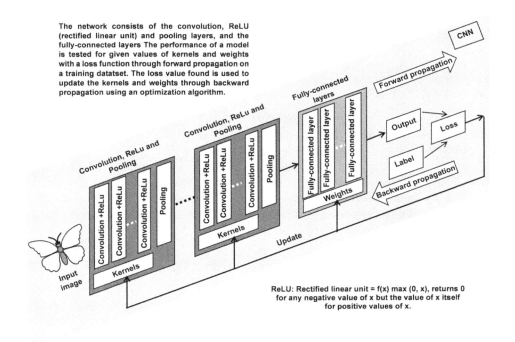

Figure 1.6. A CNN and its training procedure.

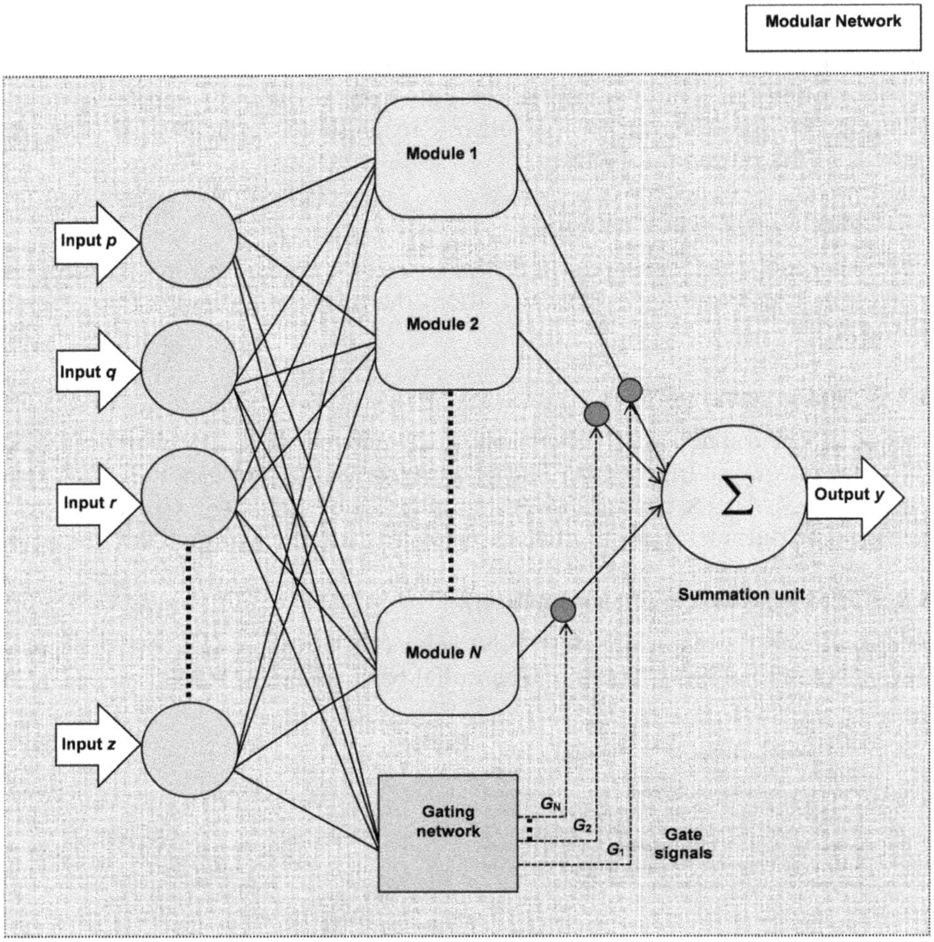

Figure 1.7. An MNN consisting of inputs p, q, r, ..., z, the constituent N modules, the gating network and gating signals G_1, G_2, ..., G_N, the summation unit and the output y.

function φ generally used to approximate a given function. The value of RBF depends only on the distance between the input and some fixed point. The distance is usually Euclidean or expressed in some specified metric. The fixed point is either the origin or some other point called a center.

1.5.9 Liquid state machine

This is an NN with randomly connected nodes. A liquid state machine (LSM) is based on the spiking NN. The spiking network incorporates the concept of time into its operating model. The LSM comprises a collection of units called nodes or neurons. Time-varying inputs from exterior sources and from other nodes are received by the nodes of the LSM. The nodes are arbitrarily connected to each other. Therefore, their interconnections behave like an RNN. Consequently, the time-varying input is transformed into a spatio-temporal pattern of activations in the network nodes. Linear discriminant units are used to read out these activations. Linear discrimination analysis is a statistical technique to determine a linear combination of features separating two classes of objects.

In this fashion, the broth of recurrently connected nodes calculates an assortment of nonlinear functions on the input. A large variety of such nonlinear functions helps in obtaining linear combinations through read-out units. These linear combinations are applied to perform mathematical operations for computer vision or speech recognition tasks. The word 'liquid' in the name 'liquid state machine' originates from its similarity to throwing a stone into a calm water or other liquid pond. The impact of the stone causes the generation of ripples in the liquid pond. The falling stone acts as the input. This input gives rise to an output which is a spatio-temporal pattern of liquid displacement, observed as ripples. These ripples resemble the spatio-temporal pattern of activations in the LSM nodes.

1.5.10 Residual neural network

The residual neural network (ResNet) is a neural network which combines the output of early layers with that of later layers.

Feature extraction is considerably improved by deepening a CNN. However, the problem of vanishing gradient degrades the network performance. To avoid this problem, a deep residual learning framework is devised (Kido *et al* 2022). It is known as the ResNet. A conventional CNN with input x and output $H(x)$ is shown in figure 1.8(a). The ResNet depicted in figure 1.8(b) has a skip connection structure. Here, the input is trained using the equation

$$H(x) = F(x) + x \qquad (1.1)$$

instead of passing through the convolutional layer. In this equation, $F(x)$ denotes the residual function. The 'plus' operation is implemented via a skip connection. In the skip connection, an identity mapping is performed to link the input of the subnetwork with its output. This connection is referred to as a residual connection. An identity map or identity transformation is defined as a function possessing a characteristic property. The characteristic property is that it always returns the same value that was used as its argument in an untouched form. This means that when $F(x)$ is the identity function, the equality $F(x) = x$ is valid for all the values of x to which $F(x)$ is applicable.

Eventually, the training of two convolutional layers in the network is done by applying the equation

$$F(x) = H(x) - x \qquad (1.2)$$

Thus, $F(x)$ is easily learned even for a small difference between x and $H(x)$. The ResNet does not allow the gradient to decrease even in the deepest layers of the model. Hence, the image features can be trained correctly.

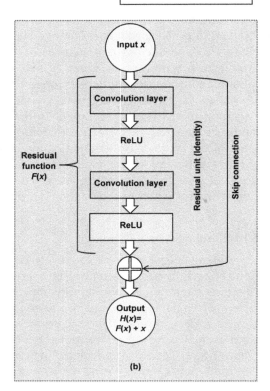

Figure 1.8. A ResNet: (a) conventional NN and (b) ResNet with a skip connection construction. The part shown as residual function $F(x)$ contains two convolution layers and their ReLU layers. The residual unit, the skip connection and the ' + operation' are indicated.

1.5.11 Generative adversarial network

The generative adversarial network (GAN) is a network that learns to generate new data having the same statistics as the training data. It is composed of two parts (figure 1.9):

    (i) A generator NN: It creates new fake data by taking random noise as input. The fake data is similar to the real dataset.

    (ii) A discriminator NN: It differentiates the real data from the generated fake data by taking both real and fake data as inputs. It determines the probability that a sample has come from the real dataset rather than the fake dataset produced by the generator.

The discriminator tries to maximize the frequency of correct classification of received data as input. Conversely, the generator tries to make the classification correctness less accurate (Remya *et al* 2021). Thus, both the NNs are engaged in a mutual competition with each other, throwing dust in each other's eyes. Each tries to

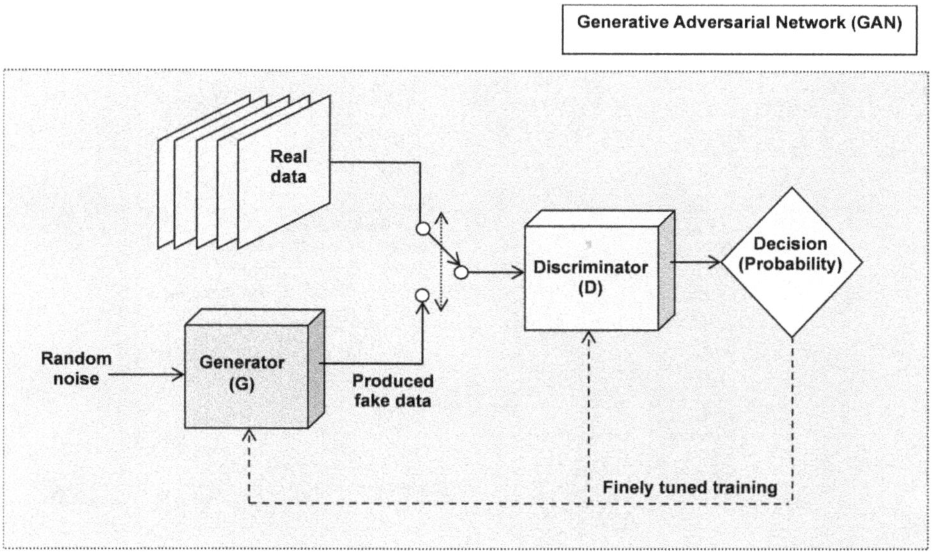

Figure 1.9. A GAN showing the generator (G) supplied with random noise for fake data production, and the discriminator (D) supplied with both real and fake data inputs giving a decision in the form of a probability value about whether the data is received from the real or fake dataset. Both the discriminator and generator undergo finely tuned training. The discriminator is trained to predict and classify images correctly. The generator is trained to deceive the discriminator by convincing it that the samples produced by it are real.

establish its superiority over the other in its respective aim. In this competition, the generator seeks to minimize the probability that the fake data produced by it are recognized by the discriminator as fake data. On the opposite side, the discriminator tries to maximize the probability of proper classification of data into real and fake categories. As the generator and discriminator improve in their own performances in their assigned roles, the generated data more closely matches with the training data in quality.

1.6 AI computing

1.6.1 Procedures and applications

It is a procedure involving gigantic mathematical calculations. The procedure uses computers and software. AI computing professionals, generally data scientists, collect and compile huge sets of data based on past incidents or happenings. Data is also congregated through experiments run to solve an assigned problem. For solving the problem, the AI professionals either choose from a library of AI models or design and test entirely new AI models that are suitable for their specific applications. Finally, the stakeholder companies run their data through the chosen AI models in a process known as inference. These exercises provide valuable insights into the problem and succour to achieve better outcomes by improving upon traditional practices. New possibilities are thus generated in the respective field under study (Howard 2024).

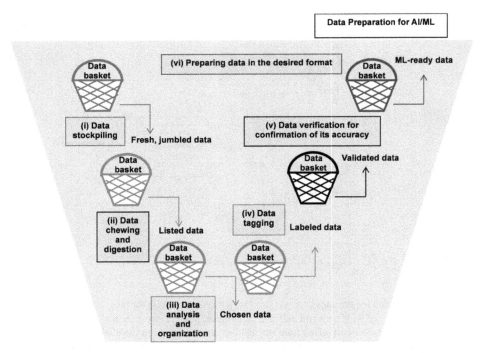

Figure 1.10. Gathering data, systematically compiling it and making it acceptable for AI/ML work.

Figure 1.10 shows the methodology of preparing data for AI/ML computing. We begin with a data basket. It is a database in which all data will be stored. Then we move through the stages below:

   (i) Accumulation of data from various sources: this activity gives us raw, mixed-up data.

  (ii) Data munching and assimilation: this process yields listed data.

 (iii) Data scrutiny and systematic arrangement. This is done to produce the chosen data for the problem to be solved.

 (iv) Data tagging: It results in labeled data.

  (v) Data authentication and endorsement of its correctness: It gives certified data.

 (vi) Data preparation by putting it in the required format: It is essential to make it ready for ML work: this gives ML-ready data.

An AI cycle comprises multiple stages. Figure 1.11 shows the stages of a typical AI cycle:

   (i) Understanding the problem and defining the strategy to obtain a solution.

  (ii) Identifying the data needed, planning for arranging and collecting the data.

 (iii) Formulating a new model or applying an existing one.

 (iv) Model corroboration.

  (v) Model deployment in actual situation, i.e., field trial.

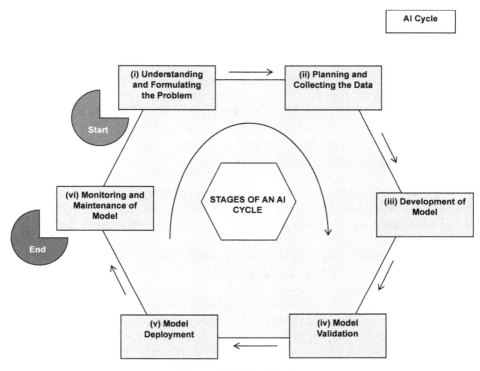

Figure 1.11. The AI cycle.

(vi) Monitoring, updating and maintaining the model by including new inputs and improvements.

Manufacturing companies are striving to make intelligent production lines and achieve real-time inspection and management of inventory, logistics, and distribution. Agricultural scientists are using AI computing for automatic control of crop planting, irrigation, and fertilization; as well as for detection and diagnosis of plant diseases in their farms through image recognition. Transport companies are harnessing AI computing for automation of route planning, navigation and parking. In healthcare, AI facilitates faster genome sequencing for diagnosing and treating genetic diseases. In smart electrical power distribution grids, AI computing is used for optimizing the energy distribution.

1.6.2 AI-focused processor design considerations

To meet the computational challenges outlined above, AI processors enhance performance not only by dimensional shrinkage of transistors in the chip. They also seek the help of AI-focused design features. These features arise from the fact that the calculations involved in AI algorithms are often independent, predictable and similar in nature. The distinct characteristics of calculations are identified and exploited to accelerate the speed of computations drastically. Some of the improvements made in AI chips are:

(i) Extensive parallel processing: Concomitant parallel processing of a large number of calculations is done in place of the sequential processing done by CPUs. Parallel processing is implemented by constructing a processor with a plurality of functional units. These units perform dissimilar or similar operations simultaneously. The data is distributed among these units for synchronized execution. Parallelism is achieved by:

  (a) Performing arithmetic or logical operations while fetching the next instruction.
  (b) Executing several instructions at the same time.
  (c) Performing arithmetic or logical operations on multiple sets of operands.

(ii) Choosing low-precision calculations whenever permissible: In low-precision floating-point arithmetic, computations are performed by reducing numerical accuracy with narrow bit-width. This modification improves the performance of the numerical programs. The performance is enhanced by virtue of small memory footprint required, faster computing speed observed, and significant energy saving achieved (Sakamoto *et al* 2020).

(iii) Adopting schemes for decreasing memory access and storage times: This is done with the help of registers, cache memory and availing the benefits of non-uniform memory access. Registers are temporary memory units of data storage. They are located in the processor, instead of the RAM. Their proximity to the processor enables speedy accessing and storing of data.

Cache memory is a small-sized, volatile, and extremely fast computer memory. It is built into the central processing unit of a computer to provide high-speed data access to the processor. It is able to hasten the data access by storing frequently used computer programs, applications and data. It is the fastest memory in a computer.

Non-uniform memory access (NUMA) scheme is a computer memory design strategy. This strategy is followed in multiprocessing systems. In NUMA, the memory is partitioned into local and non-local or remote segments based on their vicinity to processors. Memory access time is determined by the location of a memory segment with respect to the processor accessing it. Prompt accesses to local memory segments than remote memory segments bring down latency.

(iv) Increasing the use of domain-specific architectures (DSAs): DSA constitutes a type of heterogeneous architecture. It efficiently unifies general-purpose cores with specialized hardware accelerators to elevate energy efficiency. It also provides programming flexibility using specially-constructed programming languages for translating the computer codes for execution by the AI processor. The hardware, software, and systems aspects in DSAs are meticulously tailored. During tailoring, the designer keeps an eye on maximization of the overall efficiency of applications in a stipulated target domain (Krishnakumar *et al* 2023).

As a result, a mammoth collection of specialized processors has been developed. These processors cater to different tasks and applications. Some of these processors are useful for training. Others perform well in inference tasks. Training is teaching a model to execute a given task. Inference is the application of knowledge acquired by the model on new data.

The term 'AI processor' pertains to a broad class of integrated circuits or microchips. This class of microchips encompasses various integrated circuits designed for easily handling the uniquely complex computational requirements of AI algorithms with great dexterity, innovativeness and efficiency.

AI chips fall under the group of specialized computing hardware. This hardware is intended for the development and deployment of AI systems. These systems are dedicated to operations on analysis of data, ML, computer vision and natural language processing (NLP). The difference between a traditional and an AI processor is apparent from the quantity and type of data they can process and the number of calculations they can do in the same time duration.

1.6.3 Glimpses of state-of-the art in AI computing

From sources across the web, we find information about the AI chip announcements and demonstrations. The web updates reveal that many of them belong to a system-on-chip (SoC) class. In an SoC, multiple dies are assembled and encapsulated inside one package. Some of these SoCs are designed as network acceleration services (Dilmegani 2024). The performances achieved by AI chips are liable to frequent changes as progress continues unabated at a fast pace and websites are accordingly restructured. We cite a few examples below for sake of discussion, in alphabetical order. The readers are recommended to visit the web sources for notifications. Some of the computer terms will be difficult to understand but we shall learn their meanings as we progress through the text. See notes 1 and 2 below.

1.6.3.1 Alibaba cloud global accelerator (GA)
It is a high-availability and high-performance network acceleration service. This service is provided by Alibaba Cloud. The service offers a high security and low latency. The GA enables nearby access to the internet. It also provides cross-region deployment of applications. These are implemented by taking advantage of the high-quality border gateway protocol (BGP) bandwidth and global network infrastructure of Alibaba Cloud (Alibaba Cloud 2024).

1.6.3.2 Amazon Web Services Trainium
This is a purpose-built ML chip. The chip is used by Amazon Web Services for training of models in DL. Up to 16 Trainium accelerators are deployed by each Amazon Elastic Compute Cloud2 (Amazon EC2) for DL training in the cloud. These accelerators provide faster training with up to 50% cost-to-train savings over comparable EC2 instances. Trainium is optimized for training NLP, computer vision, and recommender models. These models are used in text summarization,

code generation, question answering, image and video generation, recommendation, and fraud detection. AWS Neuron SDK (software development kit) helps the developers to train models on Trainium accelerators and install them on AWS Inferentia accelerators (AWS 2024).

1.6.3.3 AMD MI300X discrete GPU and AMD MI300A APU (accelerated processing unit)

The discrete GPU incorporates 8 accelerator complex dies (XCDs). For reduced precision data, the chip offers peak theoretical FP8 performance of 2.6 PFLOPs^{-1}. For classic HPC workloads using single- and double-precision, it offers 163.4 TFLOPs^{-1} FP64 matrix performance for a solitary processor. The AMD high-performance, data center APU contains the CPU, GPU, and memory on one package. The accelerator computational capacity of MI300X is decreased by 25% to accommodate three CPU dies. The CPU dies are tightly coupled with six GPU dies. They share a single pool of virtual and physical memory with low latency (AMD 2023).

1.6.3.4 Apple M4 chip

It is an SoC fabricated in 3-nanometer technology. It has up to 10-core CPU and 10-core GPU. It also has a fast memory bandwidth. The chip contains 28 billion transistors. Dynamic caching in GPU allocates local memory efficiently in hardware and in real time. The dynamic memory allocation increases the average utilization of the GPU. Hardware-accelerated ray tracing and mesh shading help in geometry processing. Graphics-intensive applications are thereby assisted. The neural engine of this chip can execute up to 38 trillion (10^{12}) operations per second. The display engine of the chip enables precision, color accuracy, and brightness uniformity. The advanced media engine of M4 gives high-resolution video. Its energy efficiency delivers all-day battery life (Newsroom 2024).

1.6.3.5 Cerebra's wafer scale engine (WSE-3)

Presently, the wafer scale engine (WSE-3) of Cerebras Systems Inc., an American AI company, is the fastest AI processor. It is 57 times larger than the largest GPU with 52 times more compute cores, and 880 times more high-performance on-chip memory. Its chip size is 46 225 mm^2 with 4×10^{12} transistors. It is fabricated with 5 nm process and has 900 000 AI-optimized cores. Each core on the WSE-3 is independently programmable. The cores are optimized for sparse linear algebra operations based on tensors. The 44GB of super-fast on-chip SRAM of WS-3 is spread uniformly across the entire surface of the chip. The spreading of SRAM gives every core single-clock-cycle access to fast memory at extremely high bandwidth of 21 petabytes s^{-1}. It has a high 214 petabytes s^{-1} processor–processor interconnect bandwidth resulting in low latency. It is a cluster-scale AI compute resource. It is easily programmable like a single desktop machine. Stock PyTorch is used for this work. The WSE-3 excels over all other processors in AI-optimized cores, memory speed and on-chip fabric bandwidth (Cerebras WS-3 Datasheet 2024).

1.6.3.6 Etched Sohu chip

It is an application-specific integrated circuit. It is designed specifically for running transformer models. The transformer models are used in NLP and Gen AI. They are DL models. They are designed for sequence transduction or neural machine translation for understanding, interpretation and generation of human language. They learn context and hence meaning by tracking relationships in sequential data. They have a sequence-to-sequence architecture. This architecture alters an input sequence of vectors into an output sequence by passage through several encoder layers. The objective is to extract features from a sequence and many decoder layers in order to apply these features for production of another sequence. The transformer models involve complex operations, e.g., attention mechanisms and matrix multi-plications. For carrying out these operations, a high memory bandwidth is essential along with parallel processing. 144 GB of HBM3 memory is included per chip (Wassim 2024, Arbisoft 2024).

1.6.3.7 Google Sycamore quantum processor

It is created by Google's AI division. Its updated version is a 70 qubits transmon superconducting quantum processor. It completed a task in 200 s that would take a state-of-the-art classical supercomputer a very long time (10 000 years) to execute (Arute *et al* 2019, Swayne 2024). The transmon is a superconducting qubit. This qubit is designed for diminished sensitivity to charge noise. Google's quantum chip Willow reduces errors exponentially with scaling up (Google 2024).

1.6.3.8 Google Trillium TPU (tensor processing unit)

This sixth generation TPU shows 4.7× increase in peak compute performance per chip than the previous version TPU v5e. Its high bandwidth memory (HBM) capacity, and interchip interconnect (ICI) bandwidth are twice that of TPU v5e. The size of matrix multiply units (MXUs) is also larger than that of TPU v5e. Also higher is the clock speed. Trillium is equipped with third generation SparseCore. The SparseCore is an accelerator used to accelerate embedding-heavy advanced ranking and recommendation workloads (Chole *et al* 2018). The sparsity in AI refers to models that are designed to use a small number of significant features. These models pay no heed to the less relevant attributes. Scaling up to 256 TPUs is possible in a single high-bandwidth, low-latency pod. On the far side of pod-level scaling, Trillium TPUs can scale to hundreds of pods. They are able to connect tens of thousands of chips in a supercomputer. Trillium TPUs constitute a part of Google Cloud's supercomputing architecture, viz., the AI Hypercomputer (Vahdat 2024).

1.6.3.9 Graphcore's Colossus™ MK2 GC200 IPU (intelligence processing unit)

It is a massively parallel processor. It is co-designed from the ground up with the Poplar® SDK for machine intelligence. It is fabricated with TSMC 7 nm process. It has 59.4 billion transistors. Each IPU has 1472 processor cores. In these cores ∼9000 independent parallel program threads are run. Further, each IPU holds a 900MB in-processor-memory™. 250 teraFLOPS of computing are done at FP16.16 and FP16. SR (stochastic rounding) (Graphcore 2024).

1.6.3.10 Groq LPU™ (language processing unit) inference engine

It is a processing system with single core architecture. It is designed to handle computationally intensive applications with a sequential component. Its exceptional sequential performance quickens the processing of large language models (LLMs). It displays the ability to auto-compile >50B LLMs. It has a high compute and memory bandwidth. These features reduce the amount of time per word calculated. Faster generation of text sequences is thus allowed. Hence, it can deliver orders of magnitude better performance than a graphics processor. It provides instant memory access and high accuracy even at low precision levels. Synchronous networking is maintained even at large-scale deployments (Groq 2024).

1.6.3.11 IBM NorthPole

It is a low-precision, massively parallel, densely interconnected, energy-efficient chip. It is built in a spatial computing architecture. It also has a co-optimized, high-utilization programming model. The chip is fabricated with 12 nm node process. It contains 22 billion transistors in 800 square millimeters area. It has 256 cores. The chip can perform 2048 operations per core per cycle at 8-bit precision. It is possible to double the number of operations with 4-bit precision. Aspiring still more, the number of operations can be quadrupled with 2-bit precision.

NorthPole outperforms the prevalent architecture. Superior performance is achieved because all the memory for the device is on the chip itself, not connected separately. This integration eliminates the need to access external memory. Thus, latency and energy consumption are reduced. Consequently, the von Neumann bottleneck is eliminated (Murphy 2023, Software Engineers Blog 2023).

1.6.3.12 IBM quantum processors

IBM Heron is a 156-superconducting qubit tunable-coupler quantum processor. It is free from crosstalk errors encountered in the preceding quantum processors of IBM. IBM's quantum system two with three Heron processors is an innovatory quantum-centric supercomputing architecture (IBM 2023).

IBM Condor is a 1121qubit quantum processor. It is based on cross-resonance gate technology. The cross-resonance gate is a microwave-triggered two-qubit gate. This gate is connected through effective capacitors in a superconducting circuit. The circuit works by driving one qubit (control) at the frequency of the other qubit (target).

IBM Osprey is a 433qubit processor. IBM Eagle is a 127-qubit processor (IBM Quantum Documentation 2024).

1.6.3.13 Intel® Gaudi® 3 AI accelerator

It has a heterogeneous architecture. The architecture includes two similar compute dies. The two computer dies together contain 8 MMEs (Matrix multiplication engines), a 64 TPC (Tensor processor core) engines cluster and 24 × 200 Gbps (giga bits per second) RDMA NICs (remote direct memory access network interface cards) ports. The TPC cluster is fully programable. Further, 8 HBM2e (2nd generation high-bandwidth memory) chips comprise a 128 GB (gigabyte) unified high bandwidth memory (HBM). The compute dies are connected through a high-bandwidth, low-latency interconnect over an interposer (thin substrate) bridge. This connection is transparent to software. Therefore, it provides the performance of a large unified single die. The MMEs are able to do all the matrix multiplication operations. The operations performed by them embrace the fully-connected layers, convolutions and batched-GEMMs (general matrix multiplications) in neural networks. The TPC is a very long instruction word (VLIW) single-instruction multiple-data (SIMD) processor; SIMD is defined in section 4.2.2. The TPC is specifically made for DL applications. It seeks to accelerate all non-GEMM operations. This AI accelerator performs training and inference with 1.8 PFLOPs of FP8 and BF16 or bfloat 16 (brain floating point) compute. The memory capacity is 128 GB of HBM2e, and the bandwidth is 3.7 TB s^{-1} of HBM (Intel 2024). The BF16 is a computer number format occupying 16 bits in its memory.

1.6.3.14 Meta training and inference accelerator (MTIA), Artemis

It is a custom-made chip. It is designed for Meta's AI workloads. It is fabricated with 5 nm technology of Taiwan Semiconductor Manufacturing Company (TSMC) Limited. The accelerator has an 8 × 8 grid of processing elements (PEs). It has a high-bandwidth, low latency network-on-chip (NoC) architecture. Its operational frequency is 1.35 GHz. The thermal design power (TDP) is 90 W. Other parameters are: GEMM 177 TFLOPSs^{-1} (FP16/BF16); SIMD 2.76 TFLOPSs^{-1} (FP32); Local memory: 384 kB per PE with 1 TB s^{-1} per PE bandwidth; and On-chip memory: 256 MB with 2.7 TB s^{-1} bandwidth (Tal *et al* 2024).

1.6.3.15 Microsoft Azure Maia 100

It is designed to run cloud-based AI workloads. This chip is made on 5 nm node using advanced packaging technology from TSMC. It has a dedicated companion to match the thermal profile of the chip. It includes rack-level, closed-loop liquid cooling. Maia 100 servers are designed with an Ethernet-based network protocol. The aggregate bandwidth is 4.8 terabits per accelerator (Borkar *et al* 2024).

1.6.3.16 Mythic M1076 AMP™ (Analog Matrix Processor)

It integrates 76 AMP tiles to store an exceedingly large number of weight parameters. It delivers up to 25 TOPS (trillion operations per second). It executes matrix multi-plication operations without any external memory. The operations are performed at higher resolution and lower latency. Thus, it delivers the AI compute performance of a desktop GPU. However, it has substantailly lower~ 1/10 power consumption. Power used in running complex models is 3–4 W. It has a 4-lane PCIe 2.1 interface. Up to 2GBs^{-1} of bandwidth is offered for inference processing (Mythic 2024).

1.6.3.17 NVIDIA GH200 Grace Hopper CPU-GPU superchip for giant-scale AI and HPC workloads

It combines a CPU and a GPU into a single superchip. The combination is enabled via a 900GBs^{-1} total bandwidth chip-to-chip interconnect. Memory coherency enables programming of both the CPU and the GPU. The programming is done with a unified programming model. The CPU uses 72 CPU cores with top per-thread performance. Higher energy efficiency is endowed than traditional CPUs. Up to 480 GB of memory, and up to 500 GBs^{-1} of memory bandwidth per CPU are available. Up to 3.2 TB s^{-1} of total bisection bandwidth is provided. The bisection bandwidth is the minimum bandwidth on hand when a network is partitioned into two sub-networks of equal size. The ninth-generation data center GPU gives an order-of-magnitude improvement compared to previous generations. Thread block clusters and thread block reconfiguration improve spatial and temporal data locality. They provide assistance to applications for always keeping all the units busy with asynchronous execution engines (NVIDIA 2024).

1.6.3.18 SambaNova Reconfigurable Dataflow Unit™ (RDU)

The RDU is a processor providing native dataflow processing and programmable acceleration. It consists of a tiled array of reconfigurable processing and memory units. These units are connected through a high-speed, three-dimensional on-chip switching fabric The components of an RDU are:

(a) A pattern compute unit (PCU) to execute a single, innermost-parallel operation in an application.

(b) A pattern memory unit (PMU) consisting of specialized scratchpads that provide on-chip memory capacity.

(c) A high-speed switching fabric connecting PCUs and PMUs and composed of three switching networks: scalar, vector and control.

(d) An address generator unit (AGU) and a coalescing unit (CU) for interconnections between RDUs.

(e) The rest of the system (off-chip DRAM, other RDUs and the host processor).

The RDU architecture enables a broad set of highly parallelizable patterns. These patterns are contained within dataflow graphs. This is done in order that they are efficiently programmed as a combination of compute, memory and communication networks. A spatial programming model achieves high hardware utilization. The utilization is made possible by allowing the optimization of compute layout and minimization of data movement. When an application is started, SambaFlow configures the RDU elements. During configuration, custom processing pipelines are created. These pipelines permit data to flow through the complete computation graph. In such manner, execution of an optimized dataflow graph for that application takes place (SambaNova Systems Whitepaper 2021).

Note 1: HPC (high-performance computing) refers to high-speed parallel processing of complex computations on multiple servers. A thread is a single unit of execution within a process that carries out instructions. A thread block is a group of threads working together at the same time to execute a function. Memory bandwidth is the rate at which a processor reads data from the memory or writes data into the memory of the computer. Memory bandwidth is measured in gigabytes s^{-1} implying 10^9 bytes s$^{-1} = 8 \times 10^9$ bits s^{-1}, bandwidth is the amount of data (in gigabytes, megabytes or some other unit) transferred between the processor and the memory in a given period of time (in seconds).

Note 2: TFLOPs^{-1} and PFLOPs^{-1} stand for tera (10^{12}) and peta (10^{15}) floating point-operations per second, respectively. They are units of measurement of computational performance. Other units used are gigaFLOPs^{-1} for giga (10^9) floating point-operations per second and exaFLOPs^{-1} for exa (10^{18}) floating point-operations per second. Floating-point numbers are non-integer fractional numbers. They have a decimal point inside them that moves in an extensive range. The operations refer to mathematical operations of addition, subtraction, multiplication and division. FP8 (8-bit floating point), FP16 (16-bit floating point) or half precision, FP32 (32-bit floating point) or single precision and FP64 (64-bit floating point) or double precision are binary floating-point computer number formats. They are used for representation of numbers with varying degrees of precision. Different number of bits are occupied by the number in the computer memory during representation of numbers by these formats. The precision of a format is determined by the number of bits assigned to its exponent and mantissa. As an example, the number 3.3×10^3, the decimal scientific notation for 3300 has a mantissa of 3.3, a base of 10 and an exponent of 3.

1.7 Aims, scope and organization of the book

This book seeks to provide a comprehensive survey of the physics of operation of a few representative AI processors among diverse chips available, along with the related technologies, to highlight the key concepts and innovations underlying their development. This will help to get a wide-sweeping perspective of the underlying principles, technologies, the available opportunities and future prospects. The book is divided into 15 chapters as follows:

1.7.1 Chapter 1: The AI, ML, DL and Gen AI

This chapter introduces AI and its key technical terminology. AI is a blanket term or hypernym circumscribing computer algorithms and software for building smart machines which imitate human capabilities of perception, understanding, reasoning and decision-making. The imitation is demonstrated by the ability of machines to learn from training data and perform complicated tasks with imagination and ingenuity. ML is a subcategory of AI in which machines learn from supplied data and contrive to frame adaptable models for execution of defined tasks. DL is a subsection of ML. Its distinctive feature is the use of deep NNs for analyzing data. Gen AI is a subcomponent of DL. It exclusively exhibits the ability of creating new content resembling real-world data, e.g., text, images, music, audio, video and more.

Knowing the technical jargon of these fields will smoothen further study.

1.7.2 Chapter 2: Electronic computing fundamentals

This chapter offers a quick revision of the binary number system. Combinational and sequential logic circuits are delineated. The esteemed CMOS logic family is discussed. Preliminary arithmetic circuits are surveyed. These include different types of adders and subtractors, such as half and full adders, subtractors, multipliers, etc.

1.7.3 Chapter 3: Central processing unit, and the von Neumann bottleneck

The central processing unit is often hallmarked as the mainstay or the lynchpin of conventional computers. It was developed prior to the ML revolution. It is a general-purpose processor based on the von Neumann architecture. It provides great flexibility in computing. But its workflow of fetching data from memory, performing calculations on it and storing the results back in memory limits its throughput. The reason is slower memory access than calculation speed. This obstacle is spotlighted as the von Neumann bottleneck.

The chapter demystifies the working of CPU. It explains the special tasks for which a CPU is a mandatory requirement. It brings out the reasons for inadequacy of the CPU for AI work.

1.7.4 Chapter 4: Parallel computing architecture

Also called high-performance computing or supercomputing, it is a leading-edge technology. It utilizes a group of processing units to solve large computational problems quickly. This accomplishment is rendered possible through inter-communication and mutual collaboration. Thus, plenty of time is saved and the efficiency is improved. Parallel computing implements multiple computational tasks simultaneously or in an overlapping fashion, which is antagonistic to the 'one by one' style of serial computing. It is contrasted from concurrency too. The concurrency practice involves execution of multiple tasks on a single processor by rapidly switching between them to give the impression of simultaneity. But these tasks are not necessarily done simultaneously. Parallelism is the truly simultaneous execution of different parts of a program on multiple processors. This chapter explores the salient features of parallel computing.

1.7.5 Chapter 5: Optimized AI-computing within physical limits of transistors

Harnessing the full potential of AI-driven workloads demands purpose-built hardware, designed in concordance with the requirements of AI algorithms. AI chips have unique design features to speed up the identical, predictable, independent calculations by:

(i) Carrying out a large number of calculations in parallel instead of sequential processing.

(ii) Sacrificing precision in a manner that successfully implements AI algorithms using a smaller number of transistors for the same calculation.

(iii) Speeding up memory access and with the aid of specially-built programming languages.

Hardware/software provisions to meet this goal are surveyed in this chapter for realization of AI-specific accelerators.

1.7.6 Chapter 6: Graphical processing unit

In this chapter, the advantages and limitations of GPUs relative to CPUs are discussed. Originally developed for high-speed graphics processing, the primary advantage of GPUs is massive parallelism due to use of 2500–5000 ALUs enabling the simultaneous execution of thousands of mathematical operations. It is a co-processor to shoulder some of CPU's workload. Whereas CPUs have multi-core processors, functioning with MIMD (multiple-instruction, multiple-data) architecture, GPUs use SIMD (single-instruction, multiple-data) architecture, section 4.2. The SIMD architecture is more suited to ML problems requiring the implementation of the same process on numerous data items. NVIDIA GPUs have compute unified device architecture (CUDA) or tensor cores or both types of cores. The CUDA core GPUs are excellent for general-purpose computing tasks in scientific research and ML. Tensor core GPUs optimize matrix multiplication operations frequently used in NNs, providing hardware acceleration to computations involving matrices or tensors.

1.7.7 Chapter 7: Tensor processing unit

This chapter delves into the prominent features of TPU, a neural processing unit. It is a custom-designed application-specific integrated circuit developed by Google to solve matrix and vector operations for DL through multiply and accumulate functions at ultra-high speeds using Google's tensor flow platform and by pairing with a CPU to supply and execute instructions. The heart of the matrix multiplier unit of TPU is the 2D systolic array. The systolic array is a monolithic network of processing elements called cells or nodes. Each cell rhythmically performs a sequence of operations on the data received from its upstream neighbor. This operational sequence may be computation of a partial result, storing it within itself and passing it to its downstream neighbor.

1.7.8 Chapter 8: Neural processing unit

Hardware implementations of NNs have been customarily restricted to specific types of learning. However, they have promising capabilities for acceleration of general-purpose code that is tolerant to small errors. Neural processing units constitute a class of trainable accelerators having potential implementations in the digital and analog domains. Several types of NPUs such as RENO, AXNet and DianNao family are described. Also discussed is the Ascend AI processor chip of Huawei Technologies containing three computing units dedicated to matrix, vector and scalar computations.

1.7.9 Chapter 9: Convolutional neural network processor, and the vision processing unit

Traditional processors display low energy efficiency in the execution of deep CNNs algorithms owing to the extensive amount of computation required. A CNN processor is described. This chapter also looks at the main features of VPU (vision processing unit) with respect to GPU. The VPU is a specialized hardware optimized to work as a co-processor with CPU to take the load from CPU and assign it to an ASIC (application-specific integrated circuit). This ASIC can efficiently process the gargantuan amounts of data associated with ML tasks, and particularly in image processing and machine vision for object and facial recognition. It comes under the family of NPUs, which speed up the processing of ML algorithms by operating on predictive models such as ANNs. Intel VPUs handle demanding computer vision and AI workloads efficiently.

1.7.10 Chapter 10: Compressed and sparse neural network processors

Deep NNs are not only computationally intensive, they also consume a great amount of digital storage. The term 'sparsity' in AI inference and ML refers to a matrix of numbers including many nulls or zero values. The nulls will not appreciably influence a calculation. The sparse NN processor performs inference on a compressed sparse network model. It hastens the resulting sparse matrix-vector multiplication. By working on the compressed sparse model, the processor is able to yield momentous speedup and energy saving.

1.7.11 Chapter 11: Graph analytics processor for graph algorithm computations

Conventional processor architectures are deficient in dealing with the throughput and memory requirements of graph algorithms. Such algorithms are being increasingly used in applications exploiting large databases. To overcome their shortcomings, a fundamental rethinking of parallel architectures for graph problems is done. The graph-processor architecture utilizes several innovative features. The exclusive features include sparse matrix-based graph instruction set and a cacheless memory system. These are backed by accelerator-based architecture and a systolic sorter. A high-bandwidth multidimensional toroidal communication network is used with randomized communications. A field-programmable gate array (FPGA) prototype of the graph processor has been developed by Lincoln Laboratory. It has shown noteworthy enhancement in performance over conventional processors in graph computational throughput. The chapter underscores the capabilities and opportunities offered by the graph processor.

1.7.12 Chapter 12: Associative memory processor

Associative processor is a non-von Neumann architecture. It uses associative memory supplemented by an induction-like approach for computing. It is based on data-centric processing, also called processing-in-memory. This processing mode minimizes the movement of data by performing computations at the most appropriate location in the memory hierarchy. Several problem arise during fetching all data to a processor situated far away from the data storage. The processor stumbles across latency (access time), bandwidth (throughput of data transfer), and energy constraints accompanying data transfers. As opposed to this widely followed practice, processing-in-memory facilitates processing of data directly at its place of residence. It greatly uplifts the performance and energy efficiency of processing of colossal amounts of data, theoretically by orders of magnitude

1.7.13 Chapter 13: Quantum computing principles and devices

The principles of quantum computing have their roots firmly implanted in quantum mechanics. Quantum mechanics is a famous revolutionary theoretical framework in physics. It has accurately modeled the behavior of particles at the microscopic and nano-dimensional scales. Quantum computing achieves breakthroughs in computational capabilities by reaping the benefits of the many unconventional features inherent to quantum mechanics. Here, all knowable information about a physical system is represented by a quantum state. Silicon spins (leveraging the intrinsic spins of electrons in a Si-based structure), trapped ions, and superconducting transmons represent three of the leading approaches for quantum computing. Unlike the bits of classical computer which acquire only two values, 0 and 1, the qubits utilize the principle of quantum superposition. According to superposition principle, any two or more distinct quantum states can be added together to yield a valid quantum state. Conversely, any quantum state can be expressed as the sum of two or more states. As a result, the qubits can exist in several states at the same time. Properties of quantum bits or qubits are described, and devices for making quantum bits are briefly discussed.

1.7.14 Chapter 14: Quantum logic gates and circuits

The development of quantum computing technologies builds on the distinctive features of quantum physics while borrowing familiar principles from the design of conventional electronic circuits. The fundamental concepts required for designing and operating quantum computing circuits are introduced. The state-of-the-art efforts to fabricate and demonstrate quantum gates and circuits for performing arithmetic calculations are reviewed. This appraisal is followed by outlining the generation and synthesis of quantum circuits for higher-order logic using quantum computing devices.

1.7.15 Chapter 15: Quantum processing unit

This chapter provides an overview of QPU (quantum processing unit). It is a chip created on the physics of sub-atomic particles called quantum physics. It contains interconnected quantum bits known as qubits in place of bits in a classical CPU. The supporting infrastructure of a QPU is vastly more complex and variable than the classical computer hardware. In future, quantum computers will be used in conjunction with traditional computers to solve intricate problems that fall outside the capabilities of these old computers. But 'old is gold' and these old-fashioned computers are treasures. They are real assets that have no substitutes for many tasks, and therefore will not perish.

1.8 Summary and the way forward

1.8.1 Highlights of chapter 1 at a glance

(i) The terms 'artificial intelligence', 'machine learning', 'deep learning' and 'generative AI' were defined. Their specific domains and usage areas were mentioned.

(ii) AI is the ability of machines to impersonate human intelligent behavior. ML deals with machines learning from training data without being explicitly programmed for the purpose. DL utilizes deep neural networks to learn hierarchical representations of data. Thereby it automatically extracts distinctive features from data for pattern recognition. Gen AI focuses on creating models for producing new content resembling existing data that is indistinguishable from human-made content.

(iii) AI is an extensive, all-inclusive concept. ML learns patterns from data, DL utilizes the capabilities of deep NNs for intricate pattern recognition. Gen AI creates new content.

(iv) AI is the overarching field, ML is a subset of AI, DL is a further specialized subset of ML and Gen AI is a category within AI for art and new content creation, e.g., visual images and speech to enable facial and voice recognition and natural language understanding.

(v) Different types of ML, namely, supervised, unsupervised, semi-supervised and reinforcement learning were introduced.

(vi) Amongst ANNs, feed-forward neural network, perceptron, multilayer perceptron, recurrent neural network, long short-term memory network, convolutional neural network, modular neural network, radial basis function network, liquid state machine, residual neural network and generative adversarial network were briefly touched upon.

(vii) Procedures and applications of AI computing were examined. AI-focused processor design considerations were pointed out.

(viii) The state-of-the-art scenario of AI processors was sketched by discussion of salient features of a few of the vast variety of AI chips in the market or new chip announcements.

(ix) Aims, scope and organization of the book were laid out by summarizing the contents of the different chapters.

(x) A few key words describing the contents of this chapter are: AI, ML, DL, Gen AI, ANN, FFNN, perceptron, MLP, RNN, LSTM, CNN, MNN, RBFNN, GAN, LSM, ResNet, AI chips.

1.8.2 Getting ready to begin chapter 2

Whether it is AI, ML, DL or Gen AI, the core issue is that all of them are concerned with solving mathematical problems. The problem-solving strategies are formulated as algorithms to arrive at numerical solutions, comments and decisions. For solving these problems, they seek the help of computers. The computers themselves rely on their processors, the central controlling and calculating engine. The processors are made by mastering and systematically applying digital electronics principles and engineering, which we shall undertake in the next chapter.

References

Alibaba Cloud 2024 Global accelerator, 2009–2024 Copyright by Alibaba Cloud https://alibabacloud.com/en/product/ga?_p_lc=1

AMD 2023 AMD CDNA™ 3 architecture, the all new AMD GPU architecture for the modern era of HPC and AI 2023 Advanced Micro Devices, Inc. pp 1–27 https://amd.com/content/dam/amd/en/documents/instinct-tech-docs/white-papers/amd-cdna-3-white-paper.pdf

Arbisoft 2024 A closer look at etched and the world's first transformer ASIC, Linkedin 2024, https://linkedin.com/pulse/closer-look-etched-worlds-first-transformer-asic-arbisoft-eyplf?trk=organization_guest_main-feed-card_feed-article-content

Arute F, Arya K, Babbush R, Bacon D, Bardin J C, Barends R, Biswas R *et al* 2019 Quantum supremacy using a programmable superconducting processor *Nature* **574** 505–10

AWS 2024 AWS Trainium 2024 Amazon Web services https://aws.amazon.com/machine-learning/trainium/

Borkar R, Wall A, Pulavarthi P and Yu Y 2024 Microsoft Azure: Azure Maia for the era of AI: from silicon to software to systems, Microsoft 2024 https://azure.microsoft.com/en-us/blog/azure-maia-for-the-era-of-ai-from-silicon-to-software-to-systems/

Cerebras WS-3 Datasheet 2024 Wafer-scale engine 3: the largest chip ever built, 2024 Cerebras Systems Inc. https://8968533.fs1.hubspotusercontent-na1.net/hubfs/8968533/Datasheets/WSE-3%20Datasheet.pdf

Chahal H 2022 *Ethics of AI: principles, rules and the way forward* (Observer Research Foundation) pp 1–15 ORF Issue Brief No. 589, November 2022

Chole S, Tadishetti R and Reddy S 2018 SparseCore: an accelerator for structurally sparse CNNs *Proc. SysML Conf, the Conf. on Systems and Machine Learning (February 15–16) (Stanford, CA)* pp 1–3

Dilmegani C 2024 Top 20 + AI Chip Makers of 2024: In-depth Guide Copyright © 2024 AIMultiple https://research.aimultiple.com/ai-chip-makers/

Emerson J 2023 Ripples of Generative AI: How Generative AI Impacts, Informs, and Transforms Our Lives, IngramSpark, Tennessee, USA p 118

Goodfellow I, Bengio Y and Courville A 2016 *Deep Learning, Adaptive Computation and Machine Learning Series* (Cambridge, MA: MIT Press)

Google 2024 Meet Willow, our state-of-the-art quantum chip https://blog.google/technology/research/google-willow-quantum-chip/

Graphcore 2024 Designed for AI: intelligence processing unit, introducing the Colossus™ MK2 GC200IPU Copyright Graphcore https://graphcore.ai/products/ipu

Groq 2024 The Groq LPU™ inference engine: Purpose-built for inference performance and precision, all in a simple, efficient design Groq, Inc. https://wow.groq.com/lpu-inference-engine/

Howard 2024 Understanding AI computing: an introduction to artificial intelligence, feature solutions (FS) https://community.fs.com/article/understanding-ai-computing-an-introduction-to-artificial-intelligence.html

IBM 2023 IBM debuts next-generation quantum processor & IBM quantum system two, extends roadmap to advance era of quantum utility https://newsroom.ibm.com/2023-12-04-IBM-Debuts-Next-Generation-Quantum-Processor-IBM-Quantum-System-Two,-Extends-Roadmap-to-Advance-Era-of-Quantum-Utility

IBM Quantum Documentation 2024 Processor types, Heron, Revisions July 2024 https://docs.quantum.ibm.com/guides/processor-types

Intel 2024 Intel® Gaudi® 3 AI accelerator white paper, pp.1–24, 2024 Intel Corporation https://intel.com/content/www/us/en/content-details/817486/intel-gaudi-3-ai-accelerator-white-paper.html

Kido S, Kidera S, Hirano Y, Mabu S, Kamiya T, Tanaka N, Suzuki Y, Yanagawa M and Tomiyama N 2022 Segmentation of lung nodules on CT images using a nested three-dimensional fully-connected convolutional network *Front. Artif. Intell.* **5** 782225 1–9

Krishnakumar A, Ogras U, Marculescu R, Kishinevsky M and Mudge T 2023 Domain-specific architectures: research problems and promising approaches *ACM Trans. Embed. Comput. Syst.* **22** 1–26

Murphy M 2023 IBM: A new chip architecture points to faster, more energy-efficient AI https://research.ibm.com/blog/northpole-ibm-ai-chip

Mythic 2024 M1076 analog matrix processor Mythic® https://mythic.ai/products/m1076-analog-matrix-processor/

Newsroom 2024 Apple introduces M4 chip, Copyright 2024 Apple Inc. https://apple.com/in/newsroom/2024/05/apple-introduces-m4-chip/

NVIDIA 2024 NVIDIA Grace Hopper Superchip Architecture Whitepaper 2024 NVIDIA Corporation, p 39 https://resources.nvidia.com/en-us-grace-cpu/nvidia-grace-hopper?ncid=no-ncid

Oliver T 2023 *AI for Absolute Beginners: A Clear Guide to Tomorrow (AI, Data Science, Python & Statistics for Beginners)* (London: Scatterplot Press Ltd) p 162

Remya R K, Vidya K R and Wilscy M 2021 Detection of deepfake images created using generative adversarial networks—a review ed M Palesi, L Trajkovic, J Jayakumari and J Jose *2nd Int. Conf. on Networks and Advances in Computational Technologies, NetACT19* (Cham: Springer Nature) pp 25–35

Rothman D 2020 *Artificial Intelligence by Example* (Birmingham: Packt Publishing) p 578

Sakamoto R, Kondo M, Fujita K, Ichimura T and Nakajima K 2020 The effectiveness of low-precision floating arithmetic on numerical codes: a case study on power consumption *HPCAsia'20: Proc. of the Int. Conf. on High Performance Computing in Asia-Pacific Region (Fukuoka, Japan, 17 January)* (New York: Association for Computing Machinery) pp 199–206

SambaNova Systems Whitepaper 2021 Accelerated Computing with a Reconfigurable Dataflow Architecture 2021 SambaNova Systems, Inc., p 10 https://sambanova.ai/hubfs/23945802/SambaNova_Accelerated-Computing-with-a

Software Engineers Blog 2023 (TheGeneralAICo.) https://medium.com/@ibrahimmukherjee/ibms-new-north-pole-chip-is-a-big-deal-here-s-why-6ff6c1e7acf1

Swayne M 2024 Google claims latest quantum experiment would take decades on classical computer *The Quantum Insider* Copyright Resonance Alliance Inc. https://thequantuminsider.com/2023/07/04/google-claims-latest-quantum-experiment-would-take-decades-on-classical-computer/

Tal E, Viljoen N, Coburn J and Levenstein R 2024 Meta: our next-generation meta training and inference accelerator https://ai.meta.com/blog/next-generation-meta-training-inference-accelerator-AI-MTIA/

Theobald S 2021 *Machine Learning for Absolute Beginners: A Plain English Introduction* (London: Scatterplot Press Ltd) p 194

Turing A M 1950 Computing machinery and intelligence , *Mind* **LIX** 433–60

Vahdat A 2024 Compute, Google Cloud: announcing Trillium, the sixth generation of Google Cloud TPU https://cloud.google.com/blog/products/compute/introducing-trillium-6th-gen-tpus

Wassim 2024 Etched Sohu: revolutionizing AI with transformer-specific chips, Medium, https://medium.com/@maxel333/etched-sohu-revolutionizing-ai-with-transformer-specific-chips-4a8661394f49

IOP Publishing

AI-Processor Electronics
Basic technology of artificial intelligence
Vinod Kumar Khanna

Chapter 2

Computing electronics fundamentals

The terminology of 'electronics', 'digital electronics', 'digital and analog signals', 'the electronic computer and its processor' is reviewed. The binary number system and Boolean algebra are introduced. Combinational and sequential logic are distinguished. The seven basic logic gates (NOT, AND, OR, NAND, NOR, XOR, and XNOR) are explained. The MOSFET and CMOS technologies are described. Operations of CMOS-based NOT gate and the universal logic gates (NAND and NOR) are elucidated. Circuits for realization of all the logic gates from the two universal gates are enquired into. The S-R and D flip flops used as memory elements, are made from NAND gates. Half-adder and full-adder circuits for binary addition as well as half-subtractor and full-subtractor circuits for binary subtraction are discussed. Subtraction is exemplified through addition using 1's and 2's complement of the subtrahend. Steps of binary multiplication are illustrated, and a binary multiplier circuit is presented. Binary division process is also explained. Addition, subtraction, multiplication and division are the four fundamental operations serving as the primary structural units of all mathematical computation processes.

2.1 Introduction

This chapter presents the basic concepts of digital computing electronics in a nutshell. The topics covered include the logic circuits used for performing AND, OR, NOT, and combinations of these decision-making operations on relational statements. Arithmetic circuits for carrying out addition, subtraction, multiplication and division on binary numbers are synopsized (Ward 2023). Recalling these ideas and principles will be helpful in understanding the contents of the succeeding chapters.

2.2 Electronic signals and related terms

To start the ball rolling, let us begin with definitions of a few commonly used terms.

2.2.1 Electronics

Electronics is a panoramic scientific and engineering discipline with a monumental scope and far-reaching applications.

On the scientific flank, it is a subfield of physics. It explores and applies the principles of physics to design, fabricate and operate devices that work by manipulating the flow of electrons and other electrically charged particles in vacuum and solids.

On the engineering side, it falls under electrical engineering. It uses passive components such as resistors, capacitors, inductors, transformers and standard PN diodes in conjunction with active devices such as Zener and tunnel diodes with negative differential resistance; photodiodes, LEDs and Schottky diodes; transistors (metal-oxide semiconductor field-effect or bipolar junction variants, insulated gate bipolar transistors), thyristors, triacs, integrated circuits, and vacuum tubes like traveling wave tubes (TVTs), magnetrons, klystrons and gyrotrons:

   (i) To control and amplify electric current.
  (ii) To convert it from direct to alternating current and vice versa.
 (iii) To carry out transformations from analog to digital signals and conversely.
 (iv) To perform a multiplicity of operations on electrical signals of different waveforms. These operations encircle an encyclopedic, thorough and in-depth range of tasks from information storage, processing and communication to power conversion and conditioning. Time-varying voltages and current signals are used in electronics for visualization, registration and transmission of information.

2.2.2 Digital and analog electronics

Digital electronics is a branch of electronics dealing with the study of digital signals involving the design and fabrication of devices, circuits and systems to produce such signals and apply them to make a variety of equipment for domestic and industrial services. Similarly, analog electronics is concerned with continuously variable signals.

2.2.3 Digital signal

A digital signal is a signal used for representing data as a sequence of discrete values (figure 2.1(a)). At any instant of time, it can acquire only one of a finite number of distinct values. A two-level discrete signal representing the OFF and ON states of a switch is widely used in digital computing.

2.2.4 Analog signal

An analog signal is a representation of data using continuous values (figure 2.1(b)). At any given time instant, its value is a real number within a continuous range of values.

2.2.5 Electronic computer and its processor

A digital computer is an electronic machine which accepts raw data, stores and processes it according to the instructions of the user and provides output results as meaningful information. The electronic device used in the computer for processing

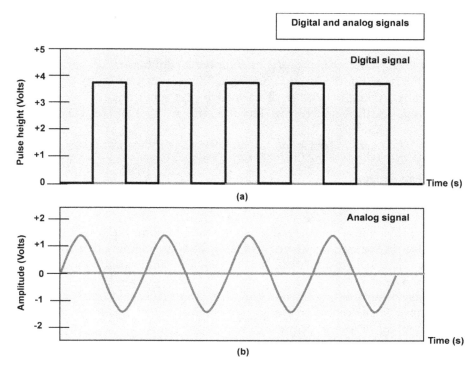

Figure 2.1. Electronic signals: (a) digital and (b) analog. Analog signals are continuous signals, e.g., sine waves while digital signals are non-continuous signals, e.g., square waves. Analog signals are more susceptible to noise while digital signals enable noise-free data processing.

input data is called the processor of the computer. This book is about processor electronics, and particularly about processors that are specially made for execution of artificial intelligence (AI) tasks, i.e., AI processors. Therefore, the success of AI depends primarily on the AI processor. It is no exaggeration to accept that AI processor electronics is the basic technology of AI because the progress of AI is interwoven with developments in AI processing capabilities. Indeed, it is the foundation of AI. We know that AI is a game-changing or disruptive technology that can alter how the people think and work, significantly impacting the way established industries and markets, businesses and society operate (Păvăloaia and Necula 2023). Since AI electronics influences the development and dissemination of AI resources, its relevance cannot be underestimated. Above all, it needs deep appreciation and utmost attention.

Analog AI using analog computing systems for AI is an emerging approach promising faster training speeds and lower energy consumption. It is touted as the ideal approach for AI processing, not only because of its power efficiency but also owing to its potential to unfasten new features in the small edge devices. Analog AI processors can transform the field of AI by removing the limitations of digital processors. They are highly useful in areas in which digital processors encounter problems.

2.3 Binary numbers and Boolean algebra

2.3.1 The binary number system

It is the number system widely used in computers (Bourdillon 1985, Malvino and Brown 1999). This number system has base 2, and uses two digits only, 0 and 1 called the bits. A binary number is formed in the same manner as the decimal number except that there are only two bits: 0, 1, 10, 11, 100, 101, 110, 111..., 1000, ..., 10000, ..., 100000, ...

The two bits 0 and 1 correspond to the OFF and ON states of a switch.

2.3.2 Boolean algebra

This is a branch of algebra which differs from elementary algebra in two primary respects:

   (i) In Boolean algebra, the values of the variables are the truth values. These values are designated as true and false. They are denoted by 1 for true and 0 for false. In elementary algebra, the values of the variables are continuous numbers.

   (ii) Boolean algebra carries out logical operations instead of arithmetic operations. These logical operations are done with the help of logical operators. Examples of logical operators are:

      (a) Negation (NOT, denoted by ←),
      (b) Conjunction (AND, denoted by ∧),
      (c) Disjunction (OR, denoted by ∨), and
      (d) Equivalence (≡).

Contrastingly, elementary algebra performs arithmetic operations. Here, the operators used are arithmetic operators such as addition (+), subtraction (−), multiplication (×), and division (÷). Therefore, Boolean algebra is a formal way of representation of logical operations, in exactly the same way that elementary algebra presents a description of numerical operations.

A Boolean function is also called a switching or truth function. It is defined as a function whose input and output values can acquire values from a set of two elements, viz., true and false. It takes arguments and produces a result from a set of two elements only.

A truth table is a structured representation of all possible combinations of the truth values that are conceivable for the input variables of a Boolean function, and their corresponding output values. It consists of columns for the various possible combinations of the truth values of input variables vis-à-vis rows showing their related output values.

2.4 Binary logic

2.4.1 Logical operations

A logical operation is a defined as a symbol or word connecting two or more phrases (groups of words) of information. Its purpose is to test whether a certain relationship between the phrases is true or false, YES or NO, valid or invalid. Logic operations are used for modelling the flow of information in an electrical circuit. We define the four common logical operations mentioned in section 2.3.2, namely, negation, disjunction, conjunction, implication or equivalence, as follows:

    (i) Negation (Logical Not or Logical Complement): It is represented by the symbol ($\leftarrow$), a dash with a tail ($\leftarrow$A), or an overhead bar sign on a letter ($\bar{A}$). It reverses the truth of a statement, e.g. $\leftarrow$A or $\bar{A}$ = not A means that: if A is true, then $\leftarrow$A or $\bar{A}$ is false. Conversely, if $\leftarrow$A or $\bar{A}$ is true, then A is false.

   (ii) Logical Conjunction or Logical 'AND' Operation: When applied on a set of operands, it produces a value of true if and only if all its operands are true, e.g., A AND B AND C is true if A, B and C are all true.

  (iii) Disjunction (Logical Disjunction or Logical 'OR') Operation: When performed on a set of operands, it produces a value of true if and only if one or more of its operands is true, e.g., A OR B OR C is true if either A is true or B is true or C is true.

  (iv) Implication (Equivalence or Bientailment): The statement 'X implies Y' means that 'if X is true, then B must also be true'. X is the antecedent and Y is the consequent.

2.4.2 Logical operators

A logical operator describes the execution of a specific logical operation between two or more expressions. Typical examples of logic operators are NOT, AND, OR, etc. This operator is represented by a symbol or word connecting two or more expressions. By insertion of this symbol/word, the expressions are linked in such a manner that the value of the resulting compound expression is determined only by the connotation of the inserted word representing the operator, and the values of the original expressions. The NOT, NAND, NOR, AND, OR, XOR, and XNOR are a few logical operators in everyday use.

2.4.3 Combinational and sequential logic circuits

Based on the circuit arrangement and the method adopted for determination of the output of the circuit, digital logic circuits are subdivided into two main classes, viz.,

    (i) Combinational logic circuits, and
   (ii) Sequential logic circuits.

The main distinguishing feature between the two types of circuits is that the output of a combinational circuit is dependent only on its present inputs but the same is not valid for a sequential circuit. A sequential circuit has some of its outputs fed back as inputs to the circuit. Hence, the output of a sequential circuit may depend on its past inputs as well as the present inputs that are supplied for performing a sequence of operations. To re-emphasize, only the present inputs play decisive roles for the output of a combinational circuit. But the output of a sequential circuit is influenced by both present as well as past inputs. Essentially, the sequential circuit is a combinational logic circuit. It consists of input variables (X), computational circuits in the form of logic gates, and output variable (Z).

Sequential circuits are further divided into two genera, namely synchronous and asynchronous types. In a sequential circuit, all changes in state variables are synchronized by the clock signal. As a result, all changes in the logical levels of its components take place simultaneously.

In an asynchronous sequential circuit, the state variables may change at any time. This happens because they are not synchronized by the clock signal.

Most digital logic circuits work in a synchronous fashion. The underlying reason is that synchronous circuit designs are easily created and verified. Nonetheless, asynchronous logic circuits benefit from the non-restriction of their speed by the clock signal. They run at the maximum speed of their logic gates. Although practically all computers are synchronous in nature, asynchronous computers have also been constructed.

An important sequential circuit is a flip-flop or latch. It is a bistable multivibrator with two stable states, SET and RESET. These states conform to binary numbers 0 and 1. The flip-flop is used as a register circuit in a computer. The register is used as a memory device to store information. It has one or more control inputs, and one or more outputs. Its output remains constant until a pulse is applied to its clock input for triggering it to switch to the opposite state.

2.4.4 Basic logic gates

A logic gate is a device or circuit that performs a pre-defined logical operation on one or more binary inputs. The execution of this operation results in returning a single binary output in accordance with its operational definition.

2.4.4.1 NOT gate
This is also called an inverter. The NOT gate (figure 2.2(a)) is a single input/single output gate. It reverses the polarity of the binary input to produce a high output (1) when given a low input (0). Contrarywise, a low output (0) is obtained when given a high input (1).

2.4.4.2 AND gate
This is a logic gate having two or more inputs and one output (figure 2.2(b)). Its output is true (1) only when all inputs are true (1). But when any one of the inputs is false (0), its output is false (0). It acts like a gate with several locks. This gate opens when all the locks are simultaneously opened. However, it remains closed if any one of the locks is closed. The gate implements the logical conjunction.

2.4.4.3 OR gate
The OR gate implements the logical disjunction. It gives true as output if any of its inputs is true (figure 2.2(c)). In other situations, it outputs false.

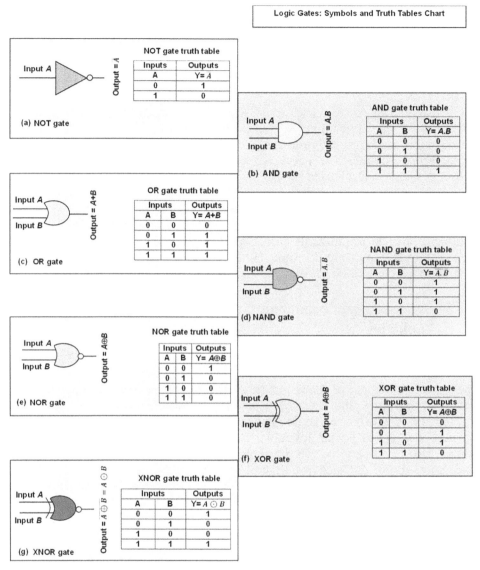

Figure 2.2. Circuit symbols and truth tables of different logic gates: (a) NOT gate, (b) AND gate, (c) OR gate, (d) NAND gate, (e) NOR gate (f) XOR gate and (g) XNOR gate.

2.4.4.4 NAND gate

A NAND gate is the shortened form of NOT-AND gate (figure 2.2(d)). It is an AND gate followed by a NOT gate. Its output is false (0) only if all of its inputs are true (1). So, the graphic symbol for a NAND gate is an AND symbol with a bubble on the output. The bubble indicates that a complement operation has been performed on the output of the AND gate.

2.4.4.5 NOR gate

A NOR gate or NOT-OR gate is a combination of the OR gate with NOT gate (figure 2.2(e)). When all its inputs are 0, the output is 1. When any one of its inputs is 1, the output is 0.

2.4.4.6 XOR gate

This is also referred to as an exclusive OR gate. It is a two-input gate that outputs a logical 1 if any one of its inputs is 1 but both inputs are not simultaneously 1; hence the adjective 'exclusive' (figure 2.2(f)). It is a fundamental building block of cryptographic circuits. The encrypted ciphertext is created by carrying out an XOR operation on a digitized message with a binary key.

2.4.4.7 XNOR gate

It is a combination of two gates, viz., the XOR gate coupled with NOT gate. Its action is equivalent to (an XOR gate + a NOT gate) (figure 2.2(g)). Its output is 1 when both of its inputs are the same, either 0 or 1. Its symbol is the same as that of XOR gate. The only difference is that the complement sign is added.

2.5 Metal-oxide semiconductor field-effect transistor and related technologies

2.5.1 MOSFETs and integrated circuits (ICs)

The MOSFET or the MOS transistor is the key device used to make digital circuits, either as an N-channel or P-channel transistor, in enhancement or depletion-mode configuration, superseding the bipolar junction transistor in many fields. It is a four-terminal device with source, drain, gate and substrate connections (figure 2.3). Halo implants, also called pocket implants, place the dopants a little below the active channel, close to the source and drain regions. These dopant placements are meant to accurately tailor the background doping in the concerned regions. Thereby, they suppress the punch-through breakdown between the source and drain through the bulk substrate. Punch-through breakdown is a failure mechanism that is likely to occur in the short-channel devices. Gate dielectric spacers provide insulation between the gate of the transistor and the source/drain metal contacts. They are also known as the sidewalls.

Silicidation is the process used to create a layer of silicide on top of silicon to reduce the contact resistance. Silicidation is done with metals such Ti, Ni or Co. It prevents penetration of metal into silicon by acting as a barrier between the metal and silicon. Further, it improves thermal stability of the devices.

The fast on–off switching speed of the MOSFET makes it the ideal device for generating pulse trains. These pulse trains constitute the electronic digital signals. Other noteworthy advantages of the MOSFET include high scalability, extraordinary transistor density, easy affordability and low power consumption.

The ICs (figure 2.4), also known as chips or microchips, are low-cost complex electronic circuits. They consist of MOSFETs and associated devices, fabricated on a small substrate of semiconductor material, generally silicon. Technological

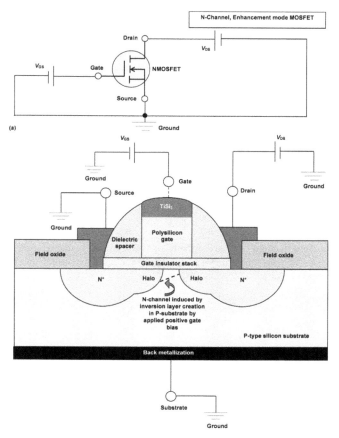

Figure 2.3. The MOS transistor, a ubiquitous component of modern integrated circuits: (a) circuit diagram symbol of N-channel, enhancement mode device, and (b) cross-sectional diagram and biasing voltages V_{GS} (gate–source voltage) and V_{DS} (drain–source voltage).

innovations have led to the placement of an unprecedentedly increasing number of transistors on a single chip. The ICs evolved through several phases of transistor count in a chip representing the integration level (Yeap *et al* 2020). The phases are presented in table 2.1.

2.5.2 Complementary metal-oxide semiconductor (CMOS) technology

The semiconductor technology used in present-day ICs is predominantly CMOS based on the two types of MOS transistors: P-channel MOS (PMOS) and N-channel MOS (NMOS) devices (figure 2.5). In a PMOS transistor, the substrate is an N-type semiconductor with P^+ source and P^+ drain regions. An NMOS transistor has the substrate of a P-type semiconductor with N^+ source and N^+ drain regions. The combination of PMOS and NMOS transistors epitomizes a great flexibility in circuit design. Static power consumption is nearly zero. Power is only consumed during the

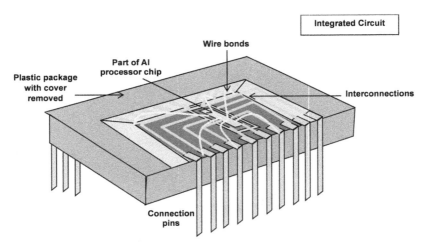

Figure 2.4. Representative diagram of a part of the IC as visible on partly removing the cover. By offering a multitude of advantages, namely, miniaturization, reliability, and energy efficiency, ICs have become the sine qua non for digital infrastructure.

Table 2.1. Integration levels and transistor counts per IC chip.

Sl. No.	Acronym for level of integration	Full form	Number of transistors per chip
1.	SSI	Small-scale integration	$10–10^2$
2.	MSI	Medium-scale integration	$10^2–10^3$
3.	LSI	Large-scale integration	$10^3–10^4$
4.	VLSI	Very large-scale integration	$10^4–10^5$
5.	ULSI	Ultra large-scale integration	$10^5–10^6$
6.	SLSI	Super large-scale integration	$10^6–10^7$
7.	ELSI	Extra large-scale integration	$10^7–10^8$
8.	GSI	Giant scale integration	$>10^8$

short time interval of switching between ON and OFF states. Low power consumption ensures less heat generation during operation of a CMOS circuit.

2.6 CMOS-based NOT gate and universal logic gates

The NAND and NOR gates are considered universal gates. They are so-called because all the logic operations can be implemented by using any one of these two gates. In other words, any logic gate can be created using either NAND gates only or with NOR gates only. Keeping this in view, we shall describe the working of NOT,

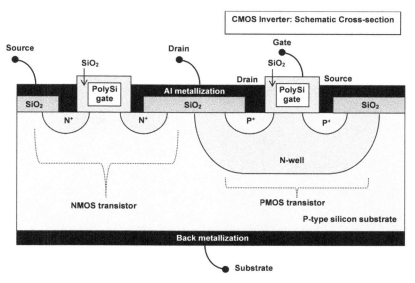

Figure 2.5. A CMOS device. Besides being less power hungry, this dual-transistor technology provides a high noise immunity and allows a high packing density.

NAND and NOR gates in CMOS technology. Then we will discuss the realization of all other gates using these universal gates.

2.6.1 NOT gate

A CMOS NOT gate is shown in figure 2.6. It consists of a P-channel MOSFET Q_1 at the top and N-channel MOSFET Q_2 at the bottom. The gate terminals of transistors Q_1 and Q_2 are tied together. Their drain terminals too are connected together. The gate terminals of the transistors Q_1, Q_2 constitute the input side of the circuit while their drain terminals form its output side. The source terminal of transistor Q_1 is connected to $+V_{DD}$ supply and the source terminal of transistor Q_2 is grounded. The working of the CMOS NOT gate circuit is described in the diagram and its truth table is also shown.

2.6.2 NAND gate

The CMOS NAND gate is shown in figure 2.7. It consists of P-channel MOSFETs Q_1 and Q_2 connected in parallel between $+V_{DD}$ and the output terminal; and N-channel MOSFETs Q_3 and Q_4 connected in series between the output terminal and ground. The operation of the CMOS NAND gate circuit is explained in the diagram, and its truth table is presented alongside.

2.6.3 NOR gate

Figure 2.8 shows a CMOS NOR gate. This gate consists of P-channel MOSFETs Q_1 and Q_2 connected in series between $+V_{DD}$ and the output terminal; and N-channel MOSFETs Q_3 and Q_4 connected in parallel between the output terminal and

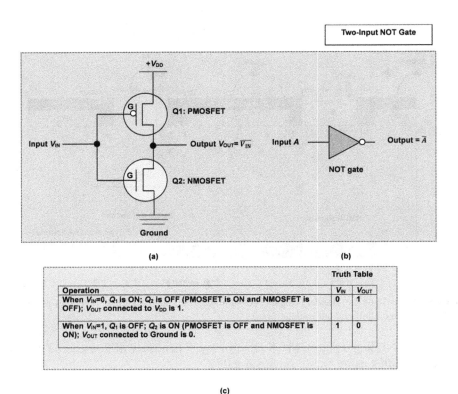

(a)

(b)

(c)

Figure 2.6. A NOT gate fabricated with CMOS technology: (a) circuit diagram, (b) symbol and (c) truth table.

ground. Circuit operation and truth table for the CMOS NOR gate are given in the diagram.

2.7 Realization of all the logic gates from universal gates

2.7.1 All gates from NAND gates

Figure 2.9 shows how the various gates are made using the NAND gate as a basic building block.

2.7.2 All gates from NOR gates

NOR gate forms the rudimentary structural element of the different gates shown in figure 2.10.

2.8 Flip-flops made with NAND gates

2.8.1 S-R flip flop

The set-reset flip-flop or the S-R flip-flop block diagram (figure 2.11(a)) has two inputs: Input 1 = Set input $\bar{S}$ and Input 2 = Reset input $\bar{R}$; and two outputs: Output 1 = Q and Output 2 = $\bar{Q}$.

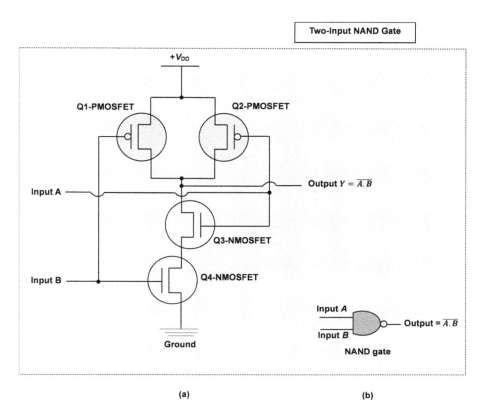

| | | | Two-Input NAND Gate |

(a)

(b)

	Truth Table		
Operation	A	B	Y
When $A=B=0$, Q_1, Q_2 are ON; Q_3, Q_4 are OFF (PMOSFETs are ON and NMOSFETs are OFF); Y connected to V_{DD} is 1.	0	0	1
When $A=0$, $B=1$, Q_2 is ON, Q_3 is OFF; Q_1 is OFF, Q_4 is ON; Y connected to V_{DD} is 1.	0	1	1
When $A=1$, $B=0$, Q_2 is OFF, Q_3 is ON; Q_1 is ON, Q_4 is OFF; Y connected to V_{DD} is 1.	1	0	1
When $A=B=1$, Q_1, Q_2 are OFF; Q_3, Q_4 are ON (PMOSFETs are OFF and NMOSFETs are ON); Y connected to Ground is 0.	1	1	0

(c)

Figure 2.7. A NAND gate in CMOS technology: (a) circuit diagram, (b) symbol and (c) truth table.

The S-R flip-flop truth table (figure 2.11(b)) shows that:
  Sl. No. 1: On applying logic 0 to the $\overline{S}$ input, the Q output is set to logic 1.
  Sl. No. 2: When logic 1 is applied to the $\overline{S}$ input, the Q output still remains at logic 1, portending that logic 1 on Q is remembered.
  Sl. No. 3: On applying logic 0 to the $\overline{R}$ input, the output Q is reset to logic 0.

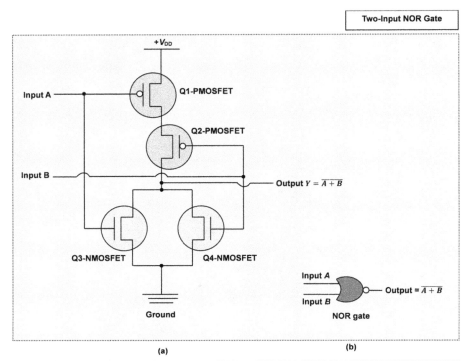

Figure 2.8. A NOR gate in CMOS technology: (a) circuit diagram, (b) symbol and (c) truth table.

Sl. No. 4: When logic 1 is applied to the $\bar{R}$ input, the Q output still remains at logic 0, which heralds that logic 0 on Q is remembered.

Sl. No. 5: When $\bar{S} = \bar{R} = 0$, then $Q = \bar{Q} = 1$. This state is disallowed in practical circuits.

Sl. No. 6: When $\bar{S} = \bar{R} = 1$, then $Q = \bar{Q} = $ Indeterminate. This uncertain state must be avoided.

The S-R flip-flop circuit (figure 2.11(c)) consists of two NAND gates. The output of each NAND gate is connected to one input of the other NAND gate. Hence, a positive feedback or cross-coupling mechanism is established between the gates.

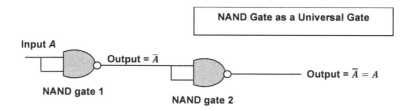

(a) Buffer from NAND gate.

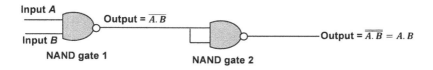

(b) AND gate from NAND gate.

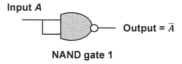

(c) NOT gate from NAND gate.

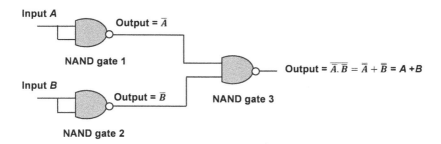

(d) OR gate from NAND gate.

Figure 2.9. Construction of all the logic gates from NAND gates: (a) buffer, (b) AND, (c) NOT, (d) OR, (e) NOR, (f) XOR and (g) XNOR.

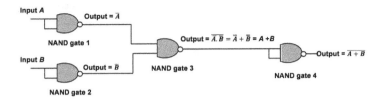

(e) NOR gate from NAND gate.

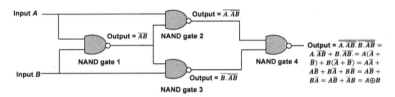

(f) XOR gate from NAND gate.

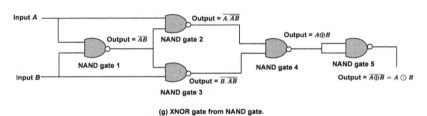

(g) XNOR gate from NAND gate.

Figure 2.9. (Continued.)

2.8.2 D flip-flop

The delay flip-flop or the D flip-flop block diagram (figure 2.12(a)) has two inputs: Input 1= Data input (D) and Input 2 = Clock (CLK); and two outputs: Output 1 = Q and Output 2 = $\bar{Q}$.

The D flip-flop truth table (figure 2.12(b)) shows that:

Sl. No. 1: On applying logic 0 to the D input, and CLK pulse falling towards logic 0, the Q output is set to logic 0 and $\bar{Q}$ output to logic 1.

Sl. No. 2: With logic 0 applied to the D input and CLK pulse rising towards logic 1, the Q output still persists at logic 0.

Sl. No. 3: On applying logic 1 to the D input, and CLK pulse falling towards logic 0, the output Q still lingers at logic 0.

Sl. No. 4: With logic 1 to the D input and CLK pulse rising towards logic 1, the Q output changes state to logic 1. This is the only step involving change of state. It is different from other steps because the D input has changed from logic 0 to logic 1. At the same time, the CLK pulse has risen towards logic 1.

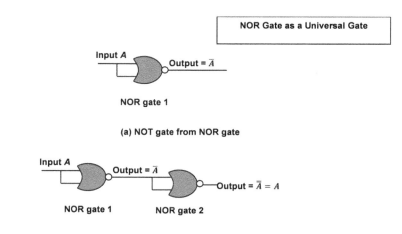

(a) NOT gate from NOR gate

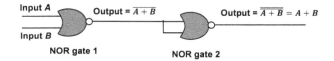

(b) Buffer from NOR gate

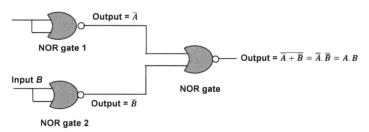

(c) OR gate from NOR gate.

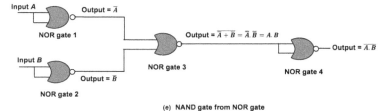

(d) AND gate from NOR gate

(e) NAND gate from NOR gate

Figure 2.10. Construction of all the logic gates from NOR gates: (a) NOT, (b) buffer, (c) OR, (d) AND, (e) NAND, (f) XOR and (g) XNOR.

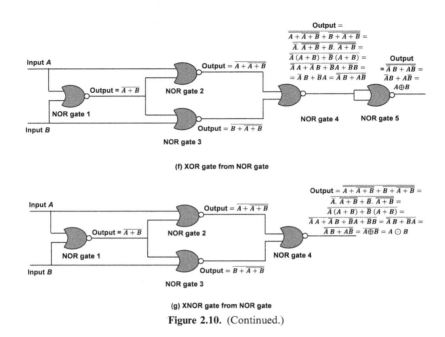

(f) XOR gate from NOR gate

(g) XNOR gate from NOR gate

Figure 2.10. (Continued.)

These steps mean that the output Q always remains the same as the D input and changes state only if the D input changes at the rising edge of the clock. Thus, in a D flip-flop, the output can only be changed if the D input changes at the clock edge. If the D input changes at other times, the output is not altered.

The D flip-flop circuit (figure 2.12(c)) consists of an S-R flip flop with two additional NAND gates, and one NOT gate. It is used in shift registers and counters.

2.9 Binary addition

This is done with an adder. The adder is a digital circuit for addition of numbers. Adders are of two types: half adder, and full adder. They are utilized in making the arithmetic and logic units (ALUs) of a computer.

2.9.1 Half-adder

The half-adder (HA) is a circuit which adds together two binary numbers A and B. It produces two outputs S and C (figure 2.13). Output S is the sum while output C is the carry. It consists of an XOR gate and an AND gate. The XOR gate produces the sum while the AND gate gives the carry. Its shortcoming is that it can add only two inputs. This deficiency limits its use to only those cases in which there is no carry to be added. Existence of a carry will require a third input.

2-18

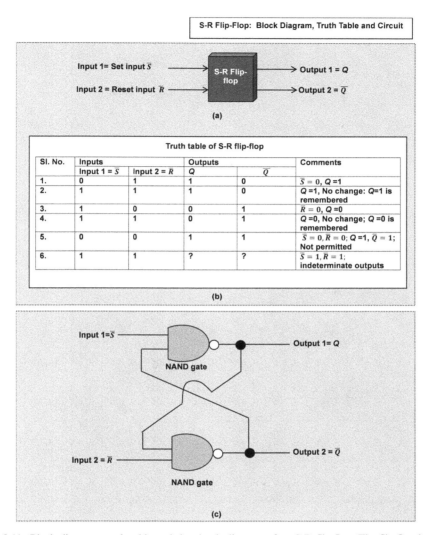

S-R Flip-Flop: Block Diagram, Truth Table and Circuit

(a)

Truth table of S-R flip-flop

Sl. No.	Inputs		Outputs		Comments
	Input 1 = $\bar{S}$	Input 2 = $\bar{R}$	Q	$\bar{Q}$	
1.	0	1	1	0	$\bar{S} = 0$, Q =1
2.	1	1	1	0	Q =1, No change: Q=1 is remembered
3.	1	0	0	1	$\bar{R} = 0$, Q =0
4.	1	1	0	1	Q =0, No change; Q =0 is remembered
5.	0	0	1	1	$\bar{S} = 0, \bar{R} = 0$; Q =1, $\bar{Q} = 1$; Not permitted
6.	1	1	?	?	$\bar{S} = 1, \bar{R} = 1$; indeterminate outputs

(b)

(c)

Figure 2.11. Block diagram, truth table and the circuit diagram of an S-R flip-flop. The flip-flop is used in memory devices such as registers for temporarily storing data.

2.9.2 Full-adder

The full-adder (FA) performs an addition operation on three binary digits (figure 2.14). It consists of three inputs and two outputs. It has two inputs A and B, and a third input, namely, the carry from the previous stage. The carry input is designated as C_{IN}. Its two outputs are the sum S and carry C_{OUT}.

It consists of two HAs and one OR gate. The first HA adds the binary numbers A and B. A partial sum S_P is thus produced. The second HA adds input carry C_{IN} to the partial sum S_P produced by the first HA. This addition yields the final sum S as

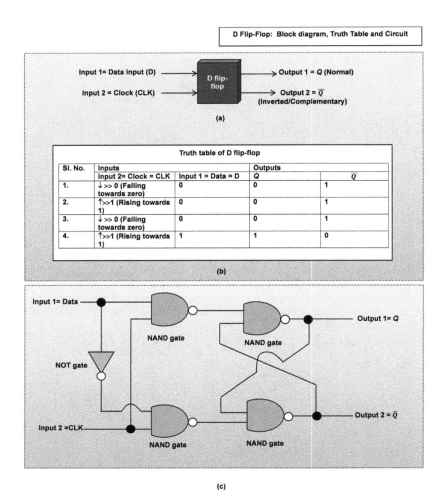

Figure 2.12. Block diagram, truth table and circuit diagram of a D flip-flop. It can store a single bit of data per clock cycle.

the output. If any of the HAs produces a carry, there will be an output carry C_{OUT}. The C_{OUT} will be an OR function of the two HA carry outputs.

2.10 Binary subtraction

2.10.1 Half subtractor

The half-subtractor (HS) block diagram is given in figure 2.15(a). The diagram shows that it has two inputs: input 1 = A, and input 2 = B; and two outputs: output 1 = Difference D and output 2 = Borrow.

The truth table of the HS is displayed in figure 2.15(b). It displays the results of all combinations of 0 and 1 that can possibly occur between the inputs and outputs.

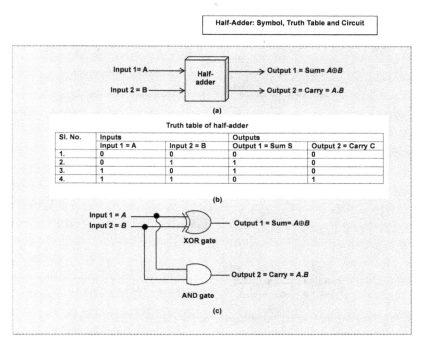

Figure 2.13. A HA. It is so named because it realizes half the function of a full adder.

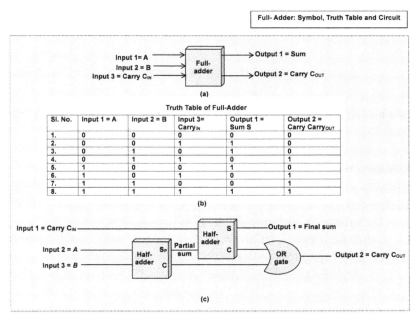

Figure 2.14. An FA. It is a basic component of the arithmetic-logic circuit of the CPU in a digital computer.

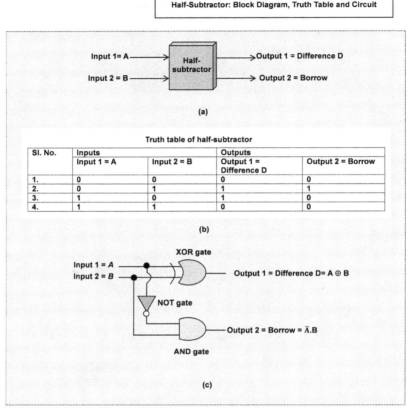

Figure 2.15. Half binary subtractor.

A single-bit full binary subtractor (figure 2.15(c)) is made using XOR, AND with NOT gates.

2.10.2 Full subtractor

The block diagram of a full-subtractor (FS) is shown in figure 2.16(a). The FS has three inputs: input 1 = A, input 2 = B and input 3 = Borrow B_{IN}; and two outputs: output 1 = Difference D and output 2 = Borrow B_{OUT}.

The FS truth table (figure 2.16(b)) presents the results of all possible combinations of 0 and 1 that can take place between the inputs and outputs.

A single-bit full binary subtractor (figure 2.16(c)) is made using two HSs and an OR gate.

2.10.3 Subtraction through addition

Subtraction means addition of a positive number with a negative number. So, it is easy to surmise that subtraction can also be done using adder circuits. Subtraction

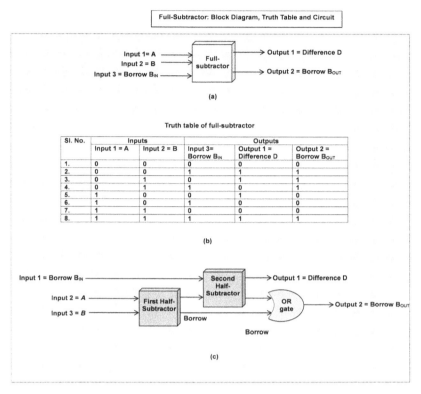

Figure 2.16. Full binary subtractor.

involves two numbers: the subtrahend and the minuend. The subtrahend is the number that is being subtracted from the given number. The minuend is the given number from which subtraction is being done.

2.10.3.1 Subtraction scheme using 1's complement of the subtrahend

The 1's complement of a binary number is the binary number which is obtained by changing every 0 bit to 1 and every 1 bit to 0 in that number. The steps of subtraction procedure are (figure 2.17):

    (i) The minuend A and subtrahend B are identified.

   (ii) The 1's complement of the subtrahend B is found.

  (iii) The determined 1's complement of the subtrahend B is added to the minuend A giving the sum.

  (iv) If there is no carry over the sum, the 1's complement of the sum is calculated. This is the answer, and it is negative.

   (v) If the carry over the sum is 1, then it is dropped and 1 is added to the least significant bit (LSB) of the sum to get the answer.

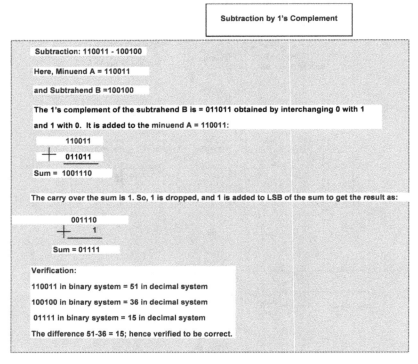

Figure 2.17. Subtraction using 1's complement.

2.10.3.2 Subtraction scheme using 2's complement of the subtrahend
The 2's complement of a given binary number is the binary number which is obtained by taking 1's complement of the binary number followed by adding 1 to the least significant bit of the 1's complement. The steps of subtraction are (figure 2.18):
  (i) The minuend A and subtrahend B are recognized.
  (ii) The 2's complement of the subtrahend B is found.
  (iii) The determined 2's complement of the subtrahend B is added to the minuend A giving the sum.
  (iv) If there is no carry over the sum, the 2's complement of the sum is calculated. This is the answer, and it is negative.
  (v) If the carry over the sum is 1, then it is dropped, and the result is positive.

2.11 Binary multiplication

2.11.1 Steps of binary multiplication

Multiplication involves two numbers: the multiplicand and the multiplier. The multiplicand is the number with which another number is multiplied. The multiplier is the number multiplied to the multiplicand. Multiplication of two binary numbers is done by calculating partial products; these are either 0 or the first number. These partial products are shifted left, and then added together to get the product. The steps of binary multiplication are (figure 2.19):

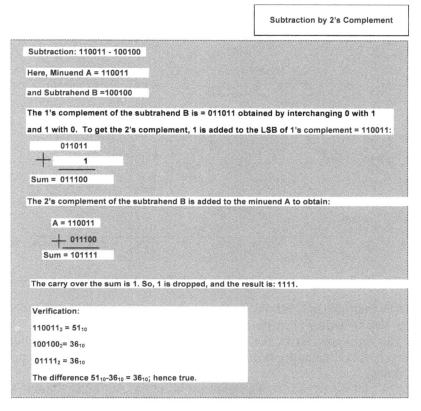

Figure 2.18. Subtraction using 2's complement. It simplifies arithmetic operations involving positive and negative numbers by avoiding the use of subtraction algorithms.

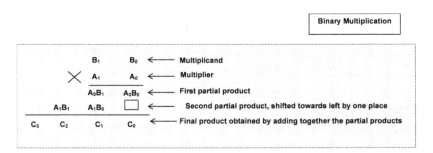

Figure 2.19. An example of binary multiplication. It involves shifting a set of partial products and adding them together.

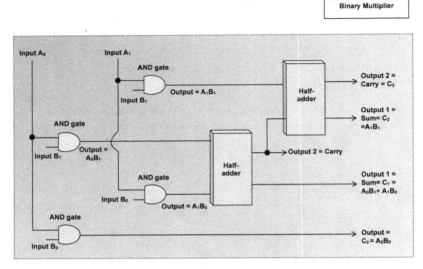

Figure 2.20. An array multiplier circuit. It works on shift-and-add multiplication principle.

(i) The multiplicand and the multiplier are written one below the other in such a way that their rightmost bits are aligned with each other.

(ii) The rightmost bit of the multiplier, which is its LSB, is multiplied with every bit of the multiplicand. The result called the partial product is written down.

(iii) The bit located next to the rightmost bit of the multiplier is multiplied with every bit of the multiplicand. The partial product obtained is written down below the previous partial product after shifting to the left by one position.

(iv) The above process is repeated with the next bit of the multiplier. The partial product is written down below that for the preceding step after shifting to the left by one position.

(v) A similar process is followed for successive bits of the multiplier. The partial product is shifted to the left side by one position each time.

(vi) All the partial products found are added together. Their sum is the product of the two binary numbers.

2.11.2 Binary multiplier

A 2-bit by 2-bit binary multiplier consists of four AND gates and two HAs (figure 2.20). In the diagram, B_1 and B_0 denote the multiplicand bits, A_1 and A_0 the multiplier bits and $C_3C_2C_1C_0$ is the product. Two AND gates are used to form the first partial product by multiplying A_0 by B_1, B_0. Another pair of AND gates is used to form the second partial product by multiplying A_1 by B_1, B_0. This partial product is shifted one position to the left. The above two partial products are added together with two HA circuits to get the final product.

2.12 Binary division

The given number being divided is the dividend. The number dividing the given number is the divisor. The number obtained as the result of division is the quotient. The division steps are (figure 2.21):

(i) The divisor is compared with the dividend. If the divisor < dividend, the quotient is set to 1. The divisor is subtracted from the first part of the dividend and the remainder is written down.

(ii) If the divisor > dividend, the quotient is set to 0. The next bit of the dividend is brought down. The divisor is subtracted from the current part of the dividend. The remainder is written down.

(iii) The process is repeated till the last bit of the dividend is brought down.

2.13 Summary and the way forward

2.13.1 Highlights of chapter 2 at a glance

(i) Electronic computing is largely based on digital electronics, a branch of electronics that performs various tasks using digital signals. The terms

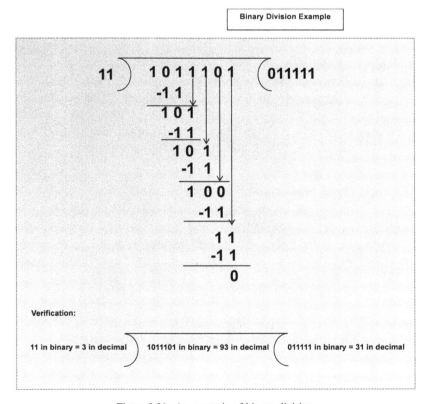

Figure 2.21. An example of binary division.

'electronics', 'digital electronics', 'digital and analog signals', 'electronic computer' and 'processor' were defined.

(ii) The binary number system and Boolean algebra were introduced.

(iii) The digital circuits are designed by using a set of seven basic binary logic gates (NOT, AND, OR, NAND, NOR, XOR, and XNOR). These gates express logical equivalence relationships between two groups of binary numbers following Boolean algebra. Operating in a combinational or sequential fashion, they carry out comparisons of data, shifting of data, data retrieval or storing and data transformation.

(iv) MOSFET and CMOS technologies were described. Operations of the CMOS-based NOT gate and the universal logic gates (NAND and NOR) were elucidated.

(v) The digital circuits for realization of all the logic gates from the two universal gates were enquired into.

(vi) The S-R and D flip flops are used as memory elements. They were constructed from NAND gates.

(vii) The digital circuits are interconnected to implement a variety of arithmetic actions, e.g., addition, subtraction, multiplication and division. The logic gates and arithmetic circuits are used to execute all the complex mathematical operations. These gates and circuits made therefrom lay down the foundations of digital computing.

(viii) HA and FA circuits for binary addition were discussed. An outline of HS and FS circuits for binary subtraction was also given. The procedures for subtraction through addition were described using 1's and 2's complement of the subtrahend. Steps of binary multiplication were illustrated, and a binary multiplier circuit was presented. Binary division process was also explained.

(ix) A few keywords for this chapter are: digital electronics, logic gates, CMOS, flip flops, binary adder, subtractor, multiplier.

2.13.2 Getting ready to begin chapter 3

Chapter 2 provided the arsenal of tools and techniques which is used to construct the processor circuits for AI. The next chapter will show how they cooperate together to perform the job of a central processing unit, the mainstay device of contemporary computers.

References

Bourdillon J F B 1985 *Computer Electronics Made Simple Computer Books* (London: Elsevier/ Heinemann) p 204

Malvino A and Brown J 1999 *Digital Computer Electronics, Glencoe* (New York: McGraw-Hill) p 544

Păvăloaia V D and Necula S C 2023 Artificial intelligence as a disruptive technology—a systematic literature review *Electronics* **12** 1102 1–37

Ward H H 2023 *Mastering Digital Electronics: An Ultimate Guide to Logic Circuits and Advanced Circuitry* (Berkeley, CA: Apress) p 488

Yeap K H, Isa M M and Loh S H 2020 *Introductory Chapter: Integrated Circuit Chip, Integrated Circuits/Microchips* (London: IntechOpen) pp 1–13

IOP Publishing

AI-Processor Electronics
Basic technology of artificial intelligence
Vinod Kumar Khanna

Chapter 3

Central processing unit, and the von Neumann bottleneck

The central processing unit (CPU) is the computing workhorse of conventional digital computers. Basic units inside a CPU are its processing elements known as cores, the two controllers (the memory controller and input/output controller) and the cache along with input/output devices. The components of a core are the control unit, clock, arithmetic-logic unit (ALU), registers (program counter, memory address and memory data registers, current instruction register, and accumulator) and on-chip Level 1 cache along with data, control and address buses. Computer memory hierarchy comprises: Level 0: CPU registers; Level 1: cache memory (static random-access memory, SRAM); Level 2: cache memory (also SRAM); Level 3: primary memory (main memory): dynamic random-access memory (DRAM); Level 4: secondary memory (magnetic hard disk drive); and Level 5: tertiary memory (magnetic tapes or optical disks). Subcategories of computer architecture are: instruction set architecture, microarchitecture and system design. The universally adopted von Neumann architecture is based on the stored program concept involving a fetch-decode-execute cycle. The von Neumann bottleneck is encountered because memory access is slower than the computation speed. Two types of processors are designed, viz., the RISC (reduced instruction set computer) and CISC (complex instruction set computer) processors. A multicore CPU operates on the multiple instructions, multiple data (MIMD) architecture. Threading, multi-threading and hyperthreading in a CPU are defined. The usage scenario of CPUs in artificial intelligence (AI) is discussed. Acceleration technologies in CPUs include Intel® advanced vector and matrix extensions. In the Harvard architecture competing with the von Neumann architecture, the storage and handling of data and instructions are explicitly isolated. The disadvantages of Harvard architecture are elucidated and reasons for the popularity of von Neumann architecture are explained.

doi:10.1088/978-0-7503-6259-7ch3

3.1 Introduction

In the incipient developmental stages, computers required physical rewiring to perform different tasks. These early computers having a program permanently wired inside were aptly called fixed-program computers. The CPU is generally defined as a device for execution of computer programs. Devices that could be called CPUs came with the advent of stored program computers whose concept was proposed by Jon von Neumann in the 1940s.

Intel® 4004 was the first general-purpose programmable commercial processor chip manufacturing launched by Intel Corporation, a flagship company in the semiconductor chip manufacturing and computer industry, with an advertisement in *Electronic News* in November 1971: announcing a new era in integrated electronics, a micro-programmable computer on a chip (Faggin *et al* 1996). Thus, a building block became available to engineers for customization with software to perform different functions in a wide variety of electronic devices.

Over the years, the form, design and size of CPU has changed but its fundamental operation remains the same. This chapter looks into the essential parts, structural organization and working of the CPU. Its success story is elaborated. Its short-comings are also highlighted.

3.2 The conventional computing workhorse

Turning the pages of computer history, we find that for a long time, prior to the revolutionary developments in the field of AI, traditional computing by a digital computer (figure 3.1) relied heavily on its workhorse that is most admired and respectfully called the 'central processing unit or CPU' (figure 3.2). It was and is even today the key processing component of a computing system. It is a general-purpose processor that serves innumerable applications. Its applications encompass a very extensive and varied empire ranging from the simple word processing on a laptop, performing online money transfers and bank transactions, directing machines in factories for industrial process control, processing of images and video to the highly complicated equipment for controlling airplane and rocket engines, and so forth. The CPU processes data and instructions, and controls the entire computer system. A system having a single CPU, is designated as a uniprocessor system while a system using many processors is a multiprocessor system (Yadin 2016, Wilson 2024).

3.3 Basic units inside the CPU

More basic units inside a CPU are its processing elements known as cores (figure 3.3). A CPU usually consists of several cores; such a CPU is a multicore CPU. A CPU can also be made with a single core; it is a single-core CPU. The core (s) are given the responsibility of execution of computer programs and assigned several other duties. Besides the processing core(s), the other components of the CPU are the two controllers (the memory controller and input/output controller) and the cache. Clearly distinguishing the CPU from the core, we note that the core comprises the control unit, the clock, the arithmetic-logic unit and memory in the

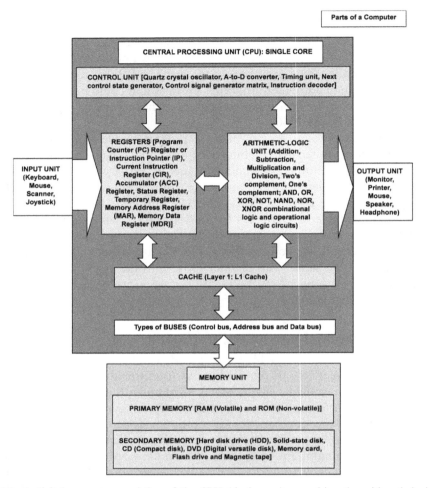

Figure 3.1. A digital computer consisting of the CPU (single core) comprising the arithmetic-logic unit, registers, control unit, L1 cache, buses; together with the external memory unit (primary and secondary memory), and input/output units. Directions of data/instruction flow are shown. The digital computer shown in figure 3.1 must be differentiated from the CPU shown in figure 3.2 and the core shown in figure 3.3. The terms 'computer', 'CPU' and 'core' must be clearly distinguished.

form of cache and registers. Thus, the processing core is a more fundamental building block of a computing system than the CPU.

3.4 The processing core

As already mentioned, the components of a core are:

(i) Control unit: The control circuits inside this unit coordinate and control the flow of data inside the CPU. They fetch data and instructions, decode the instructions and execute them on the given data. In other words, they interpret the instructions and activate the necessary circuits within the

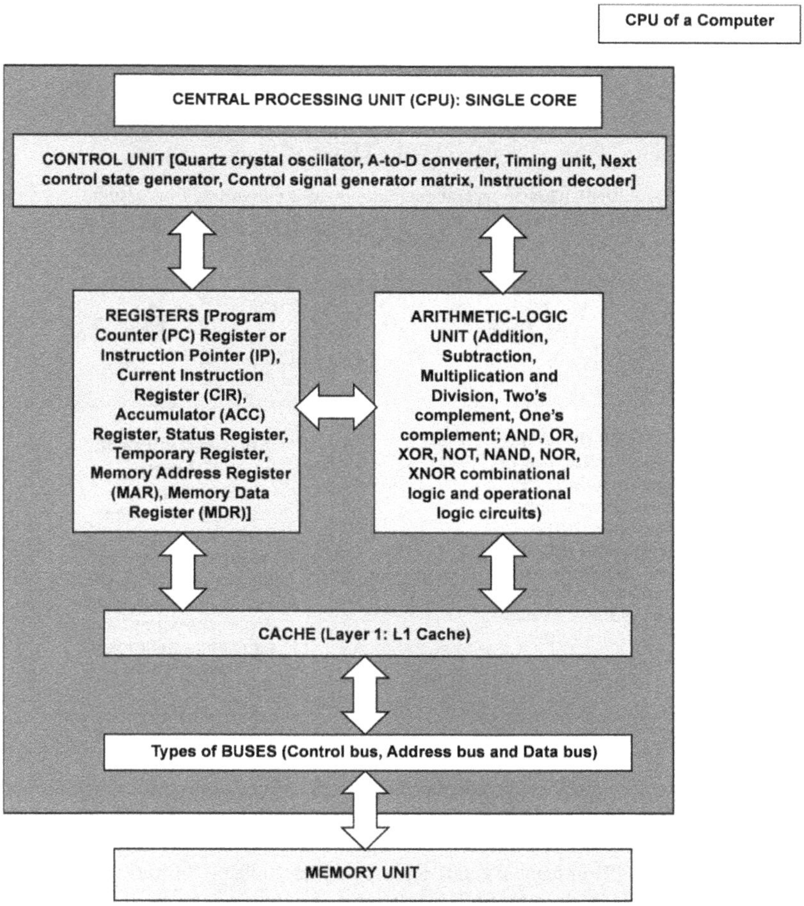

Figure 3.2. The CPU (single core) of a digital computer consisting of the arithmetic-logic unit, registers, control unit, L1 cache, and buses. External memory unit is also shown, and the directions of data/instruction flow are indicated by arrows. The reader must distinguish between the CPU shown in figure 3.2, the digital computer shown in figure 3.1 and the core shown in figure 3.3. Differences among the terms 'CPU', 'computer' and 'core' must be clearly understood.

ALU of the core to ensure their correct execution. Combinational logic in the control unit produces the appropriate control signals corresponding to the recent instruction (Harris and Harris 2015).

(ii) Clock: The clock is used along with the control unit for coordinating all the components of the core by regular transmission of an electrical pulse to keep all of them in synchronization.

(iii) Arithmetic-logic unit: This performs arithmetic (add, subtract, multiply, divide) and logic (decision-making NOT, ADD, OR, etc) operations. An ALU performs tasks like bitwise operations, addition, subtraction, and comparisons on receiving inputs. It produces outputs according to the

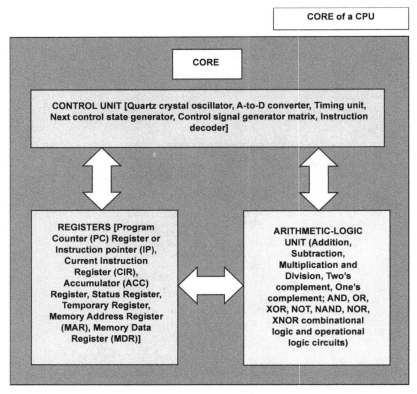

Figure 3.3. The processing core of a CPU in a digital computer consisting of the arithmetic-logic unit, registers and control unit. Two-way communication occurs between arithmetic-logic unit and registers. Bidirectional linkages also exist between the ALU/ control unit and registers/control unit. The difference between the core shown in figure 3.3 and the CPU in figure 3.2 and digital computer in figure 3.1 must be noted. Demarcation among the terms, 'core', 'CPU' and 'computer' must be made.

instructions fed. Figure 3.4 shows the circuit symbolic representation and bus organization for the ALU.

The main components of an ALU are adders, subtractors, logic and control circuits. Calculations are done using binary numbers. The adder circuits add two binary numbers bit by bit. They consider carry-over from the previous bit. For subtraction, complement circuits are used. The negative value is added to the original number and any overflow is discarded. Multiplication involves repeated addition. Division entails repeated subtraction.

The logic gates in the ALU take inputs to produce outputs based on predefined rules. Any data manipulations are carried out with logic gates.

(iv) Registers: These are miniscule memory units for storing small quantities of data of immediate use, e.g., the instruction being decoded, the address of the next instruction to be executed, and the intermediate results of

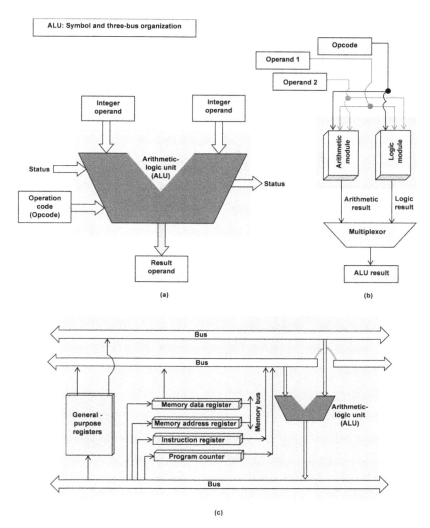

Figure 3.4. ALU and its three-bus organization: (a) symbolic diagram of ALU and its input/output signal. Opcodes are the portions of an instruction specifying the operation to be carried out, e.g., addition, subtraction, etc. Operands are the data/addresses on which opcode works by providing the memory location to be used by the processor. Status input informs about a previous operation. Status output talks about the current instruction. The data flow from top downwards. The control signal enters from left. The status signal enters from left and exits from right. (b) Flow diagram and (c) the three-bus arrangement of ALU.

calculations. Registers are named according to their roles in computing, e.g., program counter (PC), memory address register (MAR), memory data register (MDR), current instruction register (CIR), and accumulator (ACC). The contents of the different registers are: the PC register contains the address of the next instruction to be executed, MAR holds the address of the data to be accessed, MDR holds the data being transferred to/from memory, CIR contains the instruction being currently processed and the ACC holds intermediate arithmetic and logic results.

    (i) Level 1 Cache: This is a small amount of random-access memory built directly on the chip for temporary storage of data and instructions that are likely to be reused. It enables faster processing than that possible through waiting for data/instruction fetching from main memory.

    (ii) Buses: These are high-speed internal connections between different components of the core for sending address, data and control signals. They are called address, data and control buses depending on the purpose of their usage. The data bus is bidirectional while the address and control buses are unidirectional.

3.5 Components of the CPU

3.5.1 Main components

To emphasize what we have already said, the CPU has four main components:

    (i) Processing cores: Each core has all the components given in section 3.4.

    (ii) Memory controller: This is a controller for accessing memory. It serves all the cores.

    (iii) Input/output controller: This is a controller for data input or output for all the cores.

    (iv) Level 2 or Level 3 Cache: They store data and instructions that are frequently accessed.

3.5.2 Input/output devices

An input device is used for entering data into the computer system in the form of textual matter, images, video or audio signals. An output device delivers the output to the user in a friendly form, e.g., as a printed document, an image, a sound or video signal.

3.5.2.1 Input devices

Keyboard: This is used for entering text for word processing.

    Mouse: This is used for controlling the cursor on the screen of the display unit to activate various available features.

    Microphone: This is used for giving voice commands.

    Scanner: This converts the images of objects into digital signals.

3.5.2.2 Output devices

Video monitor: This displays images by converting electrical signals into optical signals. The optical signals are observed as tiny dots of light on the screen called pixels.

    Printer: This reproduces text or images on paper using a laser or ink jets.

    Speaker: This converts electrical signals into sound signals for human hearing.

3.6 CPU cache, and the computer memory hierarchy

3.6.1 CPU cache memory

Cache memory is a hardware component located closer to the processor core than the main memory. It is a faster, smaller memory that stores copies of data from frequently used locations in the main memory. These are the recently used instructions or program codes. The aim of cache is to speed up future requests for these data by minimizing the time needed for accessing data. Most CPUs have a hierarchy of cache levels: level-1, level-2, and so on. The cache is implemented with static random-access memory (SRAM). The SRAM is one of the two types of RAMs used. The other type is the dynamic RAM (DRAM).

3.6.2 Random-access memory

RAM (random-access memory) is a short-term electronic memory located close to the processor that can be read and altered in any order. It is used for storing the working data and machine code. It is a semiconductor memory usually made as an integrated circuit chip.

3.6.2.1 SRAM

SRAM is CPU cache memory. It is a chip-based component that is directly accessible by the CPU for temporarily storing the data being currently processed by it. Here, transistors are used for data storage. SRAM is a low-density device in which data are stored as voltages. SRAM is made of bistable circuits called flip-flops discussed in section 2.8. On applying a digital signal to a flip-flop, it remembers its state unless it is instructed to switch to another state. The ability to retain/recall a binary state enables its utilization as a memory device in digital circuits.

Its advantages are provision of fast and efficient data access, low power consumption and smaller heat generation. SRAM need not be periodically refreshed; hence it is referred to as 'static memory'. However, it requires a continuous power flow. The data is lost as soon as the computer is switched off.

3.6.2.2 DRAM

DRAM (dynamic random-access memory) is used for the main memory of a computer. DRAM is a high-density device in which data is stored in the form of energy. DRAM consists of a two-dimensional grid of memory cells. The memory cells comprise a metal-oxide-semiconductor (MOS) transistor and a small capacitor. Here, the data are stored as charges on capacitors. The capacitors leak away charge with time and get discharged. So, the DRAM must be periodically refreshed; hence it is a dynamic memory. The DRAM is:
   (i) less expensive than SRAM,
   (ii) slower than SRAM,
   (iii) less resistant to radiation than SRAM,
   (iv) has higher latency or delay than SRAM,
   (v) consumes more power than SRAM, and
   (vi) produces more heat than SRAM.

Both SRAM and DRAM are volatile memories. They require that power be continuously kept ON for information storage.

3.6.3 Hierarchy in memory

The memory hierarchy of a computer has the appearance of a pyramid structure with different levels of memory (figure 3.5).

Level 0: CPU registers: The registers are made inside the CPU by using flip-flops. They have the least access time. But they are the most expensive memory components and also the smallest in size (kilobytes).

Level 1: Cache memory: This memory is used for storing the segments of a program that are frequently accessed by the processor. It is a small-size memory. It is generally made by using SRAM. It is less expensive than register memory.

Level 2: Cache memory: This is also made from SRAM but is larger and slower than level 1 cache.

Level 3: Primary memory (main memory): This directly communicates with the CPU and with auxiliary memory devices through an I/O processor.

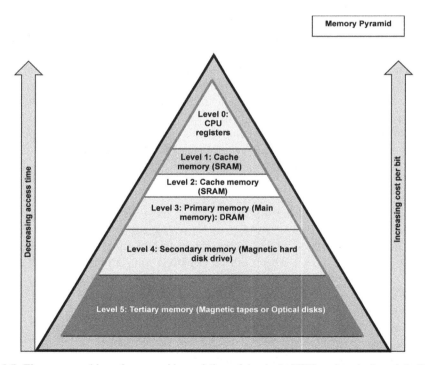

Figure 3.5. The memory hierarchy pyramid consisting of levels 0 (CPU registers), 1 and 2 (SRAM), 3 (DRAM), 4 (hard disk drives) and 5 (magnetic tapes/optical disks). On moving from level 5 to level 1, the memory access time decreases and the cost per bit increases.

This memory is less expensive than cache memory. It is larger in size (gigabytes) than cache memory, and is made with DRAM.

Level 4: Secondary memory (magnetic hard disk drive): This is used as a backup storage. This is a non-volatile memory. It is cheaper than the main memory and larger in size (a few TB). It stores/retrieves digital data using magnetic recording. In magnetic recording, different patterns of magnetization in a material are used. The hard disk drive has one or more rigid rapidly rotating platters coated with a magnetic material and paired with magnetic heads. The heads are arranged on a moving actuator arm for reading and writing data to the platter surface. Individual blocks of data are stored and retrieved in any order providing random storage and access.

Level 5: Tertiary memory (magnetic tapes or optical disks): This is used for storing removable files. It is the memory which is cheapest in price and largest in size (1–20 TB).

3.7 Computer architecture

3.7.1 Defining architecture

Computer architecture is the end-to-end structure of the system. It includes the set of rules and protocols governing the interaction between software and hardware. Compliance with these rules is critical to management of the functionality and execution of various tasks and operations on a computer. Clarifying further, architecture is a description of the structure of a computer system made from its component parts. It deals with the details of interconnectivity of the hardware devices in the computer. It also describes the modes of data transfer and processing for making arrangement of the registers, logic circuits for calculations, buses and other parts of the computer in the most effective way for an explicit objective. Frequently, a high-level description is provided. This description ignores details of particulars of the implementation.

Computer architecture defines the system in an abstract and conceptual manner. It seeks to elucidate the functionality and implementation of the system. It is engrossed in the attributes of a computer system that are visibly discernible to the user. These attributes directly impact the logical execution of a program. Noteworthy among these attributes are the addressing techniques, instruction sets, and the bits used for data.

A computer architect creates a computing system or improves its performance in order that it performs efficiently and effectively. In this specialized engineering activity, a lot of valuable time and effort are expended. This activity is undertaken to increase the speed of computation, reduce the cost, minimize power consumption, and boost immunity to programming errors.

3.7.2 Computer organization

Computer architecture is differentiated from computer organization. The computer organization deals with the realization of the abstract model to build the system in accordance with the specifications laid down by the computer architecture.

The realization is achieved by examining the way in which the system is structured and the linkages of its different operational attributes to fulfil the desired goals.

3.7.3 Subcategories of computer architecture

3.7.3.1 Instruction set architecture

The instruction set architecture (ISA) is a part of the abstract model of a computer. It defines the ways in which the CPU of the computer is controlled by the software. In a computer, the software is converted to machine instructions using a compiler/interpreter. The instructions are executed using hardware. The ISA elucidates the software/hardware interface in terms of the machine codes supported, the address and data formats used, the word size, process registers, and the hardware employed for managing the main memory. Features of interest are the memory consistency, addressing modes and virtual memory as well as the input/output models of implementation. A functional definition of storage locations (registers and memory) is given along with the arithmetic and logic operations of addition, multiplication, branching, loading, storing, etc inclusive of the explicit methods of invoking and addressing them. The ISA is classified into three types: stack, accumulator, and register-set architecture. This classification is based on the locations where operands are stored and the manner in which they are named, explicitly or implicitly.

3.7.3.2 Microarchitecture

This is also known as computer organization. It describes the method of implementation of an ISA in a particular processor. The microarchitecture explains the constituent parts of the processor by clarifying their interconnection and interoperation for implementing the ISA. The microarchitecture is represented by dataflow diagrams. These diagrams display the structural components of the system, e.g., the gates, registers, and ALUs. Generally the data and control paths are separated.

3.7.3.3 System design

This includes the entire physical hardware. The hardware consists of the data processors, and multiprocessors to define how the machine will meet the user requirements, e.g. the interfaces used, data management techniques applied, and so on.

3.8 von Neumann or Princeton architecture

3.8.1 Stored program concept

The von Neumann architecture, also called the Princeton architecture, works on the stored program concept. This concept works with a fetch-decode-execute cycle (Godfrey and Hendry 1993). A single shared memory is used for both programs and data. Programs are simply data. Hence, they should be stored in the same way as data. The data and machine code instructions are stored as binary digits in a single memory. Machine code instructions are fetched from the memory, one instruction at time. Then they are decoded and executed by the processor. Subsequently, the processor cycles around to bring the next instruction. The same procedure is repeated for this instruction. The cycle continues serially. However, it stops when all the instructions have been fetched.

3.8.2 Fetch-decode-execute cycle

The steps below (figure 3.6) are defined considering that the CPU contains only a single core.

(i) The memory address in the PC is copied into MAR.

(ii) The memory address in the PC is modified/incremented to the address of the next instruction to be fetched.

(iii) A signal is sent to the memory address in MAR. This signal is conveyed along the address bus.

(iv) The instruction or data in the memory address is sent to MDR. This signal is transmitted along the data bus.

(v) The instruction or data in MDR is copied to CIR.

(vi) The instruction or data in CIR is decoded, and executed.

(vii) The result is stored in ACC.

(viii) The cycle goes back to step (i).

A multicore processor will be discussed in section 3.10.

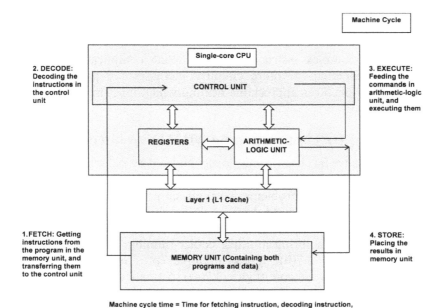

Figure 3.6. The fetch-decode-execute-store cycle running in a digital computer with a single-core CPU. The steps in this cycle include: fetching instructions form the memory to the control unit, translating them into commands in the control unit, executing the commands in arithmetic-logic unit and transferring the results to memory. The time taken in carrying out these steps is the machine cycle time, which may take several clock cycles for completion.

3.8.3 von Neumann bottleneck

Let us relook at the operation of the CPU. A CPU loads instructions and values from memory. It performs a calculation on the values. Then it stores the result of every calculation back in memory. The instructions can be executed only one at a time and in a sequential manner. Therefore, a limitation is imposed on throughput by data transfer between the CPU and the memory. The limitation is unavoidable because of the shared bus between program memory and data memory. Memory access speed is lower than the calculation speed. The bandwidth of data transfer (maximum amount of data transferred in a given time interval, measured in MBs^{-1}) between CPU and RAM is much lower than the speed at which the CPU is able to process data. The result is that the processor remains idle during the time of memory access. The CPU throughput limitation happening is this way is called the von Neumann bottleneck. It is the Achilles heel of the CPU.

3.9 RISC and CISC processors

3.9.1 RISC (reduced instruction set computer) processor

This is capable of processing a restricted number of simple instructions. Each instruction is broken down into simpler instructions. The simpler instructions can be rapidly executed using less circuitry. The process consumes low power, and generates less heat avoiding the need for dedicated cooling.

3.9.2 CISC (complex instruction set computer) processor

This can execute complex instructions to complete complicated tasks with few instructions. Requirement of more circuitry increases the power consumption. Increase in power produces appreciable heating which must be dissipated through a heat sink.

3.10 Multi-core CPU operation on MIMD architecture

Processor performance has been continuously boosted by reducing the chip size and placing more transistors on the smaller chip. Increased power consumption and heat generation are two critical problems worrying chip designers. Consequently, the performance improvement slowed down. A faster chip can be fabricated but it will occupy a larger die area. To come out of this deadlock, the semiconductor industry diverted itself to manufacturing multiple cooler-running, superior energy-efficient processing cores (figure 3.7(b)). These multiple cores replace the single highly powerful, hot-running single core (figure 3.7(a)). A dual-core processor is 1.5 times faster than a comparable single-core processor (Geer 2005). Semiconductor chip manufactures jumped on the bandwagon, and are relinquishing the single fast processor model to move towards the multicore processor paradigm (Hennessy and Patterson 2017).

A multicore CPU is one having more than one processing core on a single chip contained in a single package, e.g., a two-core processor (dual-core), a three-core processor (tri-core), a four-core processor (quad-core) and so forth. It falls under the

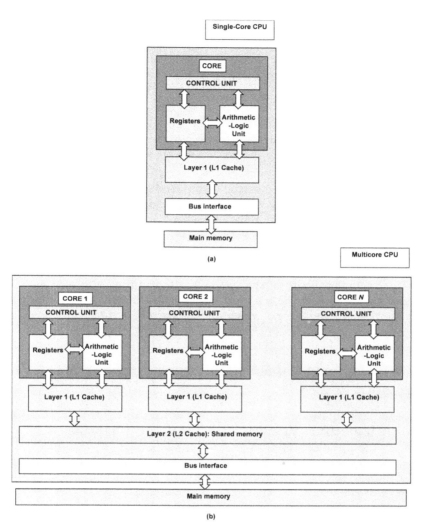

Figure 3.7. Single and multiple-core CPUs: (a) single-core and (b) multicore. In (a), the single core contains the control unit, registers and arithmetic-logic unit. The single core along with the LI cache and bus interface constitutes the CPU. In (b), there are N cores, each of which contains its individual control unit, registers and arithmetic-logic unit. Each of these N cores has its own separate L1 cache. The L1 caches of the N cores are joined to and hence share a single L2 cache. The bus interface lies between the shared L2 cache and the external memory.

category of MIMD architecture. The different cores of this processor execute different threads (multiple instructions) while operating on different parts of the memory (multiple data).

The MIMD architecture will be discussed in chapter 4, section 4.2.4 together with other architectures under Flynn's classification. It works in shared and distributed memory styles. In the shared memory MIMD architecture, all the cores can directly access the whole memory. In the distributed memory MIMD architecture, each core has its own individual memory. In the hybrid architecture, both shared and distributed memory types are combined together. The cores are similar in a homogeneous or symmetric system. They are dissimilar in a heterogeneous or asymmetric system.

A primary advantage of multicore processors is multitasking, i.e., the ability to execute several tasks simultaneously. Each core works on its separate set of instructions. This division of work enables parallel processing and thereby faster task execution. Hence, a multicore processor accomplishes more work during each clock cycle. It can therefore be designed for lower-frequency operation. A higher clock speed causes transistors to switch faster producing more heat and consuming more power. The lower-frequency multicore processor design has a relaxed cooling requirement as compared to the higher frequency single-core counterpart.

To derive the benefits of multicore processors, the programmers must create multithreaded software. This is done by subdividing the work into segments that take approximately similar execution times taking into account the speed of each core and its memory-access capability. While the multiple threads will drive the numerous cores, it must be ensured that these threads communicate among themselves with the best time sequence to assimilate the outputs of multiple simultaneous tasks together.

3.11 Threading, multithreading and hyperthreading in a CPU

3.11.1 Threading

A CPU executes lightweight sets of instructions called threads. A thread is a sequence of instructions supplied to the CPU by a program or application. Threading is the technique used to produce threads.

3.11.2 Multithreading

The larger the number of threads that a CPU can execute concurrently, the greater the number of tasks it can complete. This is done by a process called multithreading. It entails subdivision of a single process into many sub-processes that can be performed at the same time independently. Multithreading enables several threads to run simultaneously on one or more cores in a single processor. Thereby, it increases the number of tasks executed in any given period of time to achieve quicker response times, reduce latency and utilize the resources more efficiently (Nanos 2024).

Multithreading is done at the software or hardware level. Several threads can be executed by multithread programming with the help of the operating system. However, multithreading requires several physical cores. It consumes more processing power, is more expensive and may not be available in all systems.

Multithreading is done for tasks that do not require data dependencies, e.g., desktop applications such as file input/output, web browsing, networking, and image processing.

3.11.3 Hyperthreading

Hyperthreading further boosts the performance of processors. It allows them to execute two or more threads concurrently. It is done at the hardware level by dividing the single physical CPU into two or more virtualized CPU cores. The virtual CPU cores work independently to execute multiple threads simultaneously. Unlike multi-threading, which is done on a single processor or more than one processor, hyper-threading is done on a single processor. It does not yield as much processing capacity as multithreading. Contention and cache thrashing are likely to occur.

Hyperthreading is done for tasks such as scientific and mathematical simulations, video editing and 3D rendering.

3.12 Usage scenario of CPUs in AI

As long as it is necessary to process one problem at a time, CPUs are able to do complex mathematical calculations briskly. But no sooner than multitasking is required, they tend to slow down. Therefore, use of CPUs is restricted to the narrow, specialized AI workloads. Their applications include only those algorithm-intensive tasks where assistance of parallel processing is not sought. A few noteworthy instances may be mentioned:

(i) Running small-scale deep-learning tasks of neural network operations, e.g., executing inference for lightweight and efficient models. Model training is done by a few companies while inference is run by millions of users. Inference jobs obviously do not impose heavy hardware needs compared to their magnitudes during training. In some situations, CPUs can meet the inference requirements. NNs can be compressed and power-efficient models can be formulated. These models can run across unspecialized hardware platforms. While a heavy-duty processor is essential for model training, a lightweight processor can handle the inference job. CPUs can fill such slots of applications.

(ii) Running real-time inference and machine learning (ML) algorithms that are not easily amenable to parallelization.

(iii) Executing recurrent neural networks (RNNs) based on sequential data. RNNs are adroit at handling sequential data, e.g., time series or textual data enabling them to understand context and order. The context and order are crucial for applications in language processing, where the meaning of a word depends on the preceding words. RNNs capture this dependency very effectively.

3.13 CPU for AI or ML work

A powerful CPU is required having a large number of cores (32–64) with 16-core minimal and a high clock speed ~5.6 GHz. A large number peripheral component interconnect express (PCI-e) lanes ~16–32 is necessary. Advantages of a large number of cores, a high clock speed and a large number of PCI-e lanes call for more elaboration. A CPU with more cores can process more instructions in a given period of time. However, if a core has to wait for the result from another core to process the next phase of instruction, a delay may occur. The faster the clock speed, the greater the number of fetch-decode-execute cycles, and hence the number of instructions processed in a given time interval. A larger number of PCI-e lanes gives a higher data transfer rate.

3.13.1 Intel Xeon and Core processors

The Xeon and Core series of processors are powering consumer-level and industrial-grade servers and workstations assiduously. They are also powering the military and commercial systems through and through. The two series have distinct functions. Xeon CPUs are used extensively in enterprise servers and workstations and Core CPUs in mainstream consumer PCs (Techbuyer 2024).

3.13.1.1 Intel® Xeon® 6 processors

The innovative P-core and E-core microarchitectures of the Intel Xeon 6 processor adopt a two-core strategy:

    (i) the larger, performance-driven cores (P-cores) handle compute-intensive and AI workloads, while

    (ii) the smaller, energy-efficient cores (E-cores) take care of high-density and scale-out workloads in the background.

The two types of cores work with the same instruction sets, BIOS (basic input/output system), and built-in I/O accelerators supporting the same drivers and operating systems.

    The P-cores and E-cores are designed with different aims in mind:

    (i) Performance-cores (P-cores): These are hyperthreaded cores with built-in matrix and vector engines. They are proficient in handling compute-intensive AI workloads. With the flexibility to scale up to 128 cores, 12 memory channels, and 96 PCIe lanes per socket, they are able to meet varying application requirements. For alleviating memory bandwidth bottlenecks, innovative multiplexer combined rank (MCR) dual In-Line memory module (DIMM) delivers up to 8800 mega transfers per second (MTs^{-1}). Intel® advanced matrix extensions (Intel® AMX), support INT8, BF16, and FP16 (new) data types. Therefore, P-cores are suitable for several AI models ranging from object detection to midsize Gen AI (Intel® Processors-1).

    (ii) Efficient-cores (E-cores): These cores contain single-threaded cores out-spreading to parallel workloads. With up to 288 cores, they deliver a high core density and extraordinary performance per watt. These capabilities make them suitable for high-throughput cloud-scale workloads. They have high memory bandwidth and rich I/O with up to 12 channels of DDR5, up to 64 lanes of compute express link (CXL) 2.0, and up to 96 lanes of PCIe 5.0. For integrating AI inferencing, they have Intel® advanced vector extensions 2 (Intel® AVX2) with support for both vector NN instructions (VNNI)/INT8 and BF16/FP16. They omit matrix engines in favor of higher, single-threaded core density (Intel® Processors-2).

Example: Intel® Xeon® 6780E Processor: Lithography: Intel 3, cores 144, threads 144, maximum turbo frequency: 3 GHz, processor base frequency: 2.2 GHz, cache 108 MB, maximum memory size 1 TB, TDP 330 W (Intel® Processors-3). Turbo frequency is the maximum clock speed at which a processor can operate under a heavy workload.

3.13.1.2 Intel i9 core processor

14th generation Intel core i9-13900K processor having (8 P-cores + 16 E-cores = 24 total cores) with 32 processor threads, 5.8 GHz maximum turbo frequency, Intel® Smart Cache (L3) Size: 36 MB, Total L2 Cache Size: 32 MB, maximum memory bandwidth: 89.6 GB s^{-1}, lithography Intel 7, base power: 125 W, maximum turbo power: 253 W (Intel® Processors-4). Turbo power of a processor is the same as its turbo frequency; it is also called turbo boost or maximum boost clock.

3.13.2 AMD Ryzen™ 9 7950X desktop processor

This processor has 16 cores, 32 threads, base clock 4.5 GHz, maximum boost clock up to 5.7 GHz, L1 cache: 1024 kB, L2 cache: 16 MB, L3 cache: 64 MB, TSMC 5 nm FinFET technology, Default TDP: 170 W, Liquid cooler for optimal performance (AMD Ryzen Processors 2024).

3.14 Acceleration technologies in CPUs

Hardware-acceleration techniques are built into CPUs, e.g., Intel® Advanced Vector Extensions (Intel® AVX) and Intel® Advanced Matrix Extensions (Intel® AMX) (Mongkolsmai © Intel Corporation).

3.14.1 Intel® AVX

In Intel® AVX, 16 YMM registers perform a single instruction on multiple pieces of data. Each YMM register is 256 bits wide. Its lower half (128 bits) is the corresponding XMM (extended multimedia) register. Each YMM register holds and participates in simultaneous mathematical operations. These operations are done on eight 32-bit single-precision floating point numbers or four 64-bit double-precision floating point numbers (Reinders 2017). YMM stands for 'Y-mm'; 'Y' indicates a wider data register than XMM and 'mm' suggests a multiple-word register.

3.14.2 Intel® AMX

Intel® AMX is an extension to the x86 Instruction Set Architecture (ISA). It is designed to work on matrices targeting at acceleration of deep-learning training and inference processes on the CPU. It is useful for natural-language processing, image recognition and recommendation system workloads. It consists of two components:
  (i) A set of two-dimensional registers (tiles), which can hold sub-matrices from larger matrices in the memory.

(ii) An accelerator called tile matrix multiply (TMUL) which contains instructions that operate on tiles (Villarreal 2024).

3.15 Harward architecture

3.15.1 Harvard versus von Neumann architecture

This architecture evolved to overcome the von Neumann bottleneck (Wanhammar 1999). In the Harvard architecture (figure 3.8), the storage and handling of data and instructions are distinctly separated. Separate buses and memory units are used for these activities. This is in opposition to the von Neumann architecture. In the von Neumann architecture, the processing, memory, input, and output components are interlinked through a single central system bus.

The Harvard architecture:

(i) uses isolated physical addresses for data storage/accessing and instruction storage/accessing,

(ii) uses different buses (not the same bus) for data/instruction transference, and

(iii) executes any instruction in a single clock cycle.

As both the data are read/written and instructions are fetched simultaneously, considerable time is saved. So, it results in faster speed of execution than the von Neumann architecture.

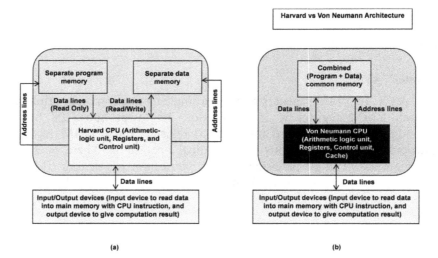

(a) (b)

Figure 3.8. Comparison of the Harvard architecture with von Neumann architecture: (a) Harvard and (b) von Neumann. In (a), there are two memory units, one program memory unit and one data memory unit. The CPU consists of ALU, registers and control unit. In (b) there is a single memory unit shared by both program and data. The address lines are unidirectional, the data lines are bidirectional between CPU and data memory as well as CPU and input/output devices but are unidirectional between CPU and program memory. The CPU consists of ALU, registers, control unit and cache. The address lines are unidirectional but the data lines are always bidirectional.

3.15.2 Disadvantages of Harvard architecture

All the advantages provided by the Harvard architecture are acquired at the cost of increased hardware requirement, more space occupation and wastage, higher price and greater power consumption. Any free portion of data memory cannot be used for instructions and conversely. So, the memory distribution must be balanced between the data and instruction needs by the manufacturers. Furthermore, several more complex controlling procedures must be built. A control unit for two buses is needed. This makes the processor design sophisticated and leaves limited processor flexibility for modifying instructions at runtime, requiring access to the separate instruction memory. Lack of flexibility makes certain types of programming more difficult. Fixed instruction length in Harvard architecture restricts the size of executable code. Applications with larger code bases are difficult to run. These shortcomings make it suitable mainly for specific real-time applications of embedded systems in signal processing and microcontrollers. In these situations, speed and efficiency are in critical demand.

3.15.3 Popularity of the von Neumann architecture

The simpler and economical von Neumann architecture operating by keeping apart the program instructions from data in a single memory unit, is therefore the widely adopted standard model for general-purpose computing. Its versatility, high adaptability, ease of programming, and widespread compatibility lay down the foundation of present-day computers. It has greatly facilitated the development of a diverse range of software applications for modern computing needs, and has profoundly shaped the evolution of computers. Quite the opposite, the Harvard architecture is less commonly seen in the prevalent across-the-board computing devices, despite its advantages of speed and efficiency

3.16 Summary and the way forward

3.16.1 Highlights of chapter 3 at a glance

(i) The CPU is the superstar computing dignitary that works tirelessly like a a busy bee in conventional digital computers. It acts as a responsible worker and dependable performer carrying all their loads (Faggin *et al* 1996, Intel® Processors-1 to 4, Reinders 2017, Techbuyer 2024, Villarreal 2024, AMD Ryzen Processors 2024).

(ii) The CPU is the big, usually square-shaped integrated circuit chip installed in a socket on the motherboard of a computer on a heatsink. The heatsink has several cooling fins for dissipation of heat produced during the operation of CPU. Sometimes a small fan is mounted nearby for further cooling.

(iii) The CPU has a flexible general-purpose architecture. The components of a CPU along with their functions and roles in CPU operation were described in this chapter.

(iv) Basic units inside a CPU are its processing elements known as cores. The cores operate along with the two controllers (the memory controller and input/output controller) and the cache using external input/output devices.

(v) The components of a core are:

    (a) The control unit to fetch, decode and execute instructions and manage operational functions.

    (b) A clock to synchronize the functions of different components for their coordinated functioning.

    (c) An ALU, which is a digital circuit consisting of adders, subtracters and logic gates for performing arithmetic and logic operations.

    (d) Registers (program counter, memory address and memory data registers, current instruction register, and accumulator) that supply operands to the ALU and store intermediate results.

    (e) On-chip level-1 cache to store copies of the data from frequently used locations in the random-access memory (RAM) for saving time compared to accessing data from level-2 RAM or main memory; along with data, control and address buses.

(vi) The CPU uses peripheral input/output devices, and auxiliary storage units to transform input data into usable information output.

(vii) In brief, the CPU is the processor of a computer while the core is a processing element of the CPU. A computer containing a single CPU is called a uniprocessor computer and a computer built with several CPUs is known as a multiprocessor computer. One core in a CPU gives a single-core CPU, whereas two or more cores in a CPU result in a multicore CPU. The parts of a CPU are the processing cores, controllers and L2-cache. The components of a core are the ALU, the registers, L1-cache, and the control unit (Fulber-Garcia 2024).

(viii) Computer memory hierarchy comprises: Level 0: CPU registers, Level 1: Cache memory (SRAM), level 2: Cache memory (also SRAM,) Level 3: Primary memory (main memory): DRAM, Level 4: Secondary memory (magnetic hard disk drive), Level 5: Tertiary memory (magnetic tapes or optical disks).

(ix) Subcategories of computer architecture are: instruction set architecture, microarchitecture and system design.

(x) The universally adopted von Neumann architecture is based on the stored program concept involving a fetch-decode-execute cycle.

(xi) The von Neumann bottleneck is encountered because memory access rate is slower than the computation speed.

(xii) Two types of processors are designed, viz., the RISC and CISC processors.

(xiii) A multicore CPU contains multiple independent processing units, known as cores in a single package. Each core can execute instructions independently and concurrently. This ability provides increased processing power and improved multitasking capabilities to the CPU.

(xiv) A multicore CPU operates on MIMD architecture. Threading, multi-threading and hyperthreading in a CPU were defined.

(xv) Single-core processors can do basic tasks in an energy-efficient manner, while multicore processors offer better performance and multitasking capabilities, albeit with increased power consumption.

(xvi) The usage scenario of CPUs in AI was discussed.

(xvii) Acceleration technologies in CPUs include Intel® advanced vector and matrix extensions.

(xviii) In the Harvard architecture competing with the von Neumann architecture, the storage and handling of data and instructions are explicitly isolated. The disadvantages of Harvard architecture were elucidated and reasons for the far-flung acceptance and popularity of von Neumann architecture were explained.

(xix) Keywords for this chapter include: CPU, core, muti-core CPU, MIMD, von Neumann architecture, Harvard architecture, SRAM, DRAM, RISC, CISC.

3.16.2 Getting ready to begin chapter 4

Although the supreme status of the CPU for general-purpose tasks remains unchallenged, the upcoming new accelerators are based on parallel execution and, hence, some form of parallel programming to overcome the von Neumann bottle-neck where the working time of CPU is wasted by idling during the waiting period for the fetching of instructions from the memory. Parallel programming means writing the code in a form to express parallelism in any algorithm for running it on an accelerator or multiple CPUs. The appearance of parallelism is evident when running of some parts of a program takes place at the same time as other parts of that program. To understand parallelism, we take up the study of parallel computing architecture in the next chapter.

References

AMD Ryzen Processors 2024 Advanced Micro Devices Inc. https://amd.com/en/products/processors/desktops/ryzen/7000-series/amd-ryzen-9-7950x.html

Faggin F, Hoff M E, Mazor S and Shima M 1996 The history of the 4004 *IEEE Micro* **16** 10–20

Fulber-Garcia V 2024 Differences between Core and CPU https://baeldung.com/cs/core-vs-cpu

Geer D 2005 Industry trends: chip makers turn to multicore processors *Computer* **38** 11–3

Godfrey M D and Hendry D F 1993 The computer as von Neumann planned it *IEEE Ann. Hist. Comput.* **15** 11–21

Harris D M and Harris S L 2015 *Digital Design and Computer Architecture* (San Francisco, CA: Morgan Kaufmann) p 584

Hennessy J L and Patterson D A 2017 *Computer Architecture: A Quantitative Approach* (Waltham, MA: Morgan Kaufmann) p 936

Intel® Processors-1 *Intel® Xeon® 6 Processors with Performance-Cores (P-Cores)* Intel Corporation https://intel.com/content/www/us/en/products/details/processors/xeon/xeon6-p-cores.html

Intel® Processors-2 *Intel® Xeon® 6 Processors with Efficient-Cores (E-Cores)* Intel Corporation https://intel.com/content/www/us/en/products/details/processors/xeon/xeon6-e-cores.html

Intel® Processors-3 *Intel® Xeon® 6780E Processor* Intel Corporation https://intel.com/content/www/us/en/products/sku/240362/intel-xeon-6780e-processor-108m-cache-2-20-ghz/specifications.html

Intel® Processors-4 *14th Generation Intel® Core™ Desktop Processors* Intel Corporation https://intel.com/content/www/us/en/products/docs/processors/core/core-14th-gen-desktop-brief.html

Mongkolsmai T An Intro to Accelerator and Parallel Programming Intel Corporation https://intel.com/content/www/us/en/developer/articles/technical/introduction-accelerator-and-parallel-programming.html#:~:text=on%20modern%20accelerators.-,Parallelism%20Overview,task%20parallelism%20and%20data%20parallelism

Nanos G 2024 Multithreading vs hyperthreading https://baeldung.com/cs/multithreading-vs-hyperthreading#:~:text=Hyperthreading%20breaks%20a%20single%20physical,distinction%20between%20the%20two%20techniques

Reinders J R 2017 *ID 672582, Intel® AVX-512 Instructions* (Intel Corporation) https://intel.com/content/www/us/en/developer/articles/technical/intel-avx-512-instructions.html

Techbuyer 2024 *Xeon and i9: Which Intel CPU Should You Choose?* 2024 Techbuyer https://techbuyer.com/us/blog/xeon-vs-i9-which-intel-cpu-should-you-choose#:~:text=Intel%20i9%20Processors,powerful%20processors%20to%20choose%20from

Villarreal A 2024 *ID 766088, Intel® Advanced Matrix Extensions (Intel® AMX)* (Intel Corporation) https://intel.com/content/www/us/en/developer/articles/code-sample/advanced-matrix-extensions-intrinsics-functions.html

Wanhammar L 1999 *8-DSP Architectures, DSP Integrated Circuits* (Academic Press Series in Engineering) (San Diego, CA: Academic) pp 357–85

Wilson K 2024 *Exploring Computer Hardware: The Illustrated Guide to Understanding Computer Hardware, Components, Peripherals & Networks (Exploring Tech)* (Widnes: Elluminet Press) p 216

Yadin A 2016 Central processing unit *Computer Systems Architecture* (Boca Raton, FL: Chapman & Hall, CRC Press) ch 4 p 88

IOP Publishing

AI-Processor Electronics
Basic technology of artificial intelligence
Vinod Kumar Khanna

Chapter 4

Parallel computing architecture

Flynn's classification of computer organization for fast computers according to the number of simultaneously manipulated instructions/data items, consists of categories named as single stream of instruction with single stream of data (SISD), single stream of instruction with multiple streams of data (SIMD), multiple streams of instructions with single stream of data (MISD) and multiple streams of instructions with multiple streams of data (MIMD). Distinctive problems facing processing units for artificial intelligence (AI) are outlined. AI computing lays a heavy burden on the processor of a computer, which can be appreciably relieved by adopting a parallel computing architecture. Parallel programming patterns include the master–slave pattern, divide-and-conquer pattern, pipelining and speculative multithreading. Among the types of parallelism, mention may be made of bit-level parallelism; data parallelism including vector processing; task parallelism including the pipelining scheme; and instruction-level parallelism (ILP) including the superscalar processor architecture, out-of-order execution, speculative execution, and multiple issue processors. Advantages and disadvantages of ILP are enumerated. Parallelism in software engineering is done by application checkpointing, automatic parallelization and through parallel programming languages. Several types of parallel processors are discussed, notably the pipelined, vector, SIMD array and attached array processors. Multicore and multiprocessor processors are demarcated. Asynchronous operation of an array processor is described.

4.1 Introduction

As transistor density increased, performance of a single processor improved. The smaller transistors operated at higher speeds, thereby increasing the speed of the processor. The more the transistor speed, the higher is the power consumption. Power is dissipated as heat. So, the processor needs to be cooled. But there is a limit to dissipation of heat. To the contrary, if the transistor size is increased to decrease

doi:10.1088/978-0-7503-6259-7ch4
4-1

the speed, the performance of the processor deteriorates. Thus, we are caught between the devil and the deep blue sea.

How can the advantage of the increase in speed by increasing transistor density be utilized if power dissipation becomes an issue? The solution to this problem is parallel computing (Pacheco 2011). The industry took a new route for escaping from the horns of the dilemma by deviating from the trend of building a single faster monolithic processor towards constructing several simple complete processors on a single chip in the form of multicore processors. This was the dawn of parallel computing. The conventional CPU was called a single-core processor to distinguish it from the multicore processors. The multicore CPU concept was extrapolated with further advancements into pipelined, vector, array and multiprocessors.

In this chapter, different types of parallelism used to increase the computational speed of a processor are described (Grama *et al* 2003, Petersen and Arbenz 2004). Machine learning algorithms involve complex mathematical calculations requiring high-performance processors with immense computing power to execute the tasks efficiently at low power consumption and in less time, especially for real-time applications.

4.2 Flynn's classification of computer organization by the number of simultaneously manipulated instructions/data items

Very high-speed computers were classified by Flynn (1966, 1972, 2011); the word 'stream' used in this classification means the sequence of data/instructions as viewed by the computer during program execution (figure 4.1).

4.2.1 Single stream of instruction with single stream of data (SISD)

This is a uniprocessor machine in which instructions are executed sequentially. A familiar example is a CPU working on one piece of data per instruction (figure 4.1(a)). An SISD processor is perfected by concurrent instruction handling, operand acquisition, and execution. Parallel processing is possible by using multiple functional units called cores or by pipeline processing. Hence, the performance is improved on a hardline basis by using more units.

4.2.2 Single stream of instruction with multiple streams of data (SIMD)

This is a multiprocessor machine working under the supervision of a common control unit like a graphical processing unit (GPU), figure 4.1(c). It uses a single instruction to compute a complete dataset such as a picture. The same instruction is sent to all the processors from the control unit. But they operate upon different items of data. Because a single copy of instruction is applied to all the data streams simultaneously, less memory is used. A single instruction decoder is needed, thus reducing the cost. The creation, interpretation and debugging of programs is simple because it is a synchronous programming technique. However, the huge register files increase chip area and raise power consumption.

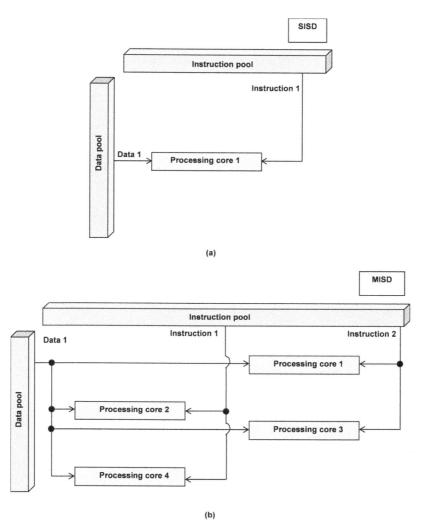

Figure 4.1. Computer architectures: (a) SISD, (b) MISD, (c) SIMD, and (d) MIMD. In (a) the instruction pool is a horizontal bar from which instruction 1 is fed to processor core 1 to act on data 1 from the data pool, which is shown as a vertical bar. This is single-instruction, single-data architecture. In (b) the instruction 1 from the instruction pool horizontal bar is fed to processor cores 2 and 4 to act on data 1 from the data pool vertical bar, while the instruction 2 from the instruction pool horizontal bar is fed to processor cores 1 and 3 to act on data 1 from the data pool vertical bar. This is multiple-instruction, single-data model. In (c) instruction 1 from the instruction pool horizontal bar is fed to processor cores 1, 2, 3 and 4 to act on data 1, 2, 3, 4 from the data pool vertical bar. This is single-instruction, multiple-data paradigm. In (d), the instruction 1 from the instruction pool horizontal bar is fed to processor cores 2, 4, 6 and 8 to act on data 2, 4, 6, 8 from the data pool vertical bar. At the same time, instruction 2 from the instruction pool horizontal bar is fed to processor cores 1, 3, 5 and 7 to act on data 1, 3, 5, 7 from the data pool vertical bar. This is a multiple-instruction, multiple data paradigm.

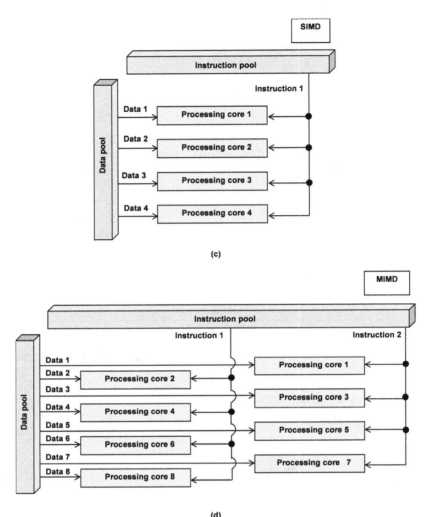

Figure 4.1. (Continued.)

Modern GPUs work in a multithreaded fashion, which is coined as single instruction, multiple thread (SIMT) embodiment. It is an extension of the SIMD model. In SIMT, the SIMD model has been augmented with multithreading, streaming memory and dynamic scheduling concepts. Multithreading entails breaking a program into smaller, executable threads. Each thread is a component of the process called a lightweight process. A process is a program under execution. It consists of text, data, heap and stack. The text is the machine code. The global, constant and variable data types are used in computer programming. Heap refers to dynamic memory allocation. The heap memory is allocated at run time. The allocation is based on a programmer's request. Stack memory is a RAM region for temporary data storage during execution of the program.

One thread is executed at a time. Multithreading is like several hands working in coordination to complete different parts of a task.

Streaming memory is a memory system optimized for handling streams of data, while dynamic scheduling involves adjustments of task assignments from real-time information about resources at one's disposal and the data flow.

4.2.3 Multiple streams of instructions with single stream of data (MISD)

This is only of theoretical significance (figure 4.1(b)).

4.2.4 Multiple streams of instructions with multiple streams of data (MIMD)

Here, numerous instructions are concurrently applied on various data streams in an asynchronous programming technique (figure 4.1(d)). No extra control unit is required. However, more decoders are used. It is more complicated and expensive than SIMD but also more efficient than SIMD.

The MIMD is used for multiprocessor and multicomputer systems in scientific computing. Each processor in the system operates independently of the other processors. Each processor works with its own program counter and instruction set.

In the shared memory MIMD model, all processors have access to a single global memory. In the distributed MIMD model, the processors communicate through an interconnection network.

The interconnection network is configured to a tree or mesh structure. Difficulties are faced in load balancing, synchronization overhead and scaling.

SIMD and MIMD are two voguish computer architectures used to enhance the performance of computing activities depending on the data and instruction streams. SIMD gives data level parallelism while MIMD yields task-level parallelism.

4.3 Challenges of AI computing, and their encumbrance on the processor of a computer

Machine learning algorithms, particularly deep neural networks, pose unique challenges. The challenges are faced regarding the resources and execution time of training/inference calculations. The burden and onus of all computations done on a computer are ultimately borne by its processor. A processor is the core component in the form of an integrated circuit that performs the calculations in a computer. In this component, different combinations of transistors and other electronic/electrical devices are wired in meticulous ways. Wirings are done to make various types of logic gates, which perform basic logical operations. Seven basic logic gates (AND, OR, XOR, NOT, NAND, NOR, and XNOR), serve as basic building blocks. They are used for creation of larger execution units, which implement any desired computation.

A weak processor will be bogged down under the load of a big problem, either slowing down or totally stopping work. The more burdensome the problem, the stronger processor required to handle it.

All-purpose processors such as the CPU of a laptop or desktop computer are deficient in dealing with such complex computations. Therefore, development of specialized processing units for the computationally-intensive AI algorithms is of paramount value. It is indeed the focus of attention of the research community worldwide.

4.4 Parallel computing

Parallel computing is essentially computing on a parallel computer with N processors (figure 4.2(b)). It strives to shorten the computing time in comparison

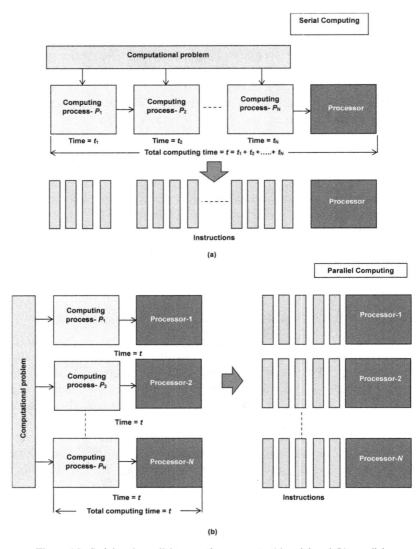

Figure 4.2. Serial and parallel computing concepts; (a) serial and (b) parallel.

to the serial computing with a single processor (figure 4.2(a)). There is a huge demand for parallelism in compute-intensive applications such as scientific and engineering calculations and simulations. Data-intensive applications, e.g., data warehousing, data mining, etc require parallelism too, and to a great extent. The necessity of parallelism exists also in network-intensive applications, e.g., remote control, telemedicine, etc (Li and Zhang 2018).

4.4.1 Parallel programming patterns

This pattern represents the path followed by programmers for parallelizing modules.

4.4.1.1 Master–slave pattern
This has two parts, namely the master and the slave (figure 4.3(a)):
  (i) The Master: This decomposes the task into subtasks, collects the results of all the subtasks and aggregates them into a final result.
  (ii) The Slave: This performs recurrent repetition of the operations: message receiving, processing of the subtask and sending back the results to the master.

4.4.1.2 Divide-and-conquer pattern
The three constituent steps of this method are problem decomposition, calculation and summarization of results. Any problem is divided into two or more sub-problems (figure 4.3(b)). Every sub-problem is solved separately to get its independent solution. These independent solutions are combined together to determine the final result. The method is organized into a virtual tree whose leaf nodes perform the calculations.

4.4.1.3 Pipelining
This will be explained under task parallelism, section 4.5.3.2.

4.4.1.4 Speculative multithreading
Speculative multithreading (SpMT) or thread-level speculation (TLS) is aggressive parallelization of sequential codes. During this parallelization, any apprehension about the success of the execution is dispelled. The hardware takes care of the success.

In a speculative multithreading execution model called Prophet, the given sequential program is partitioned into several speculative threads. Each thread executes a different part of the original program (Dong *et al* 2009). Only one thread among the parallelly executed threads is non-speculative, and delivers its results to the memory. All the remaining threads are speculative. Each thread has a spawn point (SP): thread creation instruction, and a control quasi-independent point (CQUIP): thread end instruction. In figure 4.3(c), the parallel execution of threads is compared with their sequential execution. Only thread A is the stable thread. It is safe and will verify its immediate successor. The remaining threads are speculative threads. After completion of execution by the stable, non-speculative thread, a verification of the results of its succeeding thread is carried out. If the verification shows that the successor is correct, the stable thread commits all its data to it. The stable thread also gives a stable token to its successor before quitting. On receipt of the stable token, the successor acquires the status of the new stable thread.

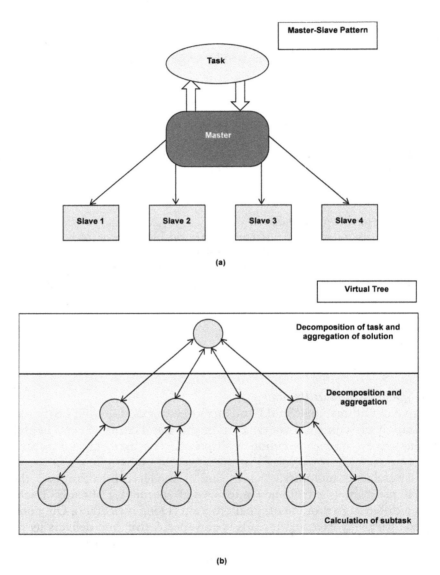

Figure 4.3. Parallel programming patterns: (a) simplified representation of the master–slave pattern, (b) organization of the divide-and-conquer strategy in the form of a virtual tree and (c) SpMT model showing the reduction of time obtained by replacing sequential execution with parallel execution.

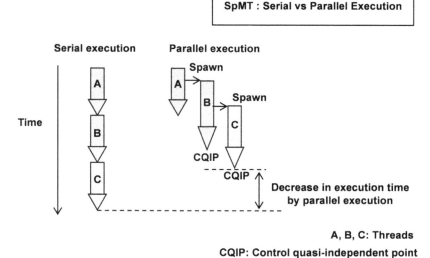

A, B, C: Threads

CQIP: Control quasi-independent point

(c)

Figure 4.3. (Continued.)

It becomes a non-speculative thread. Also, when found correct, all the values produced by the non-speculative thread are stored in memory. On failure of verification, i.e., if found incorrect, the results are discarded. All the speculative child threads are revoked by the stable, non-speculative thread. Thereafter, the stable thread continues running and its succeeding speculative threads are re-executed.

4.5 Types of parallelism

4.5.1 Bit-level parallelism

Here, the number of bits processed in a single instruction is increased. For processing more bits, the size of processor's register is enlarged. As a consequence, the number of instructions required to perform operations on large data are reduced. To exemplify, a 16-bit processor can add two 16-bit integers with a single instruction while an 8-bit processor needs two instructions to add the same.

4.5.2 Data parallelism

Here, the given data, e.g., an array of numbers or a matrix, is split up into smaller chunks or subsets of data. These chunks are named, e.g., Data-1, Data-2, Data-3, and Data-4. The chunks are assigned to multiple processors as the same task (task-1, suppose) for execution during the same period of time (figure 4.4(a)). The final output is obtained by combining the results. Thus, in data parallelism, the same

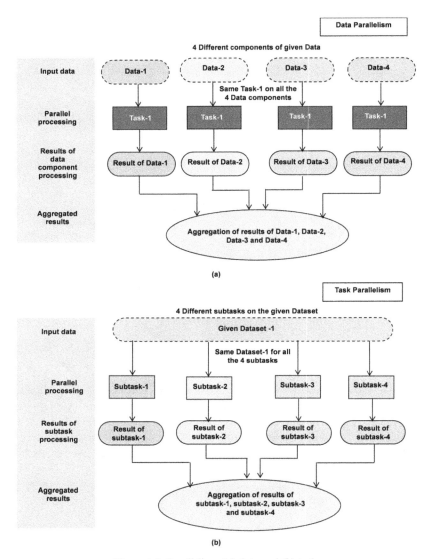

Figure 4.4. Parallelism: (a) data and (b) task.

function is simultaneously executed across the elements of a dataset. Data parallelism is a synchronous computation procedure. In this procedure, the same operation is performed on various subsets of the same data. It is designed to secure the most favorable load balance on a multiprocessor system. This is done by optimization of load balancing. In load balancing, the distribution of a set of tasks on the processors is carried out to achieve an efficient overall performance. The amount of parallelization is proportional to the size of input data. Higher speed is obtained than achieved by sequential execution. This is feasible because only one process (thread) is executed on all datasets. Data parallelism is preferred for array, matrix computations and convolutional neural network computations. A GPU uses data parallelism.

4.5.2.1 Vector processing

A vector processor is a CPU with multiple arithmetic-logic units (ALUs). This processor works with an instruction set designed for operation on large one-dimensional arrays of data called vectors. Also called super word-level parallelism, it is a type of data parallelism working on arrays or vectors of data. It needs compiler support to identify vectorization opportunities. Hardware support is required in the form of vector registers. SIMD and MIMD are the two types of vector processing.

The capability to work on multiple values differentiates a vector processor from a scalar processor. The instructions of a scalar processor operate on single data items only.

Vector processing differs from array processing. In array processing, several processors are applied to work on individual elements of the array. Antithetically, vector processing uses a single processor for execution of the same operation on many data items.

Although it is a form of data parallelism, vector processing must be distinguished from parallel processing. Parallel processing involves multiple processors working on separate tasks. In contrast, vector processing avails the services of a single processor for performing the same operation on multiple data elements.

4.5.2.2 Advantages/disadvantages of vector processing

Vector processing yields high performance by exploiting data parallelism. During this processing, memory access is reduced. Because less time is lost in waiting for data loading into memory, vector processing is more energy-efficient than conventional processing. Further, in vector processing, scaling up to handle larger datasets is possible without any sacrifice in performance.

However, vector processors have a limited instruction set that is optimized for numerical computations. For optimal performance, the data must be aligned in contiguous locations in memory. Then only efficient accessing by the processor is achievable.

4.5.3 Task parallelism

4.5.3.1 Task subdivision

Here, the total task is subdivided into subtasks termed as subtask-1, subtask-2, subtask-3, and subtask-4; a task is a unit of work performed by a processor, ranging from a simple arithmetic operation to a complex calculation.

Each of the subtasks created constitutes a small part of the workload (figure 4.4(b)). The subtasks are executed at the same time on the same dataset-1 or on different data in different parts of the same processor or in different processors, and the results are aggregated. Thus, in task parallelism, several different functions are simultaneously executed across the same dataset-1 or on different data. In other words, task parallelism is an asynchronous computation. In this computation, different operations are carried out on the same dataset-1 or on different data. Load balancing is determined by the availability of hardware and scheduling algorithms. Amount of parallelization is proportional to the number of independent tasks that are to be done. Lower speedup is possible in task parallelism than gained by data parallelism. This is so because each processor executes a different process (thread) on the same dataset-1 or different dataset(s).

Two types of task parallelism are:

    (i) Fine-grained parallelism: In this kind of parallelism, the subtasks inter-communicate among themselves very frequently delivering output in real time.

    (ii) Coarse-grained parallelism: This kind of parallelism works with infrequent communication among subtasks.

4.5.3.2 Pipelining scheme

This is a type of task parallelism which decomposes a sequential process into suboperations (figure 4.5). Each subprocess is executed in a singular dedicated segment. The dedicated segment operates at the same time with other segments. For overlapping of computations, an input register is associated with each segment in the pipeline. The register isolates the segments so that they can operate on separate data. The register is followed by a combinational circuit. The combinational circuit performs the suboperation assigned to the particular segment. During this time, the register holds the data. A clock is applied to all the registers to perform the entire segment activity.

Pipelining overlaps the execution of instructions which are independent of one another. This overlap among instructions is called instruction-level parallelism. This name is agreed since the instructions can be executed in parallel.

4.5.4 Instruction-level parallelism (ILP)

In ILP, several instructions are executed at the same time. So, the execution of the next instruction begins without waiting for completion of execution of the present instruction.

4.5.4.1 Superscalar processor architecture

In this architecture, multiple execution units are used. Each unit executes instructions independently. Several instructions are fetched together. They are sent to different execution units contingent on non-existence of data dependencies among the instructions.

4.5.4.2 Out-of-order execution

In this case, the instructions are executed irrespective of their original order as soon as their dependencies or stalls are resolved. Hence, gaps in execution due to data dependencies or stalls are filled up.

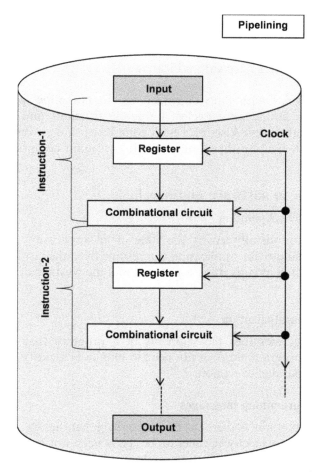

Figure 4.5. Pipeline processing.

4.5.4.3 Speculative execution
Following a conjectural approach, the outcome of conditional branches is predicted by the processor. In speculative execution, the instructions for a branch are executed before evaluation of actual condition of the branch. If found correct, the execution continues. If observed as erroneous, the results are cast-off.

4.5.4.4 Multiple issue processors
Here, the instructions are analyzed for parallel execution opportunities. After analysis, multiple instructions are distributed to execution units. As a consequence, more instructions are processed per cycle.

4.5.4.5 Advantages and disadvantages of ILP
ILP improves overall computational performance by:
    (i) reduction of execution time,

(ii) increasing throughput,
(iii) effectively exploiting memory latency,
(iv) decreasing the effect of memory bottlenecks, and
(v) providing better hardware utilization.

However, instructions with data dependencies do not allow parallel execution. Branch prediction faults waste cycles. Larger instructions and hence code sizes degrade memory efficiency. Also, processor complexity and power consumption are increased. Notwithstanding these shortfalls, it is a main component of processor design.

4.6 Parallelism in software engineering

4.6.1 Application checkpointing

This is done by periodically saving the state of an application during execution. Hence, in case of failure, the application can resume from the last saved state. In this way, the computation already done is not lost, and the wastage of associated time is prevented.

4.6.2 Automatic parallelization

Here, the compiler identifies the segments of a program that can be parallelly executed. So, the programmer does not need to manually identify such portions and prepare code for parallel execution.

4.6.3 Parallel programming languages

These languages contain constructs for expressing parallelism. These constructs allow programmers to specify parallel tasks. They need not be fretted and hassled about task scheduling, synchronization, and inter-process communication.

4.7 Parallel processors

4.7.1 Pipelined processor

Instruction pipelining is composed of overlapped and simultaneous execution of machine cycles of multiple instructions in a single processor. Therefore, the hardware of the pipelined processor is divided into several functional units (figure 4.6 (a)). Each of these functional units is called a pipeline stage. Each pipeline stage performs a dedicated and functionally independent task, and is executed in one machine cycle. The time consumed in the execution of the slowest pipeline stage determines the length of the machine cycle. All the pipeline stages are managed by the control unit. The CPU clock ensures synchronization of operation of the various stages. The synchronization guarantees coordination of all the tasks being carried out. The interface registers hold the intermediate outputs between any two stages. Each stage has a buffer-in register. From this buffer-in register, the stage receives the input data when the clock cycle starts. Also, each stage has a buffer-out register. Into this buffer-out register, the result of a stage is stored temporarily after processing the

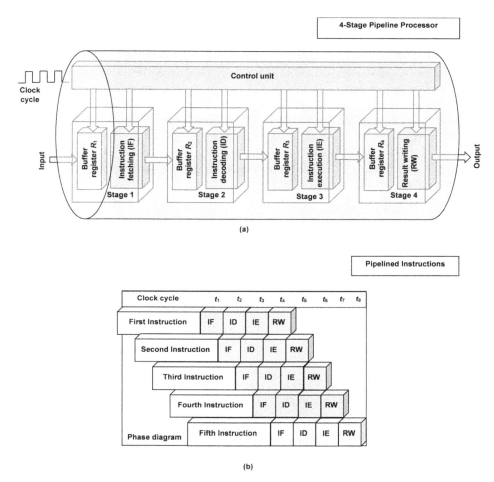

Figure 4.6. Pipelined processor: (a) structural organization of a four-stage processor and (b) its phase diagram.

input data. Obviously, the buffer-out register for a stage is the buffer-in register for the succeeding stage.

In figure 4.6(b), each clock cycle is a machine cycle. It is also referred to as a timing state. Four instructions are sent through a four-stage pipeline in which the stages are:

    (i) instruction fetching (IF) from memory,
   (ii) instruction decoding (ID) in control unit,
  (iii) instruction execution (IE) in ALU, and
  (iv) result writing (RW) in memory or register.

To completely execute the sequence of four instructions, all the instructions in the sequence pass through the four stages. The first instruction is completed in time t_4 corresponding to the 4th clock cycle. At time t_1 referring to the 1st clock cycle, when the first instruction is in instruction fetching stage, there is no possibility of its overlapping with another instruction. When the first instruction moves on to the instruction decoding stage at time t_2 for the 2nd clock cycle, the second instruction enters the pipeline at instruction fetching stage. At time t_2 of the 2nd clock cycle, two stages are busy, each with one instruction. At time t_3 pertaining to the 3rd clock cycle, the first and second instructions move on to the instruction execution and result writing stages, respectively.

The third instruction enters the instruction fetching stage. From time t_4 onwards beyond the 4th clock cycle, maximum stages are utilized. Here, five instructions are executed in eight clock cycles t_1 to t_8. In a non-pipelined architecture, these instructions will require 20 clock cycles resulting from 5 instructions × 4 cycles for each instruction.

4.7.2 Vector processor

A vector processor comprises eight main functional units. These units include: a main memory, an instruction processing unit, a vector instruction controller, a vector access controller, a vector register, a vector processor, a scalar register and a scalar processor (figure 4.7).

Both the data and instructions are stored in the memory. So, the instruction processing unit first fetches the instruction from the memory. Then it determines

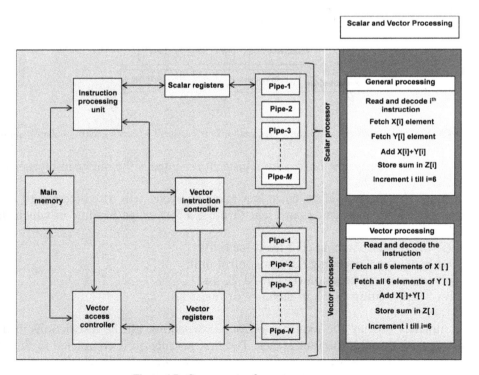

Figure 4.7. Components of a vector processor.

whether the fetched instruction is scalar or vectorial in nature, and further operations ensue according to one of the following three routes:

  (i) If the fetched instruction is a scalar in nature, then the instruction is sent to the scalar register. A scalar processor operates on it.
 (ii) If the fetched instruction has a vectorial nature, then it is supplied to the vector instruction controller. This instruction controller decodes the vector instruction to retrieve the address of the desired vector operand in the memory. The desired operand at the address thus found is fetched from the memory. It is fed to the instruction register, and the vector processor operates on it.
(iii) If the fetched instruction is a set of multiple vector instructions, then the set of multiple vector instructions is assigned by the vector instruction controller to the task system. A long vector task is divided by the processor into sub-vectors. The sub-vectors are fed to the vector processor. The vector processor uses many pipelines for execution of the instruction over the operand. The operand is fetched from the memory at the same time. The vector instruction controller schedules the various vector instructions.

Two architectures of the vector processor are commonly used:

  (i) Register-to-register architecture: In this architecture, the operand or previous results are fetched through the main memory by using registers.
 (ii) Memory-to-memory architecture: Here, the operands or the results are fetched directly from the main memory. In this case, the pipelines of the vector processor require higher startup time owing to the large memory access time.

4.7.3 Array processor

4.7.3.1 SIMD array processor
The SIMD or single instruction, multiple data stream array processor is a processor with several identical processing units. These units operate in parallel, and perform the same operation in synchronization under the supervision of the control unit (figure 4.8(a)). Each processing element has a local memory. The local memory provides extra memory to the system. Every processing element uses operands that are stored in its local memory. But the program is stored in the main memory. Array processing is basically parallel processing on a huge array of data. Therefore, it significantly increases the instruction processing speed.

4.7.3.2 Attached array processor
The array processor is attached to a general-purpose CPU. The array processor is needed to improve the performance of the host computer in numerical computational tasks (figure 4.8(b)). The array processor attachment is usually done when a large number of instructions are being executed and the host computer takes an inordinately long time for execution. Then the array processor is used to boost the performance of the host computer. The array processor is connected through an

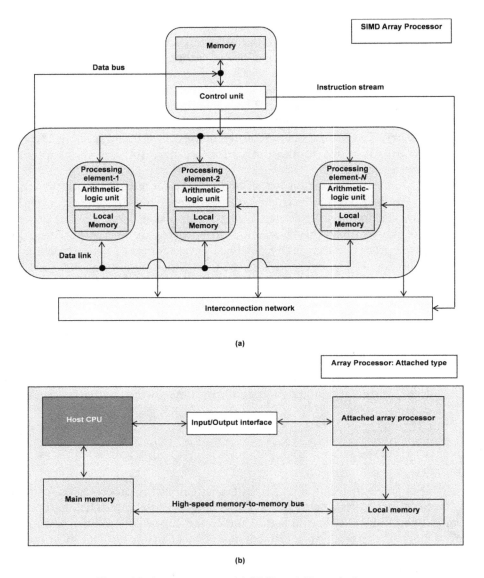

Figure 4.8. Array processors: (a) SIMD and (b) attached types.

input/output (I/O) controller to the host computer which treats it as an external interface. In this settlement, the local and main memories are connected by a high-speed memory-to-memory bus.

4.7.3.3 Asynchronous operation of array processor
A high overall capacity of the SIMD array processor system is achieved when it is run asynchronously from the host CPU. To understand this asynchronous mode of operation, we call attention to a problem afflicting synchronous processing.

The difficulty arises from the requirement of a fairly large supply current by complementary metal-oxide semiconductor logic gates during state changeover from 0 to 1. For synchronization, these changes are initialized by an active clock edge. So, they are aggregated in a synchronous circuit. Their aggregation is alarming because it leads to the occurrence of large spikes on supply current at active clock edges. These spikes cause large electromagnetic interference (EMI) and hence circuit malfunctioning.

In globally asynchronous locally synchronous (GALS) architecture, a set of locally synchronous modules communicate with each other via asynchronous wrappers. Each synchronous subsystem, i.e., clock domain runs on its own independent clock frequency. Since this design is locally synchronous, it is easier in designing than an asynchronous circuit. However, it has a globally asynchronous character owing to different, e.g. phase shifted, rising and falling active edge clock signal regimes. On account of this, the supply current spikes do not aggregate at the same time. Therefore, much lower electromagnetic interference is produced, and the EMI problem is suppressed.

4.7.4 Multiprocessor

4.7.4.1 Multiprocessor versus multicore CPU

A multiprocessor is a single processor made of two or several interactive processors. These processors work together in parallelism using a shared or distributed memory. A pathway links the various components of the system for transportation of data and instructions. It connects the I/O devices. It is called the system bus. Multiprocessors are categorized into shared and distributed memory classes.

(i) The shared memory multiprocessor: In a shared memory multiprocessor, each processor shares the main memory and I/O devices using the system bus. Such a processor is referred to as a symmetric processor (figure 4.9).

Interconnection among processors in a shared memory multiprocessor takes place in one of three possible ways:

(a) Time-shared common bus: Here, all the processors communicate through a single shared bus with the memory unit (figure 4.9(a)). When processor-1 is interacting with memory, all other processors remain idle because there is only one bus for communication. It offers simple implementation at low cost. But it provides a slow data transfer rate.

(b) Multiport memories: Here each processor interacts with a memory module (MM) through a separate bus (figure 4.9(b)). When processor-1 desires to interact with memory module 1, then port MM1 is enabled. A similar arrangement is available for other processors, allowing parallel communication among the processors. If more than one processor requests for a memory module at the same time, priority is given in the order of processor-1, processor-2, processor-3, processor-4.

(c) Crossbar switch network: A switch is installed between the memory unit and the CPU to decide about passing/not passing the request to a particular memory module based on the request made (figure 4.9(c)). The switch network provides a high data-through rate. But it is more complex and expensive in implementation because many switches are necessary.

(ii) Distributed memory multiprocessor: In a distributed memory multiprocessor, each processor has its own private memory (figure 4.10). Additionally, the processors communicate with each other and access the main memory through the system bus. As the memory is distributed with the processors, local memory accesses do not exhaust global bandwidth. Consequently, faster access times are attained. This type of multiprocessor is scalable to a large number of processors because multiple memory modules are used together with a scalable network.

The execution of a large number of processes at the same time on the different processors comprising the multiprocessor leads to the realization of parallel processing. The parallelism increases the throughput of the system considerably

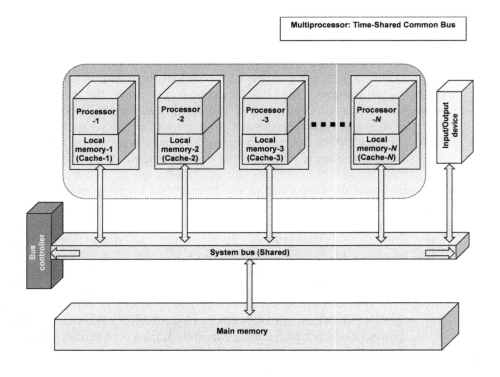

(a)

Figure 4.9. Shared memory multiprocessors: (a) shared memory, (b) multiport memories and (c) crossbar switch network.

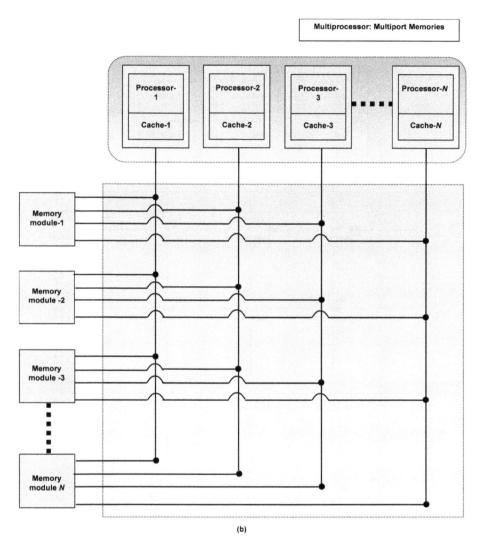

(b)

Figure 4.9. (Continued.)

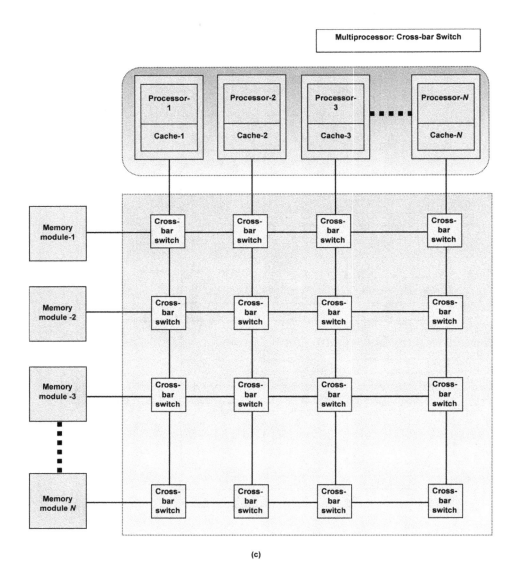

(c)

Figure 4.9. (Continued.)

relative to a single processor. However, the small parts of a given process which are taking place on different processors must be efficiently coordinated. The need of coordination adds complexity to the multiprocessor idea.

Ever since energy consumption and heat dissipation issues restricted the increase of the clock frequency and computational activities performed during one clock period within a single processor, nearly all processor manufacturers have switched to models where multiple processing units called processor cores are used in each chip to upgrade the processing power (Kirk and Hwu 2013). Therefore, most modern

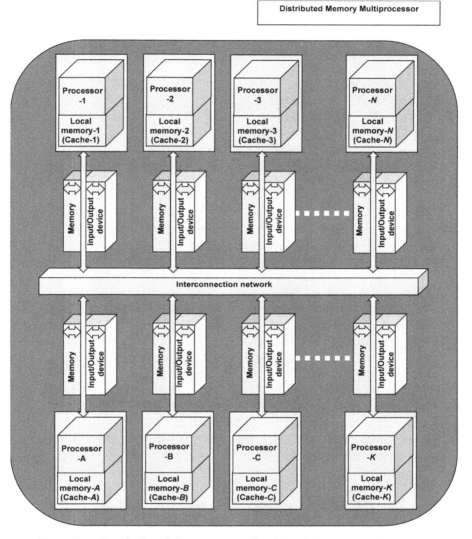

Figure 4.10. Organization of the components of a distributed memory multiprocessor.

CPUs are multicore CPUs. A multicore CPU is a processor containing two or more independent processing units. Each unit can execute instructions independently of others. The similarity between a multiprocessor and multicore CPU is that they both can do parallel computing. The dissimilarity lies in the degree of parallel processing. While a multicore CPU can run a single program faster, a multiprocessor can run several programs together at a greater pace. Furthermore, availability of several processors in a multiprocessor makes it more reliable than a multicore CPU. The reliability is achieved because failure of one processor does not affect the perform-ance of other processors. The speed and reliability advantages of multiprocessors are accompanied by disadvantages too. The drawbacks are a higher price and a more complex configuration than a multicore CPU. Present-day computers utilize both muticore CPU and multiprocessor ideas to boost the performance.

4.7.4.2 Applications of a multiprocessor

A multiprocessor is used:
  (i) As an SISD uniprocessor.
  (ii) As an SIMD multiprocessor used for vector processing.
  (iii) For hyper-threading/pipelined processing where multiple series of instructions are applied in a single view, such as MISD.
  (iv) For executing multiple, individual series of instructions in multiple views, such as MIMD.

4.7.4.3 Advantages of a multiprocessor

  (i) Parallelism enhancement: As different processors execute different tasks simultaneously, high parallelism is achieved.
  (ii) Performance improvement: Hardware sharing across multiple processors yields superior overall enhanced performance. Multitasking inside an application enables faster task execution than a single processor by distribution of responsibilities. High throughput and responsiveness are secured in several AI applications.
  (iii) Easy up-scalability: A multiprocessor is upscaled more easily than a single processor. The reason is that additional processors are included to handle the increased workloads.
  (iv) Higher reliability: As already said, even if one processor inside a multiprocessor fails, the multiprocessor operation does not stop. The remaining processors continue to execute their tasks.
  (v) Low-priced processor: This is more cost-effective than multiple single-processor systems for handling the same workload.

4.7.4.4 Disadvantages of a multiprocessor

  (i) Larger power consumption: Consumption of more power than a single processor increases the operating and maintenance costs of a multiprocessor.
  (ii) Software challenges: Specialized programming skills are essential for developing the software for effectively utilizing multiple processors.
  (iii) Inter-processor synchronization difficulties: Synchronization among processors is mandatory for correct and efficient execution of tasks inside a multiprocessor. Maintenance of synchronization introduces complexity and overhead to the system.
  (iv) High complexity: This is more complex than a single processor because additional hardware, software, and management resources are necessary.
  (v) Restricted performance benefits: Some applications may reap limited gains on running in a multiprocessor. It is mainly beneficial for heavy workloads, as frequently noticed in AI.

4.8 Summary and the way forward

4.8.1 Highlights of chapter 4 at a glance

(i) Flynn's classification of computer organization for fast computers is done in accordance with the number of simultaneously manipulated instructions/data items. It consists of categories named as SISD, SIMD, MISD and MIMD.

(ii) Distinctive problems facing processing units for AI were outlined. AI computing lays an onerous strain on the processor of a computer making it a Herculean task. The overburden can be relieved by adopting a parallel computing architecture.

(iii) Parallel programming patterns include the master–slave pattern, divide-and-conquer pattern, pipelining and speculative multithreading schemes.

(iv) Among the types of parallelism, mention may be made of bit-level parallelism, data parallelism including vector processing, task parallelism including the pipelining scheme and ILP dealing with the superscalar processor architecture, out-of-order execution, speculative execution, and multiple issue processors. Advantages and disadvantages of ILP were enumerated.

(v) Parallelism in software engineering is done by application checkpointing, automatic parallelization and through parallel programming languages.

(vi) Several types of parallel processors were discussed, notably the pipelined processor, vector processor, SIMD array processor and attached array processor. Multicore and multiprocessor processors were demarcated. Asynchronous operation of an array processor was described.

(vii) In this chapter, we came across two terms: parallel computing and concurrent processing. It must be reiterated that they have different meanings. In parallel computing, a job is broken down into discrete smaller parts called subtasks. The subtasks are run on different cores of a multicore CPU or on multiple CPUs at the exact same time. Parallelism requires hardware with multiple processors. So, a single-core CPU cannot provide parallel computing. On the other hand, concurrent processing enables a program to deal with multiple tasks even on a single CPU core for the reason that the core can switch between processes or threads without necessarily completing each one. Two subtasks can start, run, and finish in overlapping time periods. Subtask-2 can start even before the process of subtask-1 is finished. Thus, a single-core CPU, may give concurrency but not parallelism.

For parallelism, multiple cores or processors are essential. Parallelism can also be choreographed on a cluster of computers connected together to form a parallel processing cluster. Parallelism is a special kind of concurrency where subtasks are actually executed simultaneously. Therefore, parallel architectures are a quintessential cornerstone of advanced computing.

(viii) Keywords for this chapter are: parallel computing, Flynn's classification, SISD, SIMD, MISD, MIMD, bit-level parallelism, task parallelism, pipelined processor, vector processor, array processor, multiprocessor.

4.8.2 Getting ready to begin chapter 5

Despite contributing in a big way towards AI computing, parallelism alone cannot sustain the heavy load of AI data. It needs to be supported by other methods. A convenient way is to reverse the direction of thinking. We are trying to implement AI algorithms on existing processors. This is the same as if the container for a newly developed product is already made without any idea about the size of the product. Afterwards, when the product is made, it must be somehow stuffed into that container for packaging without caring about any difference between the sizes of the product and the container. Let us therefore minutely examine the subtle differences between the computation methods used in prevalent processors and the actual requirements of the AI algorithms. Let us relook at the scene from the AI algorithm side towards the processor side to know what exactly do we want the processor to do? We can redesign our processors from that angle. This approach will gear up speed and improve performance at the expense of smaller resources. Some of the steps which an already-designed processor is executing may be unnecessary or their implementational procedure may be lacking in effectiveness. These are the issues we shall pursue in the next chapter. There we will make an effort to evolve possible solutions.

References

Dong Z, Zhao Y, Wei Y, Wang X and Song S 2009 Prophet: a speculative multi-threading execution model with architectural support based on CMP *2009 Int. Conf. on Scalable Computing and Communications; Eighth Int. Conf. on Embedded Computing (25–27 September) (Dalian, China)* pp 103–8

Flynn M J 1966 Very high-speed computing systems *Proc. IEEE* **54** 1901–9

Flynn M J 1972 Some computer organizations and their effectiveness *IEEE Trans. Comput.* **C-21** 948–60

Flynn M J 2011 Flynn's taxonomy *Encyclopedia of Parallel Computing* vol **1–4** (New York: Springer) pp 689–97

Grama A, Gupta A, Karypis G and Kumar V 2003 *Introduction to Parallel Computing* 2nd edn (Harlow: Addison-Wesley) p 636

Kirk D B and Hwu W-M W 2013 *Programming Massively Parallel Processors: A Hands-on Approach* (Waltham, MA: Morgan Kaufmann) p 1

Li Y and Zhang Z 2018 Parallel computing: review and perspective *2018 5th Int. Conf. on Information Science and Control Engineering (ICISCE) (20–22 July) (Zhengzhou, China)* pp 365–9

Pacheco P S 2011 *An Introduction to Parallel Programming* (Amsterdam: Morgan Kaufmann) p 2

Petersen W P and Arbenz P 2004 *Introduction to Parallel Computing: A Practical Guide with Examples in C* (Oxford: Oxford University Press) p 278

IOP Publishing

AI-Processor Electronics
Basic technology of artificial intelligence
Vinod Kumar Khanna

Chapter 5

Optimized AI-computing within physical limits of transistors

Constraints imposed by physical downscaling limits of chip dimensions led to the evolution of domain-specific architecture (DSA). Domain-specific hardware accelerators are described including their design techniques and thumb rules for design. Artificial intelligence (AI)-optimized hardware is classified according to application and flexibility into programmable and dedicated processors. The metrics of AI-optimized processors are mentioned. Deep learning model compression is done by network pruning, quantization, knowledge distillation and low-rank matrix factorization. Advantages of network sparsity enforcement in deep learning, and sparsity-induction methods are described. Data compression for saving storage space and memory bandwidth is done by its recoding through lossless and lossy compression methods. Zero-skipping method is illustrated by the necessity of convolution neural network (CNN) workload reduction in the context of its application to CNN processing. Low-precision computing is a method applicable to situations permitting compromises with precision. The advantage gained by reduction in precision is explicated. Techniques of availing faster entry into memory via non-uniform memory access and memory interleaving are presented.

5.1 Introduction

In this chapter, we concentrate our attention on the hardware and software procedures and strategies that are specially tailored and adapted to meet AI requirements. These hardware/software plans have proved more fruitful and efficient in solving AI tasks than the general, all-purpose solutions. Many propositions made and techniques designed in recent years with this goal in mind are gaining immense popularity and wide acceptance, thereby making a strong impact on the computing scenario.

doi:10.1088/978-0-7503-6259-7ch5

A processor whose abilities and character are unsuited to the AI problem at hand looks more like a square peg in a round hole. By using it, we may be doing the absurdity of hiring an airplane to cross a road while we need to cross the road simply by walking. We may be going for an expensive solution to a problem while a cheaper one is readily available. In this context, the performance per unit price paid or per unit of power consumed becomes crucial. The saying, 'Jack of all trades and master of none' expresses the utility of such a specialist processor against the generalist processor.

5.2 Downscaling limits of chip dimensions, and evolution of domain-specific architecture

Chip design deals with the layout and structure of the electronic/electrical devices and their interconnections. The crux of the matter is to increase the computational power. Hitherto, the trend relentlessly followed for goal accomplishment has been to pack more transistors in a given area. A higher transistor count implies a more powerful processor. The chip architecture defines the basic instruction set, and the execution and memory models used by the operating system. Technology refers to the process node: 5 nm/7 nm/10 nm/16 nm/28 nm, and so on.

The well-known Moore's law states that the number of transistors per square inch on an integrated circuit increases by a factor of 2 in every 12–18 months. According to the Denard scaling law, the power consumption per unit area remains the same for every technology generation. Denard law implies that the number of transistors in a chip can be doubled with every generation without any change in power consumption. Accordingly, the supply voltage of more advanced technology chips is decreased for operating more transistors with the same power as the older technology chips. Moore's law in conjunction with Denard scaling law helped computer architects to ameliorate the performance of general-purpose processors on all-purpose programs through multitudinous technological breakthroughs, e.g., hierarchical caches, out-of-order execution, multiprocessor development, etc.

In analogy to the all-round CPUs, the AI processors are able to perform a large number of calculations per unit energy consumption. This happens because the smaller the transistors, the less the energy consumed during their operation. Further, smaller transistors are faster than larger devices.

Many processor chips have been fabricated with 5–7 nm process technology. Diameter of a silicon atom is 0.2 nm. So, transistors, the key devices of processor chips have reached a stage of miniaturization which is approaching the brink of physical limits. Beyond these limits, further downscaling places enormous economic load on semiconductor chip manufacturers. This has led to slowing down of Moore's law. Time taken to double the transistor density is therefore lengthening and a noticeable discordance with Moore's law is seen.

The slowdown of Moore's law shifted the emphasis from multifunctional, adaptive architecture towards specialized hardware. Likewise, the slackening of Denard scaling prompted engineers to divert towards multi-core processors. This happened because it was found that the performance could no longer be improved by increasing the operating frequency of a single core processor. The amendments along the above lines are not enough to meet the demands. The supremacy of the resourceful CPU is not challenged because it will have its unrivalled place in any system designed. But a heterogeneous system combining the versatile CPU with domain-specific hardware is urgently needed. The two together will provide a (flexible platform plus a computation platform with built-in specialization). Thus, DSAs evolved as a resilient platform.

Moving towards domain-specific hardware is the same thing as when we are at our wit's end, we can make a fresh start after taking stock of the situation. We can start from scratch by turning over a new leaf and remembering, 'We cannot direct the wind, but we can adjust the sails' (Dolly Parton quote). One should give due importance and respect to the customs of the people of a place or adapt ourselves to a certain situation by behaving as deemed suitable in that situation. When in Rome, do as the Romans do. Yes, the processors designed by making adjustments will give the results with the accuracy desired by the user to all intents and purposes but only for the nominated domains. Outside those domains, the usage of these processors may be less helpful.

5.3 Domain-specific hardware accelerators

5.3.1 Design techniques

As Moore's law is coming to an end, the domain-specific accelerator has become one of the few opportunities left for performance and efficiency upgradation. It is a hardware computing engine that has the specialization for solving problems of a particular domain of applications, e.g., image processing, graphics, deep learning and related tasks with orders-of-magnitude improvement in performance per unit cost paid and performance per unit power consumed. An accelerator is adjudged to be properly designed if it can cater to a broad range of applications rather than being restricted to a single or few applications. To avail the advantages of specialization, many existing applications must be refactored to decrease their bandwidth requests on global memory. Refactoring is the process of restructuring a code in such a way that its original functionality is not upset or altered in any way. The techniques used for designing these accelerators are briefly stated below (Dally *et al* 2020):

  (i) Data specialization: Specialized operations are carried out on domain-specific data types. Specialized logic is used to perform an inner-loop function to complete in one cycle the operations that normally require several cycles for execution. The inner-loop function is a function implemented by a loop nested inside another loop with the contained loop known as the inner loop.
 (ii) Parallel processing: Parallel units exploiting locality and making infrequent global memory references are highly beneficial.
(iii) Local and optimized memory: Key data structures are kept in small, local memories to increase memory bandwidth. Their compression provides bandwidth multiplication. Optimization of access patterns to global memory helps to attain a large memory bandwidth. For maximal

utilization of memory, the memory accesses are prudently scheduled and load-balanced across channels.

(iv) Overhead reduction: The overhead of program interpretation is significantly lowered by specializing the hardware.

5.3.2 Thumb rules for domain-specific architecture design

Five guidelines for DSA design are suggested following Hennessy and Patterson (2019):

(i) Minimization of distance of data movement using dedicated memories: Hardware designers, compiler writers and programmers can gainfully utilize their comprehension of the application domain for architecture design. They can select simpler and specialized memory hierarchies. These hierarchies are based on software-handled data transfers, whereas customized memories are used for particular functions inside the domain. This approach is helpful because the multi-level caches of general-purpose memory hierarchies use a large area and energy to guarantee optimal data transfer in a program.

(ii) Diversion of hardware resources to build more arithmetic units or bigger nemories: Hardware resources can be gathered by abandoning general-purpose architectural optimizations, e.g., out-of-order execution, prefetching, etc. The resources thus amassed should be used to utilize the existing parallelism by addition of more arithmetic units. Another possible course of action relates to solving any memory bandwidth problems with larger on-chip memories.

(iii) Using the easiest form of parallelism matching with the domain: The aimed domain application always exhibits an intrinsic form of parallelism. This parallelism is to be correctly recognized, advantageously used and exposed to software. The DSA is designed around the natural granularity of the parallelism. If the SIMD architecture works in the domain, the programmer will find it more convenient to use it than MIMD.

(iv) Reduction of data size and type to the simplest required for the domain: Narrower data types should be chosen to increase the effective bandwidth and utilization of memory on the chip. Adoption of these data types will lower the cost of data motion for memory-bound applications. It will also permit packing more arithmetic units in a given chip area.

(v) Using a domain-specific language for porting code to the architecture: Many such languages, e.g., TensorFlow for deep neural networks (DNNs) are gaining popularity in fostering the feasibility of porting (software adaptation) applications to DSA. In some domains, only a small, compute-intensive portion of the application must be run on the DSA. A clear-cut simplification of porting is thus provided.

5.4 AI-optimized hardware

This is a term used for computer hardware specially designed for AI purpose (Clan 2024). It can be referred to as AI-focused hardware. It is not a simple idea. The complications arise because of prevalence of a multiplicity of types, architectures, designs, and features of AI chips. Each of these is characterized by its own strengths and weaknesses. Each has its own requirements and challenges. Therefore, the AI chip choice is dictated by taking into account the targeted goals, the available budget, and user preferences. A careful comparison and evaluation of the AI chips is required based on the relevant criteria and metrics.

5.4.1 Classification according to application

AI-optimized hardware is broadly classified into two main categories:
  (i) AI training hardware: This hardware requires high computational power, memory bandwidth, and parallelism capabilities. These requirements emanate from the insistent necessity to handle the exceedingly massive amounts of data and calculations involved in training AI models. Generally, large datasets with complex algorithms and mathematical operations are used in training.
  (ii) AI inference hardware. This hardware requires low latency, power consumption, and cost. Its exigencies arise from the criticalities of running the AI models on new data using the trained parameters and weights to make predictions or decisions. The criticalities become apparent on noting that inference impels real-time or near-real-time demands to run AI models.

5.4.2 Classification according to flexibility

5.4.2.1 Programmable processors
Field-programmable gate arrays (FPGAs) are programmable processors that can be customized for AI training and inference. A vision processing unit built on FPGA for vision operation experiments is described in section 9.4. An FPGA prototype of the graph processor is discussed in section 11.6.

To tell more about FPGAs, we would like to remark that FPGAs are named as 'field programmable' because they provide the customers the ability to reconfigure the hardware inside them. They are integrated circuits that are frequently sold off-the shelf. They differ from traditional digital logic circuits constructed using discrete logic gates to provide fixed functions. They fall under the family of programmable logic devices (PLDs). The function of PLDs is undefined at the time of manufacturing. It is under the customer control and can be altered according to customer's thinking. The aforementioned extraordinary features of FPGAs make them suitable for AI tasks that require high adaptability, ingenuity and resourcefulness. Some of these tasks are machine learning, computer vision and natural language processing.

Internally, the FPGAs are reconfigurable digital logic circuits containing a matrix of configurable logic blocks (CLBs). The CLBs are surrounded by a system of programmable interconnects known as fabric (figure 5.1(a)). The CLBs and input/output blocks interconnections are made using horizontal and vertical routing channels and programmable multiplexers (PMs).

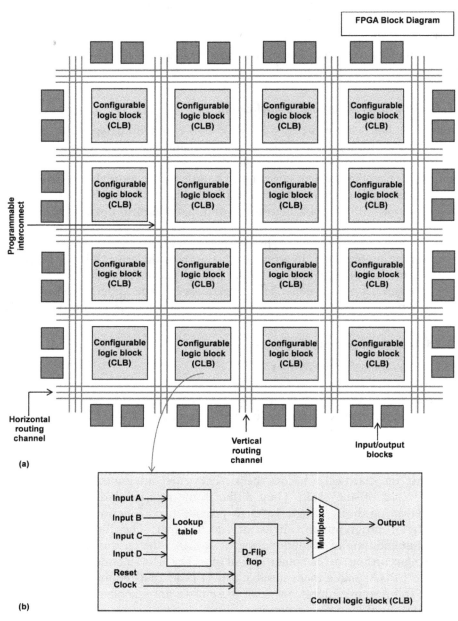

Figure 5.1. FPGA: (a) diagram showing the internal structure of FPGA and its various functional blocks, and (b) the components of a configurable logic block.

The CLB, figure 5.1(b), is a basic repeating logic resource on an FPGA. It is linked with routing resources to execute complex logic and memory functions. It synchronizes code on the FPGA. A CLB contains three components: flip-flops, look-up tables (LUTs), and multiplexers. The flip-flop (section 2.8) is a circuit capable of existing in two stable states, representing a single bit. It is a binary register used to save logic states between clock cycles on an FPGA circuit. The LUT in FPGA stores a predefined list of outputs for every combination of inputs. It thus provides a fast way to retrieve the output of a logic operation (sections 12.5–12.7 and 12.9). The multiplexer is a circuit that selects between two or more inputs.

The code for FPGA programming is written with a hardware description language, e.g., VHDL (very high-speed integrated circuit hardware description language), Verilog, etc. It is optimized for different AI tasks, algorithms, and models. So, it can provide high performance and flexibility. The FPGA does not have a fixed architecture and instruction set. It can perform the AI operations directly on the hardware without relying on software overhead, frameworks, libraries, and tools. An FPGA developer programs the physical hardware of the device. This methodology differs from the long-established method of writing the software to operate on a previously established hardwired circuit.

FPGAs are not limited in numerical representation. They can support different types of numerical representations. Fixed-point, binary, or custom representations are supported. The available choices make them useful when the AI task requires high precision and accuracy. Medical imaging, financial analysis, and scientific computing are a few such applications.

FPGAs provide hardware acceleration in cases where implementation of parts of algorithms in custom hardware helps to speed up designs. They also help in hardware prototyping. The prototyping can verify digital hardware designs, thus avoiding the high cost of application-specific integrated circuit (ASIC) manufacture. The non-recurring engineering (NRE) cost of an ASIC is very high. The NRE cost is the one-time investment into research, designing, developing and testing a new product. The exorbitantly high NRE cost of ASICs encourages device manufacturers to choose FPGAs for prototyping devices with low production volume. ASICs are preferred for devices with very large production volumes where NRE costs can be amortized across many devices.

The Virtex® Ultrascale+™ VU19P Device: This device is the world's largest FPGA. It has AMD UltraScale™ architecture. It is the first ASIC-class architecture enabling multi-hundred gigabit-per-second levels of system performance combined with efficient routing and processing data on-chip. It is co-optimized with the Vivado design suite. The Vivado is a software suite for synthesizing and analyzing hardware description language designs.

The Virtex® Ultrascale+™ VU19P is a third generation version of emulation-class devices of Xilinx, an American semiconductor company. It offers the highest logic capacity, interconnection and external memory bandwidth in an FPGA. This 16 nm device contains 35×10^9 transistors. It consists of four chips on an interposer (a thin electrical interface to spread/reroute connections). It has 9×10^6 logic cells, >2000 inputs/outputs, and 80 high-speed transceivers. Other notable features of the device include 1.5 Tb s^{-1} of DDR4 memory bandwidth and up to 4.5 Tb s^{-1} of transceiver bandwidth. It is tuned for extreme logic capacity, interconnection and bandwidth-intensive applications. Additionally, this device enables the implementation of large designs using fewer components. Advantages of software, intellectual property (IP), and tool flow maturity act as feathers in its cap (Verheyde 2019).

5.4.2.2 Dedicated processors

ASICs are dedicated processors that are customized for AI inference. They have exceptionally high performance, speed, efficiency, and reliability. These attributes impart to them compatibility for AI workloads requiring intensive processing power. On the down side, they lack flexibility and programmability. They are suitable for AI tasks that require low power consumption, cost, and latency, such as speech recognition, image recognition, and face detection.

The impact of an ASIC on AI processing is demonstrated by the achievement of extraordinarily high speeds while consuming severalfold less energy than regular chips by the tensor processing unit (TPU). The TPU is a custom-designed and developed hardware accelerator by Google. The TPU represents an illustrious example of an ASIC dedicated to AI. It is described in detail in chapter 7.

The journey from ASIC design to realization is a six-stage process involving:

(i) Application and specification finalization: The target application and performance goals are decided and documented. Accordingly, the requirements and specifications of the ASIC are laid out.

(ii) Architecture definition: The functional blocks, interconnections, and memory structures are organized and settled. These decisions are made by performing an in-depth analysis and establishing the requisite trade-offs among performance, power consumption, and area utilization.

(iii) Logic design: A detailed digital circuit representation of the ASIC is simulated and optimized for performance and functionality. The simulation and optimization are carried out using a hardware description language (HDL) like Verilog or VHDL.

(iv) Design verification: The ASIC design is rigorously tested and verified to ensure correct functionality and adherence to design specifications. Techniques like simulation, formal verification, and emulation are applied for this work.

(v) Physical design: The logical design is translated into a physical layout including geometrical topology planning, and routing. The layout is checked and finalized for manufacturing.

(vi) Manufacturing: The physical design is sent for fabrication, to a foundry. Here, the integrated circuit is manufactured using semiconductor technology.

5.4.3 Metrics of AI-optimized processors

Like regular processors, these processors are characterized for testing:

(i) Performance: This is a measure of speed and accuracy of the AI chip. It is evaluated by metrics such as teraflops, frames per second, precision, etc.

(ii) Efficiency: This is a measure of the utilization of available resources by the AI chip such as power, memory, and bandwidth. The efficiency is parametrized by watts per teraflop, gigabytes per second, joules per inference, etc.

(iii) Scalability: This measures the ease of handling of the increasing demands and complexity of the AI task. An idea of scalability is obtained by the number of cores, the amount of memory and the interconnect bandwidth.

(iv) Flexibility: This is an index of the versatility and adaptability of the AI chip for different AI tasks, applications, and environments. The number of supported AI tasks, frameworks, and numerical representations, the degree of programmability and reconfigurability constitute the flexibility criteria. The flexibility depends on AI chip instruction set, interface, and software ecosystem.

(v) Cost: This includes the acquisition, operation and maintenance cost of the chip. It varies with the AI chip quality, reliability, and durability. The chip warranty, support, and service are no less important.

5.5 Techniques of deep learning model compression

Evolution of edge AI has led to the deployment of AI algorithms and AI models directly on edge devices at the interfaces between devices with local network and the internet. The sensors or internet of things (IoT) devices made it necessary to convert large and complex models into simple models without compromising with accuracy. Ony the simplistic models can run on the low computing power and memory facilities that are at hand at the edge (Li *et al* 2023). Techniques developed for diminution of model size constitute model compression. These techniques reduce the model parameters to decrease the size of the model. Consequently, the RAM requirement during execution of the model and the storage space needed in memory are both diminished. An accompanying benefit is that the time taken by the model for making a prediction or inference is curtailed. The latency is thus significantly lowered.

5.5.1 Pruning

In pruning, redundant and inconsequential parameters are removed from the deep learning model. Generally, only the weights are removed. Biases are left untouched. In train-time pruning, the pruning process is directly integrated into the training phase of the neural network. The model is trained in such a way that the less important connections or neurons are removed. This method provides more efficient models because training is done with restricted data. In post-training pruning, the less important connections, neurons, or entire structures are identified and removed from a model which has been trained to convergence. This type of pruning offers simpler implementation. Figure 5.2 depicts the different types of pruning of NNs.

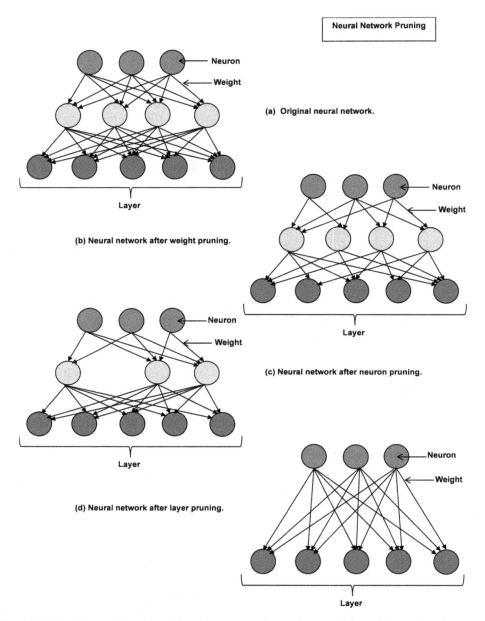

Figure 5.2. Neural network pruning: (a) starting unpruned neural network, (b) weight pruning, (c) neuron pruning, and (d) layer pruning.

5.5.2 Quantization

During quantization, precision of the weights and activations of an NN is decreased. For decreasing the precision, the weights and activations of the network are represented with fewer bits, e.g., 8-bit integers are used instead of 32-bit

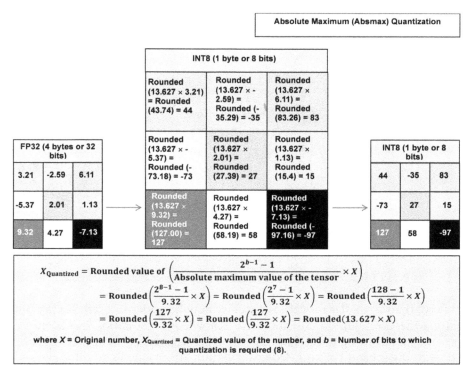

Figure 5.3. Absolute maximum method of quantization from single-precision floating-point format (FP32 or float32) numbers to INT8 data type representing signed 8-bit integers with a range between −128 and 127.

floating-point numbers (figure 5.3). As a result, the storage requirements for the model plummets. Its computational complexity too is lowered. Accuracy degradation is minimized by careful quantization. All the same, a small precision loss is unavoidable. Quantization is recommended for massive and resource-intensive models. In these models, precision loss does not adversely affect performance, e.g., in text classification and natural language processing. On the whole, quantization reduces the memory footprint requirement and computational complexity of a model. Resource-constrained devices, e.g. mobile phones, IoT and edge devices are common beneficiaries of this model compression technique.

5.5.3 Knowledge distillation

This is also called model distillation. It is a model compression and acceleration technique which involves the transference of knowledge from a large model to a model of comparatively small size without losing validity (Gou *et al* 2021). The large model is called the teacher model. It can be a DNN. The smaller size model is known as the student model. It is a shallow NN. Figure 5.4 illustrates the basic idea. The smaller model can be evaluated at a lower cost on a computationally weaker hardware device such as a mobile phone or an embedded device. Knowledge distillation is possible because although a large model has more capacity for

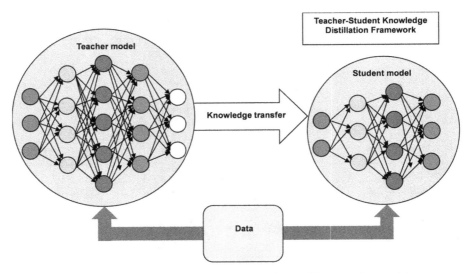

Figure 5.4. Knowledge distillation from a teacher model to a student model.

knowledge than a smaller model, its capacity may be under-utilized. It may only be escalating the computation cost incurred. Knowledge distillation has been used in visual and speech recognition, natural language processing, and other fields of AI.

The learning schemes of knowledge distillation are grouped into three classes. The classification is made on the basis of simultaneous/non-simultaneous updating of the student model with the teacher model (figure 5.5):

  (i) Offline distillation: Here the pre-trained knowledgeable teacher imparts knowledge to an untrained student.

 (ii) Online distillation: The teacher and student study and learn together with each other. So, the teacher model and the student model are updated at the same time. In this case, the knowledge distillation framework is end-to-end trainable.

(iii) Self-distillation: One model plays two roles. It acts both as teacher and student models. Thus, learning on one's own principle is followed. An example is the extraction of knowledge from the deeper sections of a network into its shallower sections.

5.5.4 Low-rank matrix factorization

A low-rank matrix is one whose rank is much smaller than the minimum number of rows and columns that the matrix can have. The rank of a matrix is the maximum number of linearly independent columns or row vectors that the matrix has.

Low-rank matrix factorization (MF) is a technique in data science which is used to factorize a given matrix into a product of two matrices with low dimension. The technique is used for model compression and acceleration. The underlying notion is that a piece of data contains latent structures. By uncovering these latent structures, it is possible to express the data in a compressed form. Here, matrix and tensor

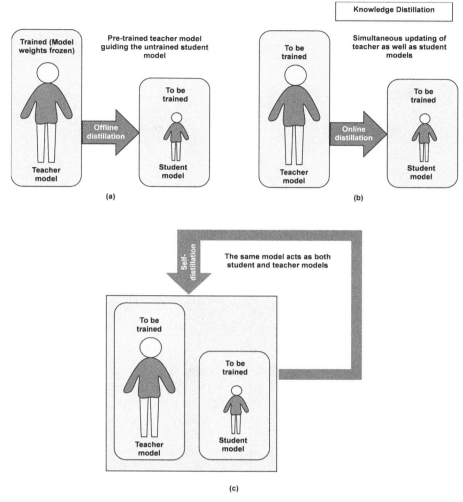

Figure 5.5. Knowledge distillation schemes: (a) offline, (b) online and (c) self.

decomposition are used to identify the redundant parameters of DNNs. Dimension reduction, clustering and matrix completion are done by matrix factorization. This allows the rapid execution of various computations (Chen *et al* 2018).

The technique is an approximation method for representation of the information in a matrix M using a matrix of rank that is smaller than the original matrix. Reduction of the rank of the matrix is accomplished by constructing it as a product of a tall but narrow left-hand matrix L_k and a short but wider right-hand matrix R_k^T as (figure 5.6)

$$M = L_k R_k^T \qquad (5.1)$$

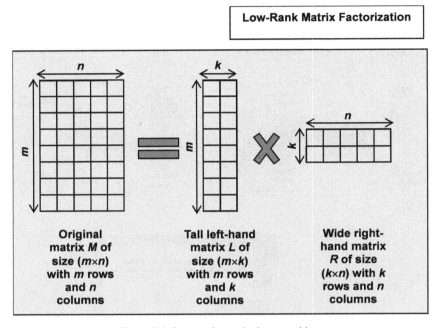

Figure 5.6. Low-rank matrix decomposition.

A matrix M of size $(m \times n)$ and rank r is decomposed into two matrices L_k and R_r. Exact reconstruction of matrix M is obtained from the decomposition when $k = r$. But when $k < r$, then the decomposition represents a low-rank approximation of M.

DNNs are used for large vocabulary continuous speech recognition (LVCSR) tasks. For achieving a good performance, the networks are trained with many output targets. Most of the training parameters lie in the final weight layer. A low-rank matrix factorization was applied to the final weight layer of DNNs. It was observed that for three different LVCSR tasks ranging between 50 and 400 h, the number of parameters of the network decreased by 30%–50% by such factorization. The diminution in the number of parameters yields an equivalent reduction in training time. The time reduction is achieved without sustaining any extra loss in final recognition accuracy than a full-rank representation (Sainath *et al* 2013).

5.6 Sparsity enforcement in deep learning

5.6.1 Advantages of sparsity

Sparsity-enforcing algorithms seek to design models using a small number of significant features or neurons at a time. This type of model designing is done by disregarding or ignoring the less relevant features. This is a reasonable idea because many of the network parameters, either weights or connections, are zero or close to zero. On that account, only a small subset of elements that contribute significantly to the model are useful for network representation. A majority of elements, being inactive, have minimal and hence negligible impact. They can therefore be ignored (Hoefler *et al* 2021).

The smaller number of non-zero parameters in sparse networks decreases the size of the model. The memory requirement is commensurately reduced. Its training time is shortened too. A sparse model shows higher computational efficiency during inference. Moreover, the introduction of sparsity improves generalization of the model to unseen data. It thus prevents overfitting. On these grounds, the memory footprint of regular networks is decreased to fit mobile devices.

5.6.2 Sparsity-induction methods

L1 regularization (Lasso or least absolute shrinkage and selection operator) is a technique used for sparsity-induction. This technique works by addition of a penalty term to the loss function. This penalty term is proportional to the absolute values of the weights associated with each feature variable. The amount of regularization applied is controlled by multiplying the penalty term with a regularization parameter λ. Regularization is performed to improve the NN generalization by decreasing overfitting. The penalty term in the loss function encourages sparsity by forcing many weights to zero value. However, other weights shrink less dramatically or remain unaffected. This optimization process helps in considerable simplification of the model.

Sparsity is introduced in activation by using dropout or dropout variant techniques during training. Here, a percentage of activations are randomly set to zero. Dropout alleviates overfitting. Structured sparsity is enforced on a group of weights. Other sparsity-induction techniques include pruning and quantization.

5.7 Data compression for saving storage space and memory bandwidth

5.7.1 Need for data compression

Memory bandwidth is a crucial restraint to system performance in many AI workloads. Data compression helps to mitigate the data storage and transmission issues (Jayasankar *et al* 2021). It has been explored as a method to save storage capacity and memory bandwidth with reference to on-chip caches, interconnections, and main memory. The upshot is that less data needs to be transferred from/to DRAM and the interconnections.

It is found that JPEG compression improves the image classification accuracy of DL (Yang *et al* 2021). Also, accuracy of machine translation is increased by directly learning from the compressed training data which are supplied as pairs of sentences in the target and source languages.

5.7.2 Data recoding

Data compression involves recoding or restructuring the information for reducing its size. This is done by using fewer bits as compared to the original representation of the information. Hence, it is also called source coding, or bit-rate reduction. The device producing data compression is called an encoder. The device reversing data compression, i.e., undertaking data decompression is known as a decoder. Data compression is done by a program using an algorithm. In this algorithm, a larger string of bits is represented with a smaller string. To this end, a reference dictionary is applied for interconversion between the two representations. Sometimes a formula is used. In this case, a pointer is inserted to a string of data already seen by the program. Whenever this data string is observed, the formula converts the concerned data string into a single bit while the primary information is preserved.

5.7.3 Lossless compression

Lossless data compression causes no loss in information. It is possible to do so because most real-world data exhibit statistical redundancy. This happens because the same piece of data exists in multiple places. For instance, the colored regions in an image may not change over several pixels. Data is compressed in a lossless manner by identification and elimination of such redundancies.

5.7.4 Lossy compression

In lossy data compression schemes, the limitations of perception of data by humans are exploited. The human eye is more sensitive to subtle luminance variations than to the color changes. These perceptual differences are utilized in compressing visuals for video and images. Non-essential bits of information are thereby rounded off.

5.8 Zero-skipping method

5.8.1 Necessity of CNN workload reduction

Deep NN technology has permeated into edge devices. Notable examples are autonomous driving, medical equipment, and robotics. CNN models are well-established for image classification, object detection, and segmentation (Kitayama *et al* 2021). With the deepening of its layer structure, the recognition accuracy of a CNN increases. In concert, the amount of computation also increases. Each image requires ~10–100 giga-operations (GOPs). Low power consumption is highly desirable for edge devices using CNNs. So, network compression without degradation of recognition accuracy is essential. Nevertheless, 70%–90% of the total convolution operation time is devoted to multiplication and accumulation (MAC) operations. A strategy to bring down the power consumption and increase the speed of operation is the need of the hour. The zero-skipping technique is one such method.

5.8.2 Application to CNN processing

Reasonably low power consumption and high speed are achieved during CNNs processing by adopting the zero-skipping method. In this method, multiplication and accumulation (MAC) are terminated as soon as the multiplication results of the input data and weight become zero. Conventional zero-skipping employs several OR circuits. Hence, the circuit overhead is increased. By combining two methods, viz., zero-skipping and active data-skipping, power consumption reduction -52.3% and low circuit overhead -45.9% are obtained. Only a small accuracy degradation (-0.29%) takes place compared with the conventional zero-skipping method (Kitayama *et al* 2021).

5.9 Low-precision calculations

5.9.1 A method for situations permitting compromising with precision

It is observed that NNs put a load on the memory system during hardware optimization for increasing computation speed. This happens when fetching the model parameters and data from the memory. Memory capacity and bandwidth are pressurized during to-and-fro data movement between the compute engine and the memory. The pressurization arises because the training sets and model parameters are lodged in the local memory. Such lodging is necessary to circumvent the slower data transfers over the PCIe connections. Computation latency and power consumption are important performance metrics. Applying reduced precision to improve these performance metrics is increasingly being used in high performance computing (HPC). Especially error-tolerant applications are selected for this mission (Cherubin and Agosta 2020).

5.9.2 Advantage gained by reduction in precision

Reduced-precision computing is a method of increasing power efficiency and decreasing memory bandwidth demand for NNs. Often 16, 8, and in some cases fewer bits of precision are sufficient in place of computing with 32-bit floating-point numbers. That is why the AI algorithms are implemented successfully by using a smaller number of transistors. Here, the higher tolerance of NNs to reduced precision is utilized. Frequently substantial power saving is obtained if computing on the least significant bit is avoided.

5.10 Faster memory access

5.10.1 Non-uniform memory access

This involves provision of faster memory ingress by non-uniform memory access. The non-uniform access style requires building multiprocessor architecture systems. In these systems, each processor is provided with its own local memory. Here access time depends on the location of the memory with respect to the processor. This provision reduces memory access times and improves general system performance. Non-uniform memory access processors are highly scalable. They can handle large workloads by including several processors and local memory nodes. However, if the same is done in a uniform memory-access system, there is serious contention of memory bus. The contention degrades the performance of the system. In this case, the additional cores or multiple processors used have to wait in queue for accessing memory.

5.10.2 Memory interleaving

For memory interleaving, the memory is divided into multiple banks or modules. These banks are organized in such a way that consecutive memory locations are stored in different banks. Hence, multiple memory locations are accessible in parallel. As a result, the time required for fetching data is decreased. Apart from shortening the memory access time, memory interleaving also improves bandwidth and reduces memory contention. Thus, it upgrades system responsiveness and overall performance.

Memory interleaving is of three types:
   (i) Byte interleaving: Here consecutive bytes are stored in different modules.
   (ii) Word interleaving: Here consecutive words are stored in different segments.
   (iii) Cache interleaving: Here consecutive cache lines are stored in different cache sets.

On the other hand, in a regular system without such interleaving, the memory is a single block. In this system, access to consecutive memory locations is obtained in a sequential fashion only. An unacceptable, annoyingly high latency is noticed in systems where memory access time is slower than processor speed.

5.11 Summary and the way forward

5.11.1 Highlights of chapter 5 at a glance

   (i) Constraints imposed by physical downscaling limits of chip dimensions led to the evolution of DSA. Domain-specific hardware accelerators were described including their design techniques by summarizing thumb rules for their design.
   (ii) AI-optimized hardware was classified according to application and flexibility into programmable and dedicated processors. The benchmarks of AI-optimized processors were mentioned.
   (iii) DL model compression was done by network pruning, quantization, knowledge distillation and low-rank matrix factorization.
   (iv) Advantages of network sparsity enforcement in DL and sparsity-induction methods were divulged.
   (v) Data compression for saving storage space and memory bandwidth was done by data recoding through lossless and lossy compression methods.
   (vi) Zero-skipping method was illustrated by the necessity of CNN workload reduction in in the context of its application to CNN processing.
   (vii) 'Low-precision calculations' is a method applicable to situations permitting compromises with precision. The advantage gained by reduction in precision was explicated.
   (viii) Techniques of availing faster memory access via non-uniform memory access and memory interleaving were presented.

(ix) The essence of this chapter is that AI hardware must be dictated by an engineer's preferences about the size and the type of datasets to be worked upon. Due consideration must be given to the AI algorithms that are to be implemented. In step with the ever-growing data in AI, efforts must be made for hardware redesign, tailoring and optimization with the aim to suit the particular needs of AI workloads for enhancement of performance and throughput of the processors. It may be noted both the training and inference phases of AI models are served by the hardware. During training, the hardware oversees iterative calculations. In the inference stage, new input data are fed to the trained model to make predictions on the hardware.

(x) Hardware optimization with respect to AI needs can greatly speed up the training of more complex models. The same was hitherto done on inappropriate computer architectures. Also, once a specialized hardware has been developed, the AI models are able to perform complicated calculations more rapidly. Engendered improvements in real-time decision-making skills enable enterprises to confront more formidable situations, resulting in breakthroughs in various AI applications.

(xi) Key words for this chapter are: DSA, AI-optimized hardware, model compression, network pruning, quantization, knowledge distillation, sparsity, low-rank matrix factorization, zero-skipping, data compression, low-precision computing, non-uniform memory access, memory interleaving.

5.11.2 Getting ready to begin chapter 6

(i) With the inexorable advances in architectural designs, it became abundantly clear to computer engineers that graphical processing units (GPUs) are significantly influencing the development of AI-optimized hardware. We know that GPUs were initially accountable for rendering images, videos, and sequences of images called animations on a computer screen in order to display high-quality graphics in video games and multimedia content. Researchers discovered that GPUs can be used in an entirely new role as AI processors to speed up calculations.

(ii) Calculations on DL models extensively involve matrix calculations. These calculations can be effectively done on GPUs because of their capacity to work in massively parallel manner. Therefore, the quest for specialized hardware architectures immediately prompted scientists to use GPUs in graphics cards in the form of AI-optimized hardware. They made suitable hardware/software modifications and developed new methods to put this idea into action. This arduous task was pioneered by companies making advanced chips for future AI factories such as the large multinational corporation and technology company NVIDIA with their CUDA framework and GPUs. The AI processing ability of GPU is a nice feather in its cap. It constitutes the contents of the next chapter.

References

Chen B, Yang Z and Yang Z 2018 An algorithm for low-rank matrix factorization and its applications *Neurocomputing* **275** 1012–20

Cherubin S and Agosta G 2020 Tools for reduced precision computation: a survey *ACM Comput. Surv.* **53** 1–35

Clan 2024 *AI-Optimized Hardware* 2024 Kocoa Technologies Private Limited https://clanx.ai/glossary/ai-optimized-hardware

Dally Y W J, Turakhia Y and Han S 2020 Domain-specific hardware accelerators *Commun. ACM* **63** 48–57

Dolly Parton quotes Copyright ©2001–2024 BrainyQuote https://brainyquote.com/authors/dolly-parton-quotes

Gou J, Yu B, Maybank S J and Tao D 2021 Knowledge distillation: a survey *Int. J. Comput. Vision* **129** 1789–819

Hennessy J L and Patterson D A 2019 Domain-specific architecture *Computer Architecture: A Quantitative Approach* 6th edn (Cambridge, MA: M K Morgan Kaufmann Publishers) ch 7 p 543

Hoefler T, Alistarh D, Ben-Nun T, Dryden N and Peste A 2021 Sparsity in deep learning: pruning and growth for efficient inference and training in neural networks *J. Mach. Learn. Res.* **22** 10882–1100

Jayasankar U, Thirumal V and Ponnurangam D 2021 A survey on data compression techniques: from the perspective of data quality, coding schemes, data type and applications *J. King Saud Univ.–Comput. Inf. Sci.* **33** 119–40

Kitayama A, Ono G, Kishimoto T, Ito H and Kohmu N 2021 Low-power implementation techniques for convolutional neural networks using precise and active skipping methods *IEICE Trans. Electron.* **E104.C** 330–7

Li Z, Li H and Meng L 2023 Model compression for deep neural networks: a survey *Computers* **12** 1–22

Sainath T N, Kingsbury B, Sindhwani V, Arisoy E and Ramabhadran B 2013 Low-rank matrix factorization for deep neural network training with high-dimensional output targets *2013 IEEE Int. Conf. on Acoustics, Speech and Signal Processing(26–31 May) (Vancouver, BC, Canada)* pp 6655–9

Verheyde A 2019 Tom's Hardware, News: Xilinx introduces world's largest FPGA with 35 billion transistors https://tomshardware.com/news/xilinx-world-largest-fpga,40212.html

Yang E-H, Amer H and Jiang Y 2021 Compression helps deep learning in image classification *Entropy* **23** 1–19

Chapter 6

Graphical processing unit

The historical background and evolution of the graphical processing unit (GPU) are recalled. The compute unified device architecture (CUDA) program structure and flow are explained. The concepts of thread, thread block, synchronized functioning of threads, kernel grid, grid hierarchy and warp are enumerated. Their hardware perspectives are provided with an explanation of the Fermi architecture of a GPU. Streaming multiprocessors (SMs) are explored along with the stream processing concept and the programming model. Reasons for enhancements in computational and communication efficiencies by stream processing are clarified. A description of the applications showing expediency of stream processing is followed by the structuring the GPU hardware to utilize the parallelism features of the application. Merits of the single-instruction multiple-thread (SIMT) model are pointed out. Its comparison with single-instruction, multiple-data (SIMD) architecture is made. Parallel thread execution (PTX) and the GigaThread controller are elaborated upon. Cache and memory hierarchy consists of registers; L1 (Level 1) cache + shared memory and L2 (Level 2) cache subsystems; and the local, global, constant and texture memories. Thread-block and warp scheduling systems comprise the thread block and dual warp schedulers, context switching and warp scheduling policies. Commercial GPU examples are cited.

6.1 Introduction

The central processing unit (CPU) has carved its own prestigious niche in computing and will never lose its importance (Powell and Smalley 2024). Removal of the CPU means that we no longer have a computer, not by a long shot. The CPU is designed to handle a variety of tasks, e.g., running various software programs, text editing and browsing the internet. The saying, 'Jack of all trades is oftentimes better than a master of one' is applicable in such situations just as a person proficient in many jobs but not an expert in one particular area can do better in many circumstances than the technicalized one highly skilled in one or two jobs only. But when those highly

doi:10.1088/978-0-7503-6259-7ch6
6-1

skilled jobs are to be done, the person with deep expertise in them will do much better than the generalized one beyond a shadow of doubt. In one case, we are talking about versatility or generalization, in the other case about specialization.

The GPU, which was already highly competent and experienced in video display, added another notch in its belt by venturing into artificial intelligence (AI) processing. The GPU is playing a dominant role in high-performance computing (HPC). The HPC deals with large computations whose requirements on memory and raw computational speed cannot be met by a single CPU as used in a desktop computer (Glaskowsky 2009, Ko *et al* 2022). On the other hand, the GPU is built by prioritizing parallel computing tasks. So, it is becoming the foundation of AI computing in applications involving extensive parallelism. Unlike the CPU, the GPU is highly suited to applications in which:
  (i) the computational requirements are large,
  (ii) appreciable parallelism exists, and also,
  (iii) throughput is prioritized over latency (Owens *et al* 2008).

This chapter will delve into the capabilities of GPUs for AI work. It will lucidly explain the unrivalled, coveted position occupied by GPU in present-day AI computing, and articulate the features that make it inimitable and idiosyncratic.

6.2 The GPU evolution

The GPU reached its present form after progressing through several stages (Hello CUDA-GPU series#1 2023):
  (i) The GPU began in its primitive version as a simple graphics adapter. In its original life, it was used as a video interface for displaying scenes on a monitor by accessing the CPU memory over a bus (figure 6.1(a)).
  (ii) During toddlerhood of GPU, specific graphics memory was incorporated in GPU to offload the CPU, resulting in faster graphics display. This renovation made the CPU memory free to function in parallel for performing other operations (figure 6.1(b)).
  (iii) During the early teenage years of the GPU's life, processors were added to the GPU for performing simple operations like rendering textures. These processors were slower than CPUs but they improved the graphics experience. They further offloaded the CPU. The processors on a GPU are called stream processors, or SPs for reasons to be explained later (figure 6.1(c)).
  (iv) In the later teenage years of GPU, more processors were added to GPU, enabling the execution of complex video operations. Unlike CPUs, the emphasis in GPU processors was on performing multiple simple operations in parallel (figure 6.1(d)).
  (v) Further ahead, the CPU began adding more cores too, enabling increased parallel processing. This upgradation for CPU received wide support because it was beneficial to add more cores than to make individual cores faster. Now the CPU was able to handle more complex workloads and the GPU became more capable. Therefore, the GPU started to be used for

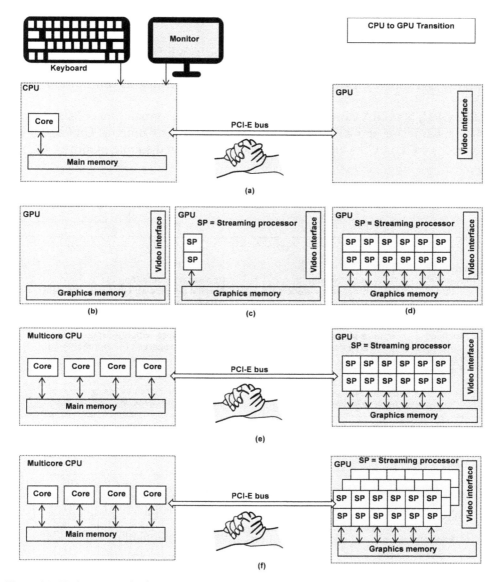

Figure 6.1. Various stages in the progress of GPU: (a) its primitive version, (b) the toddlerhood, (c) early teenage years, (d) later teenage years, (e) further ahead, and (f) attaining adolescence.

non-video processing, by offloading from CPU the compute tasks which were highly parallel (figure 6.1(e)).

(vi) Even more processors were added to the GPU as its usage for non-video tasks increased (figure 6.1(f)). The GPU matured into adolescence.

The parallel processing achieved with a GPU is easily imagined by looking at figure 6.2.

6.3 GPU launching

NVIDIA launched the GeForce 256™, the industry's first fully integrated graphics processing unit in August 1999 (NVIDIA corporate timeline 1999, Peddie 2022). NVIDIA GeForce RTX 40 series was launched for super-fast AI (Burnes and Pacelli 2024).

The GPU, NVIDIA Corporation, USA, is a high-performance parallel processor. It has a hardware and software architecture known as Compute Unified Device Architecture CUDA®, which is a proprietary and closed-source platform. CUDA

CPU vs GPU

Attribute	CPU	GPU
Purpose	General computing	Specialized computing
Skills	Serial processing proficiency with task parallelism	Parallel processing expertise with data parallelism
Latency, throughput	Low latency, small throughput	High latency, large throughput
Pipelines	Shallow pipelines with <30 stages	Deep pipelines with hundreds of stages
Distribution in Allocation of area on the chip	Large area dedicated to cache and control	Large area reserved for arithmetic-logic units
Cores	Fewer (tens or less), powerful cores with large memory	Large number (thousands or more) of less powerful cores with less memory

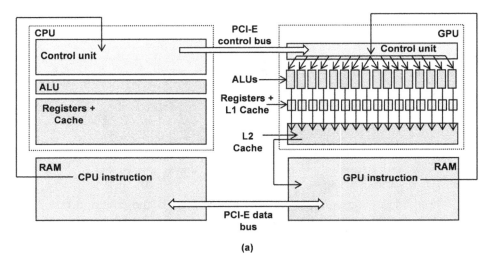

(a)

Figure 6.2. CPU, GPU and the visualization of parallelism in a GPU: (a) the CPU and GPU comparison and connection, and (b) the multiplicity of computing processes taking place in a GPU at the same time.

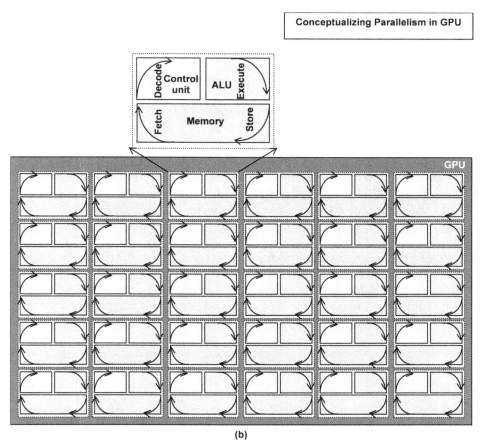

Figure 6.2. (Continued.)

enables NVIDIA GPUs to execute programs written in C, C^{++}, and other languages (Soyata 2018, NVIDIA Research 2024).

As done previously in section 6.2, it is essential to reiteratively point out the subtle distinction between a GPU and graphics or video card. We know that a CPU is inserted into a socket in the motherboard of a computer. Similarly, the graphics card is the add-in board on which a GPU is mounted. It includes components required for GPU functioning and connecting it to the remaining system. The GPU sits alongside VRAM (video memory), HDMI (high-definition multimedia interface) or display port and a component for cooling. Sometimes the GPU is built directly into the motherboard. Occasionally, it is integrated as an all-in-one chip along with other components. Interchangeable use of the terms 'graphics processing unit' and 'graphics card' frequently creates confusion, and should therefore be avoided.

High-end GPUs for machine learning generate a lot of heat, and must be properly air-cooled or liquid-cooled to avoid reduction in their lifespan. Air cooling is easier and requires little maintenance. In direct-to-chip cooling, the chilled liquid is pumped over the GPU via a cold plate attached to it. In immersive cooling, the chip is submerged in a non-conducting liquid filled in a tank for absorption of heat. The warm liquid is cooled using a heat exchanger. The cooling load is further reduced with a rear door heat exchanger (RDHx) in which a rear door with fans and liquid is affixed to the rack of the server.

6.4 CUDA program structure

Not a stand-alone computing device, the GPU is combined with a CPU. A well-defined division of labor between the two units is made (Aamodt *et al* 2018). The CUDA programs are executed in a heterogeneous environment in two separate places: the CPU and GPU. The two chips are used in a co-processing mode. In this partnership, the CPU is called the host and GPU is referred to as the device. The host and the device have their own separate memory spaces. The CPU memory is known as host memory and the GPU memory is called device memory. Both the host and device memories can be managed from the host code.

A CUDA program contains codes planned for both CPU and GPU: the host code and the device code. So, the CUDA C/C^{++} code is a blending of C (or C^{++}) and CUDA. For its compilation, a specially designed compiler, the NVCC (Nvidia CUDA Compiler), is used. This compiler can appropriately compile the C/C^{++} binaries as well as CUDA binaries (figure 6.3). The NVIDIA CUDA compiler

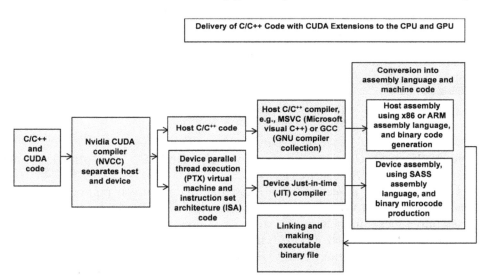

Figure 6.3. Block diagram showing the steps involved in the delivery of C/C^{++} code with CUDA extensions to the CPU and GPU. The input code is analyzed by an NVCC compiler. The host and device codes are separated. The host code is a C/C^{++} code and the device code is a PTX virtual machine instruction set architecture (ISA) code. The host code is compiled by a standard CPU compiler, MSVC or GCC. The device code is handled by a JIT compiler. An x86 assembly code is produced for the host program and SASS assembly code for device program. The respective assembly codes are converted into machine codes. From these machine codes, a binary file is constructed for regulating the operations of CPU and GPU.

(NVCC) analyzes the CUDA C/C^{++} code. Then it separates the host code from the device code by looking for special keywords for labelling data-parallel functions called kernels. The host code is compiled by GCC (GNU Compiler Collection). The device code is compiled by the CUDA compiler of NVIDIA. Compilation of the host code produces a GPU-specific machine code PTX (parallel thread execution). The PTX is an intermediate representation for the device code. The PTX code is translated in the just-in-time (JIT) compiler by the NVIDIA GPU driver. The translation generates the GPU-specific machine code SASS (scalable assembly). In just-in-time compilation, any PTX code loaded at run time by an application is compiled further to binary code by the device driver. This helps in delivering to that application the advantages offered by compiler advancements with each new device driver. However, the application load time is prolonged (NVIDIA 2024).

Further optimization is performed by NVCC personalized to the recipient GPU architecture. After the optimization, the compiled host code and the translated GPU machine code are combined together. They are properly linked to form a binary executable file for GPU operation.

Any application consists of two parts:

 (i) Sequential part: This includes management of data, transfers between memories, and configuring the GPU execution. This part is implemented on the CPU (host).
 (ii) Parallel part: The compute intensive parallel operations are implemented on the GPU.

The host code runs the C/C^{++} program on the CPU. The kernels run on the GPU. To clarify, a kernel is a function that can be called from the host and executed on the device through several parallel processes. The function is made of a sequence of statements called the function body. Values are fed in the function and the function returns a result.

6.5 CUDA program flow

A process containing parallelizable tasks is executed by heterogeneous computing (Wang and Kemao 2017). The process is addressed by programmable models and languages for heterogeneous computing (Fumero *et al* 2019). The serial workload is placed on CPU. The parallel computation is offloaded to GPU. With this planning, the program execution steps are displayed in the flowchart shown in figure 6.4.

 (i) Loading input data into host memory: The code in C or C^{++} fetches data. It stores the fetched data in host memory.
 (ii) Host-to-device transfer: The input data are copied from host memory to device memory. The data needs to be copied because the host and device have isolated memory spaces. They cannot directly access each other's memory. So, the data transference between the host and device is done via the peripheral component interconnect (PCI) bus.

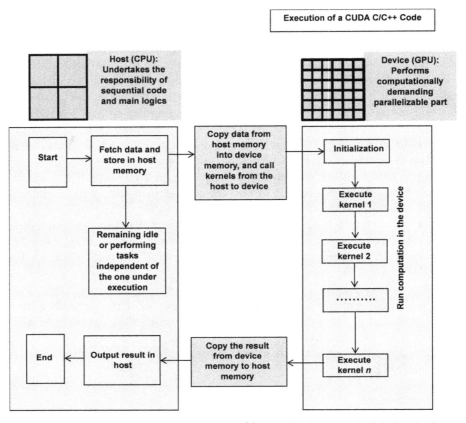

Figure 6.4. Data flow during execution of a CUDA C/C^{++} code. The flow starts by bringing the data to the host memory. Then the data is copied to the device memory. This is followed by calling the kernels from host to device, and executing the program in the device. After program execution, the result is copied to the host memory. The flow ends by outputting the result in the host.

(iii) Calling GPU kernel: From the host, a kernel is called for performing computations in the device. A kernel is a parallel-data function that runs on the GPU, written in CUDA. The different kernels are scheduled and executed on the parallel hardware of the device.

(iv) Device-to-host transfer: The output results after completion of computation are copied from the device memory to host memory.

(v) Printing the results on the host printer and using them.

6.6 Thread, thread block and grid hierarchy

6.6.1 Thread

Recall section 3.11. A thread represents the smallest series of inter-related instructions that participate in a process (figure 6.5). Each thread has a per-thread private memory or thread-private memory. This private memory is only accessible by the thread allocating it. Each thread block has a per-block shared memory. This per-block shared memory is shared by all threads in the block for reading data from and writing data to that shared memory. The per-block shared memory is utilized for inter-thread communication.

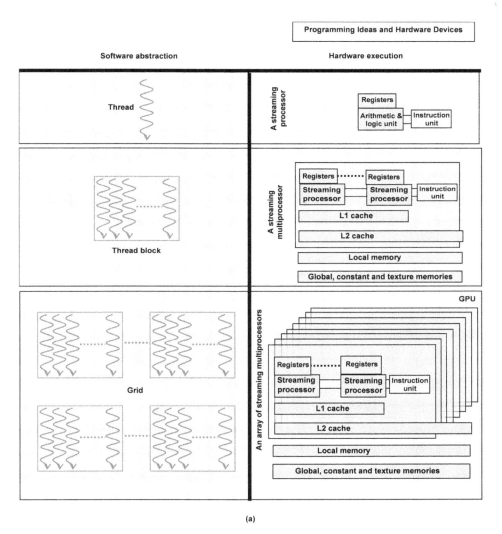

(a)

Figure 6.5. GPU conceptualization to realization: (a) Establishing a correlation between programming ideas and the hardware devices used for their practical implementation. A thread is executed by a streaming processor. A thread block is executed by a streaming multiprocessor. A grid is executed by an array of streaming multiprocessors. The array of multiprocessors constitutes the GPU. (b) Simplified drawing showing GPU chip components: SMs with individual L1-cache and shared L2-cache, global, constant and texture memories. The multiple layers displaying multiprocessors in (a) are represented by a single layer in (b).

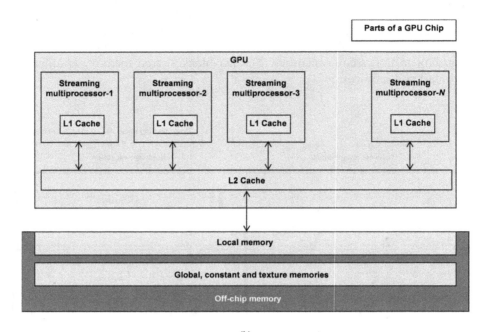

(b)

Figure 6.5. (Continued.)

The thread works as a basic processing element. It can read data from a shared global memory. This is called the 'gather operation'. The thread can also write data to the shared global memory in arbitrary locations. This is 'scatter operation'. Data exchange between threads in the same block is not allowed. However, the threads in a block can send data to threads in other blocks. Data sharing and result sharing are done in parallel algorithms.

6.6.2 Thread block and synchronized functioning of threads

A thread block is a group of concurrently executing threads. It has an x, y, z index. The x, y, z index is a three-component vector providing a unique identification of the block. The threads in the thread block cooperate among themselves. This cooperation takes place through barrier synchronization and shared memory, explained below:

   (i) Barrier synchronization: In barrier synchronization, the states of threads sharing the same resources are synchronized at two levels:

      (a) Thread block level in which a waiting period is allowed for all the threads in a thread block to attain a common synchronization point.

      (b) Block level in which the calling user thread is blocked until completion of all the work on the device.

   (ii) Shared memory synchronization: This deals with the coordination and ordering of access to shared memory when multiple threads are trying to access and modify shared memory data simultaneously. Then a synchronization mechanism is necessary to get correct and consistent results. This is done by calling a simple programming language element barrier called the barrier synchronization primitive to avert the race condition. Execution of a thread can only continue past this primitive after all threads in its block have executed the synchronization primitive.

6.6.3 Kernel grid

A kernel grid is a collection of thread blocks executing the same kernel. Blocks are organized along a line. They can also be arranged in a rectangular format. The threads inside a block have an x, y, z index within the block. Grids of thread blocks are executed on the device. They read inputs from global memory, and write the results back to global memory. They also synchronize between dependent kernel calls.

6.6.4 Warp

A warp is a bundle of threads that are scheduled together for execution of the same instructions in a lockstep fashion. The same instructions are executed in each cycle but on different data.

6.7 Hardware perspective

Threads, thread blocks and kernel grids constitute the fundamental components of CUDA compute hierarchy from the programmer's perspective. This hierarchy maps to a hierarchy of processors on the GPU. The GPU works on the kernels, which are launched in kernel grids. At a time, one kernel is executed. The execution is done via thread blocks. A thread block is executed by one of the several SMs (section 4.7.4), arranged in graphics processor clusters (GPCs). Several thread blocks run concurrently on one SM. A thread is executed by one of the large numbers of SPs or CUDA cores inside the SM. These work on warps, the groups of threads that follow the same set of instructions. Owing to execution of warps of threads in lockstep, multiple cores share the same instruction control units. This operational pattern is highly beneficial for simplification of the processor design. Briefly, a kernel is executed by the complete GPU. The thread block is executed by the SM. The warps of threads are executed by the streaming processor (SP or CUDA core). Hence, we can write:

Kernels → GPU, thread block → SM (GPC), warps of threads → CUDA core or SP.

6.8 Fermi architecture

A major advancement in GPU is the CUDA compute and graphics architecture, code-named 'Fermi architecture' (figure 6.6(a)). Its salient features are given in subsections ahead (Wittenbrink *et al* 2011).

The CUDA core used in this architecture consists of a dispatch port, an operand collector, a floating-point unit, an integer unit and a result queue (figure 6.6 (b)).

The tensor core was first introduced in NVIDIA Volta architecture. It is also used in the NVIDIA Ampere architecture. It is shown in figure 6.6(c) for the sake of comparison with figure 6.6(b) (NVIDIA Ampere architecture 2024). It performs matrix multiplication and addition using a specially designed numeric floating-point

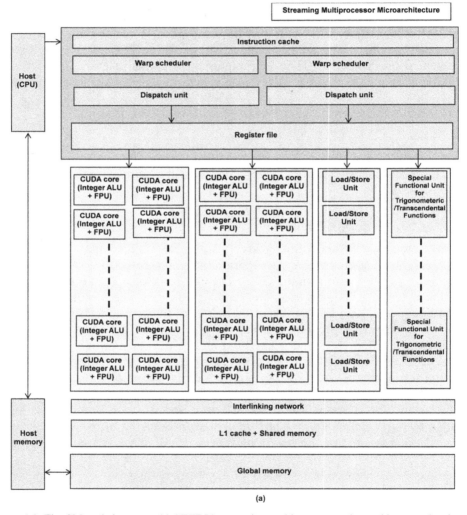

(a)

Figure 6.6. The SM and the cores: (a) NVIDIA streaming multiprocessor microarchitecture showing the middle layer of CUDA cores, the load/store units, and the special function units, flanked by the top layer of instruction cache, warp schedulers, dispatch units and register file; and the bottom layer of interlinking network, L1 cache plus shared memory and uniform cache. (b) NVIDIA CUDA core and (c) NVIDIA tensor core.

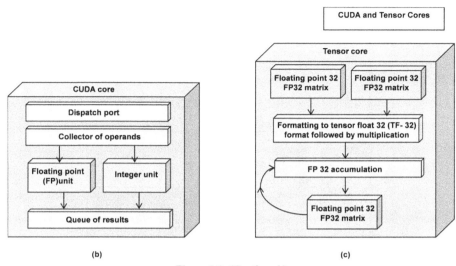

Figure 6.6. (Continued.)

format tensor float 32 or TF-32. The tensor core brings about a drastic speed-up of AI processing by providing a high acceleration to inference tasks.

6.8.1 Streaming multiprocessors (SMs)

The GPU chip is partitioned into GPCs. Each GPC is partitioned into SMs. The SMs are specialized processing units that are optimized to handle workloads involving parallel data processing tasks (figure 6.7). They work on the concept of stream processing, which is elaborated below.

6.8.1.1 Stream processing concept and programming model

A stream is an ordered set of data of the same type, e.g., integral or floating-point numerical data. Many examples of operations on streams can be cited, e.g., copying streams, extracting sub-streams from streams, calculations on streams with kernels, and so forth. The kernels operate on full streams containing multiple elements, instead of working on individual elements. They produce output streams from input streams, e.g., a coordinate transformation kernel works on a stream of points in a given coordinate system to yield a stream of points in another coordinate system.

The stream processing programming paradigm is a model which focuses on the real-time processing of continuous streams of data in motion. In this single-pass data processing model, streams or sequences of events in time are kept moving to decrease latency. The streams are viewed as the input data and output results of the computation.

6.8.1.2 Computational efficiency enhancement by the stream model

The stream prototyping of an application exposes parallelism to provide opportunities for parallel processing in various ways:

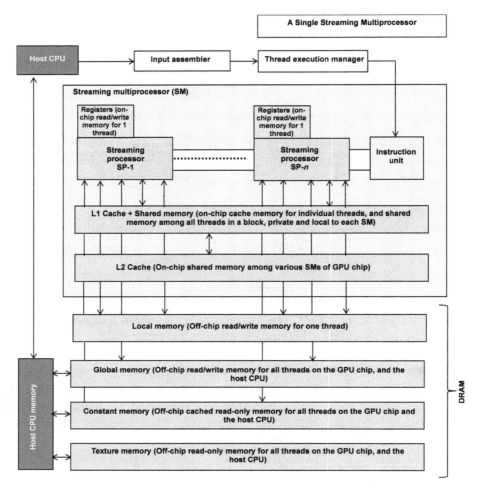

Figure 6.7. Diagram showing the main components of a single SM of a GPU. The GPU consists of a large number of such interconnected SMs: SM-1,..., SM-N. The single SM shown consists of SPs: SP-1, ..., SP-n, associated registers and instruction units. The memory system consists of (L1 cache + shared memory), L2 cache, local, global, constant, and texture memory. The (L1 cache + shared memory) and L2 cache are integrated on-chip memories. The remining memories are externally located. Each type of memory has its own defined role, purpose and usage rules. The global, constant and texture memories can read/write data from/to the host. The SM is controlled by the C^{++} program with CUDA extensions by the host through an input assembler and thread execution manager.

(i) Data-level parallelism: The kernel operation style on entire streams helps in processing of subgroups of elements in parallel. The parallelism increases in efficiency as the stream becomes longer. This is data-level parallelism.

(ii) Instruction-level parallelism: This can be exploited within the processing of a single element.

(iii) Task-level parallelism: Applications are organized from several kernels. Execution of these kernels in parallel enables task-level parallelism.

The simplicity of control flow in kernel execution diverts the attention of hardware engineers towards the data flow path from the control flow. So, the hardware has to lay more emphasis on the data-path. With accentuation on data path, more transistors are used in the data-path hardware than in the control hardware.

6.8.1.3 Communication efficiency amelioration by stream processing

Off-chip communication becomes more efficient when data streams, rather than individual elements are transferred from memory or to memory. This happens because the cost of transference is paid over the full stream instead of a single element.

By structuring applications as chains of kernels, the intermediate results of kernels are kept locally inside the kernel execution units. By keeping the results here, the data transfers to and from memory are rendered unnecessary.

Pipelining of an execution is made deeper by increasing the number of dependent steps. The deeper pipelines are known as super pipelines (sections 4.5.3.2 and 4.7.1). They allow the hardware to continue working during the time it waits for return of data from global memory. Hence the throughput is optimized, rather than latency.

6.8.1.4 Applications for which stream processing is expedient

The stream processing model greatly simplifies the software and hardware for applications which possess characteristics such as:

  (i) High compute intensity: The compute intensity is defined as

$$CI = \frac{\text{Number of computational operations}}{\text{Data accessed}} \tag{6.1}$$

expressed in units of floating-point operations per byte read, FLOP byte^{-1}.

Taking an example, let us add the contents of an array consisting of n elements. For a single-processor system, one thread will add the elements 0, ..., $(n-1)$. For a dual-processor system, the task can be shared between two threads. Thread (a) running on processor A, adds together the elements 0, ..., $(n/2-1)$. At the same time, thread (b), running on processor B, adds together the elements $n/2$, ..., $(n-1)$. So, the threads (a) and (b) run in parallel on separate processors. Increasing the number of threads to (c), (d), ..., running on processors C, D, ..., the output is produced in a much shorter time span than with a single-thread operation.

 (ii) Data parallelism: A kernel is said to exhibit data parallelism if the same function is applied to all records of an input data stream. In addition, it should be possible to process several records of data stream concurrently without waiting for results of preceding records. Data parallelism allows the same task to be simultaneously executed on multiple processors.

Data parallelism must be distinguished from task parallelism. In data parallelism, the same task is performed on different subsets of same data. On the reverse side, in task parallelism, different tasks are carried out on the same or different data. Recall discussions in section 4.5 on data and task parallelism.

(iii) Data locality: This is a temporal locality observed in signal processing applications in which the data is produced one time, read one or two times afterwards but never thereafter. This locality can be captured by intermediate streams flowing between kernels or by intermediate data within kernel functions with the help of the stream processing programming model.

6.8.1.5 Structuring the GPU hardware to utilize the parallelism features of the application

The execution of kernels in the procedural framework is assigned to self-sufficient functional units on the GPU chip. The SMs are capable of executing several thread blocks in parallel. When a kernel grid is requested by a host program, the blocks of the grid are specified. They are distributed to multiprocessors in which the necessary execution capacity is available.

As these functional units occupy different areas on the chip, the kernels can run independently on different areas of the chip at the same time to achieve task parallelism. Although separate, these functional units are proximately located on the chip with an eye toward making inter-kernel communication possible via inter-functional unit communication without accessing global memory. Consequently, the communication among kernels is greatly facilitated.

Within each stage of the procedural outline that maps to a functional unit on the chip, the independence of each constituent element of the stream is utilized. Availing the benefit of this independence, several data elements are processed parallelly to reinforce task-level parallelism with data-level parallelism. This combination of task-level and data-level parallelism allows GPUs (figure 6.8) to favorably use tens of functional units coincidentally.

In order to achieve efficient mapping of kernels to the independent functional units on the GPU chip, each SM contains scalar processor cores, e.g., CUDA cores, also called SPs. A scalar processor acts on one piece of data in one instruction while a vector processor acts on multiple data elements in one instruction (sections 4.5.2.1, 4.5.2.2, and 4,7.2).

Each SM is supported by a multithreaded instruction unit, instruction dispatch units, warp schedulers, load/store (LD/ST) units, special function units (SFUs), high-speed on-chip L1 cache plus shared memory, and interface to L2 cache.

The CUDA cores comprise integer arithmetic-logic units (ALUs) and floating-point units (FPUs). for executing SIMT instructions. The SIMT is an execution model in parallel computing. It extends the SIMD execution model by adding multithreading to it. The objective is to reduce the latency arising from memory access during instruction fetching.

The SIMD acts on multiple data with a single instruction working on all the data in exactly the same way. The SIMT operates on multiple threads with loosened restriction. Chosen threads can be activated or deactivated. Hence, the instructions

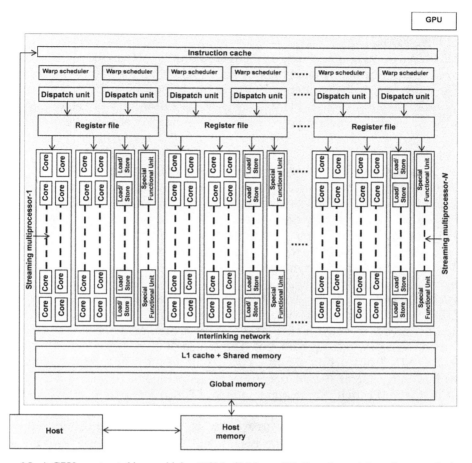

Figure 6.8. A GPU constructed by combining N SMs, SM-1, ..., SM-N, each consisting of n cores, dual warp schedulers and dispatch units. They are joined together by an interlinking network. They share a common instruction cache, the (L1 cache + shared memory) and global memory. The global memory is connected to the host memory by a high-speed PCI-express I/O bus. The instruction cache is also fed by the host.

and data are processed only on the active threads. The local data remain unchanged on inactive threads.

Each thread is mapped by the multiprocessor to one scalar processor core. Each scalar thread performs individualistically. It executes with its own instruction address and register state.

The load/store units perform data loading and storage from/to cache or DRAM. The SFUs work on the transcendental instructions such as sine, cosine, reciprocal, logarithm, exponential, power, square-root and reciprocal square root evaluation. One instruction per thread, per clock is executed by each SFU. The L1- and L2-caches will be discussed further in section 6.8.5.

6.8.2 Merits of SIMT execution model

6.8.2.1 SIMD model

SIMD is an execution model used in parallel computing. Here, the same instruction is executed and hence the same operation is performed on multiple data inputs concurrently. A SIMD compute unit is a multiprocessor or multi-core system with many processing units PU-1, PU-2, ..., PU-N. It is altogether called a vector-lane, on which a single instruction is executed on different data streams (figure 6.9(a)).

6.8.2.2 SIMT model

SIMT is an execution model used in parallel computing. In SIMT, the SIMD model is augmented by combining with multithreading. It can be visualized as a multi-core system, where each core has its own register file, its own SIMD and scalar ALUs, and its own data cache. However, there is a dissimilarity to a standard multi-core system. The standard multi-core system has multiple independent instruction caches and decoders, together with multiple independent program counter registers. In SIMT, the instructions are synchronously broadcast to all the SIMT cores. The broadcast is done from a single unit with a single instruction cache and a single instruction decoder which reads instructions using a single program counter, e.g., a GPU SM containing many streaming processors SP-1, SP-2, ..., SP-N, hosts numerous threads through a thread or a warp scheduler. Each thread carries an instruction to be executed (Instruction 1, Instruction 2).

When a SIMT instruction prescribes the execution and branching behavior of one individual independent thread operating on data input-1, data input-2, ... data input-N, the SIMT architecture applies the SIMT instruction to several independent threads: thread with instruction 1, thread with instruction 2, ... (figure 6.9(b)). These threads working in parallel are free to execute and branch independently. Processing efficiency is improved by inserting conditional break and return instructions in which threads may branch independently. Threads that execute a break or return are able to complete processing sooner than threads that do not execute the break or return (Coon *et al* 2009).

6.8.2.3 Thread divergence and reconvergence

Threads with the same instruction execute concurrently on multiple data; hence the name single-instruction, multiple-thread model. Other threads have to wait spawning thread divergence. When this happens, the threads can no longer execute in lockstep. The GPU moves to the next instruction only after all the threads in the warp have ended execution. Consequently, the running of the program slows down. Hence, thread divergence impacts the execution of parallel programs adversely.

Immediate reconvergence of threads is essential to prevent non-utilization of many hardware units. When the control flow of threads of the same warp diverges, execution of the warp is serialized for each unique path. Threads that did not take the path are disabled. The hardware tracks the activation states of the threads and schedules all the paths for execution consecutively. Once all the paths are completed, the threads reconverge and revert to moving in lockstep again.

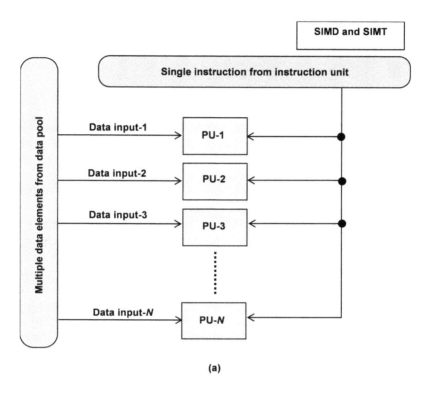

(a)

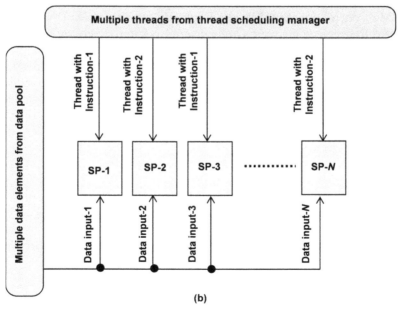

(b)

Figure 6.9. The SIMD and SIMT processing paradigms: (a) SIMD and (b) SIMT. In (a) the processing units PU-1, PU-2, …, PU-N are fed data input-1, data input-2, … from the data pool while the same instruction is given to these processing units by the instruction unit of multiprocessor or multi-core system of a CPU. In (b) the stream processing units SP-1, SP-2, ..., SP-N are fed data input-1, data input-2, ... from the data pool while the instructions (Instruction 1, Instruction 2) are given to these SPs by the thread scheduling manager of the SM of the GPU.

6.8.2.4 *Mechanisms to handle divergent control flow*

Different mechanisms are applied for management of divergent control flow on SIMT architecture. These mechanisms differ in thread scheduling and performance.

(i) Stack-based reconvergence mechanism: The Fermi architecture uses a stack-based reconvergence mechanism. The main concept of this mechanism is to store information about control flow divergence and reconvergence in a reconvergence stack (Habermaier and Knapp 2012).

Divergence is caused by the branches and loops with conditions depending on thread-specific data. Another factor responsible for divergence concerns the loops and function calls with bodies containing data-dependent break or return statements. A token is created in the reconvergence stack whenever the control flow might diverge. The token stores the information that a potentially divergent thread is continued and also the information about the participating threads in its execution. Once the control flows reconverge, the topmost token is removed from the stack. The information contained in the token is exploited by the SIMT core to continue the execution of the program.

The stack-based method can take care of thread divergence and reconvergence without intervention by the compiler.

(ii) IMT control flow SIMT cores: Implicitly-multithreaded (IMT) control flow SIMT cores execute warps in lockstep. Lockstep execution implies that each thread of a warp executes the same instruction in parallel on different operands (Park *et al* 2003). Therefore, instruction fetch and decoding costs are paid back over all threads of a warp. Further, the memory operations performed by threads of the same warp can be frequently coalesced into fewer memory accesses. Lesser memory accesses are beneficial to performance. However, the SIMT cores must be able to deal with diverging control flow that occurs when at least two threads of the same warp take different paths due to a data-dependent control flow instruction.

One strategy followed involves serializing the execution of the remaining program once the control flow diverges. Serialization without a reconvergence mechanism results in a low utilization of processing resources for branch-heavy programs. The performance degrades by a factor proportional to the warp size in the worst case. Therefore, mechanisms for reconverging threads must be applied.

6.8.2.5 *SIMT versus SIMD*

In both SIMD and SIMT models, parallelism is approached through broadcasting the same instruction to multiple execution units. Hence, the execution units are replicated. However, the fetch/decode hardware is shared among them; hence it remains the same. Briefly, the main differences between them are:

(i) SIMD operates on multiple data elements using execution units or vector units. SIMT performs the same instruction on different datasets with the help of threads, i.e., SIMT operates on multiple threads. Activation/ deactivation of selected threads allows the instructions and data to be processed only on the active threads. The SIMT is the thread equivalent of SIMD.

(ii) SIMD lanes are part of an ALU that does not know about memory. In SIMT lanes, each SIMT core may have a different stack pointer and may perform computations on different datasets.

(iii) SIMD does not provide three important features: single instruction, multiple register sets; single instruction, multiple addresses; and single-instruction, multiple-flow paths. SIMT provides these features.

(iv) The SIMT architecture shows greater flexibility than the SIMD architecture. The primary reason is that the threads in a group of threads may follow different paths through a set of instructions to process multiple data inputs.

6.8.3 Parallel thread execution (PTX)

This defines a virtual machine and ISA for execution of threads in parallel. The code written in a C^{++}-like language is translated by the compiler into an assembly language. The assembly language is written as ASCII (American Standard Code for Information Interchange) text. The ASCII is the standard code for representing alphanumeric characters. The ASCII text constitutes the PTX instructions. These instructions are translated into an executable binary code. This code is run on the CUDA processor cores.

6.8.4 The gigathread controller

This is engaged in the distribution of thread blocks to SM thread schedulers. Besides, it oversees the management of context switching between threads during execution.

It also provides two streaming data-transfer engines. One engine is used for input data transfer from the host memory to device memory to set up a GPU computation. The other engine is used for the result data transfer from the device memory to the host memory. Each engine can fully saturate Fermi's peripheral component interconnect (PCI) express, a high-speed standardized serial communication interface.

6.8.5 Cache and memory hierarchy

6.8.5.1 Registers

These comprise the fastest direct-access memory on SM for storing the operations used in computing by the SM. They are only for one thread, private to the thread, and only accessible by the thread. The register file is divided among the threads to provide read/write memory per thread. Data in registers allocated to a thread is exclusively accessible by the concerned thread. It is not accessible to other threads. It cannot be accessed even by other threads in the same block. It is also unavailable to the host. Further, it must be noted that the data is not permanently stored in registers. It is available only during the time of execution of a thread. Hence, the data in registers has the lifetime of the thread. The number of available registers depends on the workload. It decreases as the workload increases.

6.8.5.2 L1 (Level 1) cache + shared memory subsystem
It is a scratchpad memory (high-speed memory) used:
   (i) as a temporary workspace for quick data cache (register spilling/L1) by individual threads in a block, or
   (ii) for sharing data among several threads in a block (shared memory). It has the lifetime of the block.

Each SM has its individual L1 cache. This memory is very fast, working at register speeds to allow quick communication between the threads in a block. The L1 cache of a GPU is smaller than L1 cache of a CPU. But the L1 cache of the GPU has much higher bandwidth than that of a CPU.

The shared memory can only be accessed by threads in the same thread block. Low-latency access is provided but the bandwidth is very high. This memory is used for storing intermediate results in a serial calculation. Small quantities of data pertaining to one row or column of data in a matrix operation can be stored.

(L1 cache + shared memory) is read/write memory per thread (L1 cache/register spillage) and read/write shared memory per thread block. The L1 cache is controlled by hardware but shared memory is software-controlled.

6.8.5.3 L2 (Level 2) cache subsystem
The L2 cache is read/write per-grid memory, shared among all SMs. The L2 cache provides load/store services from/to global memory, and copy services from/to host CPU apart from handling texture requests. It also performs atomic operations for managing access to data to be shared among thread blocks or kernels. Atomic operations can read, modify, and write a value back to memory without interference from other threads. Thereby rat race conditions in multithreaded systems are avoided.

The L2 cache of a GPU is much smaller than the L2 or L3 cache of a CPU. However, much higher bandwidth is available in a GPU L2 cache than in its CPU counterpart. Data retrieval from L2 cache is quicker than from global memory.

6.8.5.4 Local memory
Local memory is read-only per-thread memory which is automatically used by threads during register spilling. The register spilling happens when a register file runs out of space requiring more storage than available on an SM. The local memory is used to store data that is private to a thread but cannot be accommodated in registers. Its need arises during inadequacy of space in the register file. Then it

becomes necessary to temporarily store variables for future requirement, thus preventing their erasure.

This memory stores variables that the compiler spills to local memory when a kernel needs more registers than available. Some automatic array variables are also stored by the compiler. However, the arrays that the compiler cannot determine are indexed with constant quantities. Large arrays that would consume too much register space are also not stored here.

Local memory is much larger than register memory. However, it is comparatively slower than register memory. The throughput and latency of local memory are identical to those of the global memory.

6.8.5.5 *Global memory*

It is the slow, uncached, large-size main memory of GPU. It is accessible to all threads in all blocks for reading and writing, and also to CPU. It has the lifetime of the application. It is the largest memory space offering vast amounts of data storage. Nonetheless, it is slower to access than the registers and shared memory.

This memory is allocated/de-allocated and managed by the host. It is accessible to both the host and the device. It can therefore be used for data exchange between them.

The global memory is uncached read/write per-grid memory.

6.8.5.6 *Constant memory*

It is a read-only cache accessible by all threads and the host. It has the lifetime of the application. All threads can only read the same constant memory but they cannot write to it. It is a part of the GPU's main memory with its own cache. It is unrelated to L1 and L2. The values in the constant memory are set by the CPU before a kernel is launched. The constant memory is very fast, similar to the register and shared memory.

The constant memory is cached read-only per-grid memory. It is utilized for uniformly-accessed read-only data.

6.8.5.7 *Texture memory*

The texture memory is optimized for 2D spatial locality. The texture memory is cached read-only per-grid memory. It is useful for spatially coherent random-access read-only data.

Local, global, constant and texture memories are implemented with dynamic random-access memory (DRAM). The DRAMs are error correcting code (EEC)-protected, and so also the registers, and the L1, L2 caches. Global and shared memories are the most commonly used memories, and play major roles in computations.

6.9 Thread-block and warp scheduling

6.9.1 Thread-block scheduler (TBS)

Most-room policy is adopted for allocation of a thread block to an SM. The threads of a thread block execute at the same time on one multiprocessor. Thread-block allocation is decided by the local resources offered in the SM. No sooner than one thread block has been executed, the serially next thread block is undertaken for execution. This is done by launching new blocks on the emptied multiprocessors. The next ready thread block is assigned to the SM which can host the maximum number of blocks of the present kernel.

6.9.2 Dual warp scheduler

In the dual warp scheduler, there are two warp schedulers and two instruction dispatch units for each SM. This makes possible the concurrent issuance and execution of two warps.

The SIMT unit of the multiprocessor produces, and manages the scheduling, and execution of threads in groups of parallel threads comprising warps. Discrete threads composing a SIMT warp commence together at the same program address. Otherwise, they are free to branch and execute on their own.

At the time of execution of every instruction, a ready-to-execute warp is chosen by the SIMT unit. The instruction is delivered to the active threads of the warp. At a time, a warp executes one common instruction. Realization of full efficiency is achieved when all the threads of a warp unanimously agree on their execution. In case of divergence of threads of a warp, the warp undertakes serial execution of each branch path pursued. Threads which are not on that path are disabled. After completion of all the paths, the threads again meet together to the same path of execution. Divergence takes place only within a warp. Different branches execute independently.

The execution takes place irrespective of whether they are engaged in common or disconnected code paths.

6.9.3 Context switching

If a thread block is initiated on an SM, all its warps become dwellers until their execution is completed. In case one or both its operands are not prepared, a process known as context switching ensues. This process hands over the control to another warp. During switching away, all the data of the warp persists in the register file. Quick resumption of the warp is therefore possible at the instant its operands become all set. A warp is considered to be all-set for execution when an instruction has no left-over dependence on data.

6.9.4 Warp scheduling policies

If several warps have eligibility for execution in a SM, it has to be decided which warp will get the next obtained instruction. Policies followed for making this decision are:

(i) Round-Robin (RR) warp scheduling: Each warp is given an equal share of time. The warps are placed in a circular queue. On expiry of the time for a warp, it is positioned at the end of the queue. Then the next warp is taken from the front of the queue.

(ii) Least recently fetched warp scheduling: Priority in the fetching of instruction is given to the warp for which no instruction has been retrieved for the longest time interval.

(iii) Fair warp scheduling: Equal opportunity is given to all warps about the number of instructions obtained for them. This statement means that the instruction is fetched to the warp for which the least number of instructions have been brought.

(iv) Critically aware warp scheduling (CAWS): This entails preferential provision of more time resources to the warp that will take the longest execution time.

6.10 Role of GPUs in AI context

GPUs are powerful and versatile processors. They have befitting capabilities to handle a wide range of parallel computing tasks of different intricacy levels. Such tasks are encountered in bioinformatics, drug design, and clinical trials; climate modelling and particle physics research.

(i) Applicability: GPUs are used for general-purpose AI computing tasks. These tasks are concerned with parallel processing of large data blocks of vectors and matrices.

(ii) AI/ML training and inference: GPUs provide the desired computational power for processing largely identical or unstructured data. We are aware that AI training typically involves processing mostly identical, simultaneous operations on multiple data samples. The training datasets continue to grow larger, requiring increasingly massive parallelism. Therefore, GPUs are being used more and more in data centers to quicken the processing of escalated computational requirements of deep neural network training and inference. We shall further talk about math-intensive computing in (iii).

(iii) Math-intensive scientific computing and simulation: The GPU emerged as a processor unsurpassed in parallelism. The aspiration behind GPU is to surmount the limitations of CPUs, which typically have a much smaller number of cores. As against CPU, the GPU is composed of hundreds or even thousands of cores which are capable of executing input/output operations simultaneously.

The parallelism offered by GPU is leveraged for the rapid processing of large datasets that we stumble upon very often in high-volume calculations. Such calculations deal with the execution of complex mathematical models full of matrices and vectors. These models are pervasive in machine learning algorithms. Many high-end GPUs incorporate tensor cores. These cores facilitate the execution of deep neural network algorithms.

The basic series of operations involves allocation of memory on the GPU and transference of data from CPU to GPU. Then the functions are launched that encode the core computational operations done on individual pieces of data known as kernels. The kernels operate on the threads, with a given block/grid arrangement. A kernel can access data stored on the GPU, including results from the previous kernel, and finally transfer results back to the CPU. One kernel may also launch another kernel. This is a form of dynamic parallelism in which the parent and child kernels work. The child kernel can catapult further, resulting in a nested execution hierarchy.

(iv) Optimization for floating-point computations: The floating-point calculations are those involving real numbers in contrast with integers. Advanced ALUs in GPUs endow them with enormous computational capability. The capability is measured in floating-point operations per second (FLOPS). The FLOP is a critical metric indicating the number of calculations performed every second. High-end GPUs can achieve tera-flops (trillions of FLOPS) of performance. They are able to deliver fast execution speed with accuracy.

(v) Good scalability backed up by a robust ecosystem of software tools and libraries: A comprehensive software support facilitates the use of GPUs in AI, e.g., the CUDA, TensorFlow and PyTorch frameworks allow researchers and developers to focus on building and training models. The job of developers is made comparatively easier by the diversion of their attention towards models instead of spending time in managing the hardware intricacies.

(vi) Acceleration of graph analytics with GPU: The GPUs can accelerate data-intensive graph analytics. It is a form of data analysis that helps businesses in understanding the complex relationships between data for linked entities in a network or graph. The assistance of graph specific algorithms is sought for this analysis (NVIDIA: Graph analytics 2024). The tremendous degree of parallelism and large memory access bandwidth of GPUs is a boon for graph analytics.

The thousands of small cores designed in GPUs for handling multiple tasks simultaneously are highly proficient in performing the calculation 'for every X do Y', serving as a blessing or piece of good fortune. This calculation is applicable to sets of vertices or edges within a large graph. The implementation of a few fundamental graph algorithms, viz., breadth first search, single source shortest path, and all-pairs shortest path was examined using CUDA on large graphs on the NVIDIA GPUs using the CUDA model (Harish and Narayanan 2007). Computation of single source shortest path on a 10 million vertex graph was done in 1.5 s using the NVIDIA 8800GTX GPU (Production: End-of-life).

Processing of graph algorithms will be revisited in chapter 11.

(vii) Video and graphics processing: In their traditional role, GPUs are used for processing images and videos in 3D computer graphics applications, e.g., real-time rendering of multimedia content. They allow high-resolution video streaming. They are also used for virtual and augmented reality applications.

6.11 GPU examples

(i) NVIDIA A100 and H100: The NVIDIA A100 is a GPU for data center and professional deep learning applications. It was launched in 2020 with the Ampere architecture. It has 6912 CUDA cores; 432, generation 3 tensor cores, 1.41 GHz boost clock, 78 F16 TFLOPS, 40/80 GB memory, 1.6 TB s^{-1} memory bandwidth, and 250/400 W thermal design power. The chip having 54.2×10^9 transistors is made by a 7 nm process.

While NVIDIA A100 is suited to research and production environments, NVIDIA H100 is superior for large-scale AI with 30× better inference and 9 × better training performance than NVIDIA A100. The NVIDIA H100 SXM has NVIDIA Hopper architecture, 80GB GPU memory, 3.35TB/s GPU memory bandwidth, 4th generation tensor cores, and supports FP64, FP32, BFLOAT 16, FP16 and INT8 compute tasks, with maximum thermal design power up to 700W; the NVIDIA H100 NVL has 94GB GPU memory, 3.9TB/s GPU memory bandwidth, and thermal design power 350-400W (configurable).

(ii) NVIDIA RTX A6000: This is a powerful GPU which is well-suited for deep learning tasks. It was launched in 2020 with the Ampere architecture. It has 10752 CUDA cores; 336, generation 3 tensor cores, 1.41 GHz boost clock, 38.7 F16 TFLOPS, 40 GB memory, 768 GB s^{-1} memory bandwidth, and 250 W thermal design power. Its 54.2×10^9 transistors are fabricated by a 7 nm process.

(iii) NVIDIA RTX 4090: This is a powerful consumer-grade graphics card usable for deep learning. However, it is less suited for this task than NVIDIA A100 or RTX A6000; it is exceptional for gaming and creative applications. It was launched in 2022 with Ada Lovelace architecture. It has 16384 CUDA cores; 512, generation 4 tensor cores, 2.23 GHz boost clock, 82.6 FP 16 TFLOPS, 24 GB memory, 1 TB s^{-1} memory bandwidth, and 450 W thermal design power. The chip contains 76×10^9 transistors made by a 4 nm process (GPU Mart 2023, NVIDIA H100 2024).

(iv) AMD Instinct™ MI300X discrete GPU accelerator module: This module is developed for generative AI, training, and high-performance computing. It is based on AMD CDNA™ 3 architecture. It has 304 high throughput compute units, 1216 matrix cores, peak engine clock 2100 MHz, 192 GB of HBM3 memory, 5.3 TB s^{-1} memory bandwidth (maximum peak), and 750 W total board power. A 5 nm FINFET process is used for fabrication of this chip (AMD Instinct 2023).

6.12 Summary and the way forward

6.12.1 Highlights of chapter 6 at a glance

(i) Key components of the architecture of a typical GPU were described such as the global memory, compute units and a high-bandwidth memory interface. For further reading, the reader is recommended to refer to Glaskowsky

(2009), NVIDIA Ampere Architecture (2024), NVIDIA (2024), NVIDIA
H100 (2024), NVIDIA Research (2024), AMD Instinct (2023).

(ii) The global memory stores the data processed by the compute units. The
compute units are organized into groups called SMs. Each SM contains
scalar processors, a scheduler, and a shared memory. Efficient data
sharing and synchronization among threads is established. The memory
interface ensures rapid data transfer between the GPU and the system
memory. Thus, a high throughput is maintained in AI applications.

(iii) Compute unified device architecture is a parallel computing platform
developed by NVIDIA for general-purpose computing on GPUs.
It allows developers to expand the capabilities of a GPU. Developers
can break complex problems into multiple separate tasks. They can
harness the computing power of the GPU to execute the tasks simultaneously instead of their one-by-one implementation by CPU.

(iv) For accelerating applications, the sequential part of the workload runs on
the single-threaded performance-optimized CPU. Throughout this time,
the computationally intensive portion of the application runs in parallel
on thousands of SPs of a GPU.

(v) The CUDA program structure and flow was explained followed by
enumeration of the concepts of thread, thread block and synchronized
functioning of threads; kernel grid, grid hierarchy and warp. The hardware
perspective of GPU was provided with explanation of the Fermi architecture.

(vi) SMs were explored along with the stream processing concept and
programming model. Reasons for enhancements in computational and
communication efficiencies by stream processing were clarified.
Applications were mentioned for which stream processing is expedient.
The structuring of the GPU hardware to utilize the parallelism features of
the application were described.

(vii) Merits of the SIMT model were pointed out, and its comparison with
SIMD architecture was made. PTX and the GigaThread controller were
elaborated on.

(viii) Cache and memory hierarchy consists of registers, (L1 (Level 1) cache +
shared memory) subsystem, L2 (Level 2) cache subsystem; and the local,
global, constant and texture memories.

(ix) Thread-block and warp scheduling systems comprise the TBS, dual warp
scheduler, context switching and warp scheduling policies.

(x) A few commercial GPU examples were cited.

(xi) Keywords for this chapter are: GPU, NVIDIA, AMD, SIMT, stream
processing, multiprocessor.

6.12.2 Getting ready to begin chapter 7

The GPU championed and emerged a winner in many AI scenarios. In the ensuing
chapter we look into the working of a TPU, Google's dedicated AI-optimized
hardware. The TPU is created explicitly to speed up AI workloads, particularly the

tensor-based calculations of deep learning. Its focus on a specific type of task to perform certain types of calculations makes it different from other processors, and imparts to it the capability to achieve high performance in an energy-efficient manner for training and inference workloads in AI applications.

References

Aamodt T M, Fung W W L and Rogers T G 2018 *General-Purpose Graphics Processor Architectures* (Cham: Springer Nature) p 2

AMD Instinct 2023 *Data Sheet AMD Instinct™ MI300X Accelerator Leading-Edge, Industry-Standard Accelerator Module for Generative AI, Training, and High-Performance Computing* (Advanced Micro Devices, Inc.) p 2 https://amd.com/content/dam/amd/en/documents/instinct-tech-docs/data-sheets/amd-instinct-mi300x-data-sheet.pdf

Burnes A and Pacelli C 2024 *GeForce RTX 40 Super Series Graphics Cards Launching this January, for Supercharged Gaming and Creating, with Super-Fast AI* (NVIDIA Corporation) https://nvidia.com/en-in/geforce/news/geforce-rtx-4080-4070-ti-4070-super-gpu/

Coon B W, Nickolls J R, Nyland L, Mills P C and Lindholm J E 2009 Indirect function call instructions in a synchronous parallel thread processor *U.S. Patent* US20090240931A1 pp 1–16

Fumero J *et al* 2019 Programming and architecture models ed O Terzo, K Djemame, A Scionti and C Pezuela *Heterogeneous Computing Architectures: Challenges and Vision* (Boca Raton, FL: CRC Press) pp 54–87

Glaskowsky P N 2009 NVIDIA's Fermi: the first complete GPU computing architecture a white paper by Peter N Glaskowsky prepared under contract with NVIDIA Corporation, Peter N Glaskowsky, pp 1–26

GPU Mart 2023 5 Best GPUs for AI and deep learning in July 2024, an in-depth comparison of NVIDIA A100, RTX A6000, RTX 4090, NVIDIA A40, Tesla V100 Database Mart LLC, https://gpu-mart.com/blog/best-gpus-for-ai-and-deep-learning-2024#:~:text=For%20demanding%20tasks%20requiring%20high,smaller%2Dscale%20tasks%20or%20hobbyists

Habermaier A and Knapp A 2012 On the correctness of the SIMT execution model of GPUs *Programming Languages and Systems, European Symp. on Programming (ESOP)* Lecture Notes in Computer Science vol 7211 *ed H Seidl (Berlin: Springer) pp 316–35*

Harish P and Narayanan P J 2007 Accelerating large graph lgorithms on the GPU using CUDA; S Aluru, M Parashar, R Badrinath and V K Prasanna *High Performance Computing-HiPC 2007, 14th Int. Conf.* Lecture Notes in Computer Science *(December 19–21) (Goa, India) vol 4873 (Berlin: Springer) pp 197–208*

Hello CUDA-GPU series#1 2023 http://w-uh.com/posts/230121-Hello_CUDA.html

Ko S, Zhou H, Zhou J J and Won J H 2022 High-performance statistical computing in the computing environments of the 2020s *Stat. Sci.* **37** 494–518

NVIDIA 2024 *CUDA® C++ Programming Guide Release 12.6* (NVIDIA Corporation & Affiliates) pp 1–538 https://docs.nvidia.com/cuda/pdf/CUDA_C_Programming_Guide.pdf

NVIDIA H100 2024 NVIDIA H100 Tensor Core GPU, NVIDIACorporation, https://www.nvidia.com/en-in/data-center/h100/

NVIDIA Ampere Architecture 2024 NVIDIA Corporation https://nvidia.com/en-in/data-center/ampere-architecture/

NVIDIA Corporate Timeline 1999 https://nvidia.com/content/timeline/time_99.html#:~:text=NVIDIA%20launches%20GeForce%20256%E2%84%A2,graphics%20processing%20unit%20(GPU)

NVIDIA: Graph Analytics 2024 What Is Graph Analytics? NVIDIA Corporation, https://nvidia. com/en-us/glossary/graph-analytics/#:~:text=Graph%20analytics%2C%20or%20Graph% 20algorithms,the%20graph%20as%20a%20whole

NVIDIA Research 2024 Publications, NVIDIA Corporation, https://research.nvidia.com/ publications

Owens J D, Houston M, Luebke D, Green S, Stone J E and Phillips J C 2008 GPU computing *Proc. IEEE* **96** 879–99

Park I, Babak F and Vijaykumar T N 2003 Implicitly-multithreaded processors *ACM SIGARCH Comp. Archit. News* **31** 39–51

Peddie J 2022 NVIDIA'S GeForce 256: the first fully integrated GPU, Graph. Chip Chronicles 6, https://electronicdesign.com/technologies/embedded/article/21178111/jon-peddie-research-nvidias-geforce-256-the-first-fully-integrated-gpu

Powell P and Smalley I 2024 What is a central processing unit (CPU)? https://ibm.com/think/ topics/central-processing-unit

Soyata T 2018 *GPU Parallel Program Development Using CUDA* CRC Computational Science Series (Boca Raton, FL: Chapman & Hall/CRC Press) p 476

Wang T and Kemao Q 2017 *GPU Acceleration for Optical Measurement* (Washington, DC: SPIE Press) p 4

Wittenbrink C M, Kilgariff E and Prabhu A 2011 Fermi GF100 GPU architecture *IEEE Micro* **31** 50–9

IOP Publishing

AI-Processor Electronics
Basic technology of artificial intelligence
Vinod Kumar Khanna

Chapter 7

Tensor processing unit

The terms 'tensor', 'TensorFlow' and 'tensor processing unit (TPU)' are introduced. Chronological history and features of TPU are described and its specialties for AI work are enumerated. The driving factors behind the motivation for TPU become evident by understanding the components of a perceptron and the calculation steps performed in its working. The structural units of TPU and its instruction set are described, including both off-chip and on-chip components and the complex instructions set computer (CISC) instruction set. TPU operation is explained by drawing the block diagram of the workflow of an inference task and discussion of inference implementation by TPU.

The central hub of all activities in a TPU is the matrix multiplier unit (MXU), which performs a major chunk of the computations required for machine learning (ML). The ability of MXU to perform a large number of multiply–accumulate operations in parallel is the primary cause of the high throughput of TPUs. The advantages of its systolic array architecture are mentioned. A TPU also has a unified buffer that serves as a high-speed on-chip memory, and an activation unit that applies non-linear functions to the data.

These energy-efficient application-specific integrated circuits (ASICs) provide several teraflops with a power consumption of a few watts. The TPUs are integrated into Google's cloud computing platform, Google Cloud, to allow a seamless scaling of ML workloads. Services of additional TPUs are readily availed to handle larger datasets or more complex models. In order to derive full advantage of TPUs, Google provides a software stack including TensorFlow, an open-source ML framework that is optimized for this purpose. Various aspects of TPU architecture and its operational procedures are described.

7.1 Introduction

This chapter deals with the architecture and operation of the TPU, a specialized piece of hardware developed by Google, the popular company for its unmatched

doi:10.1088/978-0-7503-6259-7ch7   7-1   © IOP Publishing Ltd 2024. All rights,

search engine technology besides many products and services. The TPU is designed and produced with a definitive purpose for acceleration of demanding artificial intelligence/machine learning (AI/ML) tasks using matrix processing: in edge computing where the data must be processed near its source such as in self-driven autonomous vehicles working in autopilot mode; in factories; and in cloud computing, e.g., chatboats simulating text or voice human conversation over the internet, generative AI, speech and vision systems.

7.2 Tensor, TensorFlow and tensor processing unit

7.2.1 Tensor

The concept of tensors is based on the generalization of scalars, vectors and matrices to n dimensions (figure 7.1). A tensor can be looked upon as a generalized matrix with n dimensions. It is an n-dimensional matrix or a multidimensional matrix. A scalar, being a single number, is a zero-dimensional tensor. A vector, being a one-dimensional array of numbers, is a one-dimensional tensor. A matrix, being a two-dimensional array of numbers, is a two-dimensional tensor. Proceeding further, an n-dimensional array of numbers is an n-dimensional tensor.

7.2.2 TensorFlow

TensorFlow is an end-to-end open-source platform created by Google for AI and ML (Ramsundar and Zadeh 2018). This platform incorporates a comprehensive, flexible software ecosystem made of tools, libraries and other resources for managing different aspects of ML. The aspects covered are predominantly training and inference of deep neural networks (DNNs) for object detection, image classification, language modeling, and speech recognition.

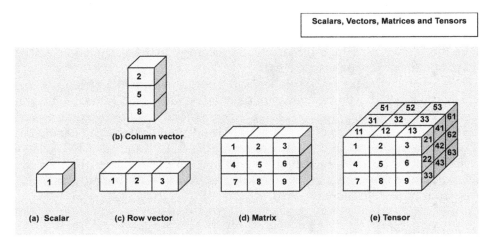

Figure 7.1. Comparison between a scalar, a vector, a matrix and a tensor: (a) scalar; (b) and (c) column and row vectors respectively; (d) matrix and (e) tensor.

7.2.3 Introduction to TPU

The TPU is a proprietary integrated circuit specially developed by Google for ML by inclusion of high-performance AI accelerator with MXU, local memory and interconnect technology on a network, and allowing communication and data transport with them to provide the necessary features and functionalities for high-volume mathematical processing workloads. The TensorFlow software of Google is used for its operation.

7.3 Chronological history and features of TPU

Although internally used in Google's data centers since 2015, the TPU was formally announced by Google in May 2016. The TPU delivers an order of magnitude higher performance per watt for ML than present-day devices. This is the same as fast-forwarding technology about seven years into the future. In turn, this is equivalent to three generations of Moore's law (Jouppi 2016). According to Sato and Young (2017), the performance achieved by TPU is 15–30× higher than state-of-the art devices like central processing units (CPUs) and graphical processing units (GPUs). Its perform-ance per watt is 30–80× higher. Augmenting performance per watt is hitting the nail on the head because it strikes the power consumption issue directly. The TPU chip can withstand reduced computational precision. Hence, it requires fewer transistors per operation and allows the cramming of more operations per second into the silicon. By using advanced and strong ML models, and also execution of these models more swiftly, the intelligent results are delivered very fast to the users.

The first version of TPU (TPUv1) was introduced in 2015 at the 28 nm process node; the process node represents the minimum feature size of the transistor gate length. The second and third versions TPUv2 and TPUv3 were introduced at 16 nm process node in 2017 and 2018, respectively. The fourth version TPUv4 appeared at the 7 nm process node in 2021 followed by fifth versions TPUv5e, TPUv5p in 2023, and the sixth generation TPU codenamed Trillium in 2024 (Jouppi et al 2017, 2020, 2021, 2023, Vahdat and Lohmeyer 2023a, 2023b, Vahdat 2024); see also section 1.6.3.4. The edge TPUv1 was launched in 2018 (Kundu 2018). The clock speed rose from 700 MHz in TPUv1 to 1750 MHz in TPUv5p, while the on-chip memory increased from 28 MB in TPUv1 to 112 MB in TPUv5p. Thus, the TPU specifications vastly improved as new generations were launched.

As compared to GPU, the TPU is aimed at handling massive low-precision computing jobs, e.g., 8-bit precision for TPUv1. TPUv5e chip can perform up to 393 trillion (tera) INT8 (8-bit integer precision) operations per second (TOPS) while TPU5p offers a peak performance of 918 trillion (tera) INT8 operations per second; tera = 10^{12}.

TPUv1 has a thermal design power (TDP) of 70 W while TPUv4 has TDP = 170 W. Thermal design power is a parameter that measures the power dissipation capability of the processor chip. It is the maximum amount of heat which a processor can dissipate during its normal operating conditions. It is expressed in units of Watts. It indicates the power consumption of the processor. So, it gives us an idea about the cooling requirements of the device. The TPU IC is mounted in a heat sink assembly. This assembly is housed in a hard drive slot of the data center rack.

7.4 Specialties of TPUs for AI work

(i) ASIC for neural network training and inference: As already mentioned in section 7.2.3, TPUs provide exclusive features for AI because they are custom-designed, specialized, ASICs developed by Google. They have the well-defined targeted goal of accelerating ML training and inference workloads. Unlike GPUs, initially designed for graphics processing and later adapted for AI, the TPUs were conceived with a purpose-built approach from the ground up keeping the demands of ML in view.

Therefore, the research endeavors culminated in a hardware solution excelling in executing tensor operations. The TPU design reflects a focused approach to build an AI hardware with an architecture that streamlines the execution of tensor operations. Such operations are the prime components of neural network algorithms. We shall elaborate on this topic in (ii).

(ii) Suitability for the mathematical operations done in neural networks: TPUs are optimized to efficiently perform the essential mathematical operations that predominate neural network training and inference with superior performance and energy efficiency. The operations of interest are the high-throughput matrix multiplications and tensor operations, the fundamental ingredients found time and again in a vast majority of AI algorithms (Cloud TPU 2024). Looking behind the scenes of matrix processing, it is essentially a combination of multiply and accumulate operations in which the first step is finding the product of two numbers, and the second step is addition of the product obtained to an accumulator. Bearing this calculation procedure in mind, the TPUs are designed with thousands of multiply-accumulators (MACs). These MACs are directly connected to each other to form a large physical matrix constituting a systolic array architecture (see section 7.6.2 (ii) for more details). The input data is streamed by the TPU host into an infeed queue. From the infeed queue, the TPU loads the data and stores it in HBM (high-bandwidth memory). Then the TPU loads the coefficient or weight values from HBM memory into the MXU. It also loads the data from HBM memory into MXU. As each multiplication is executed, its result is forwarded to the ensuing multiply–accumulator unit. Eventually, the sum of all the multiplication operation results performed between the data and the coefficient or weight values is delivered as the output. TPUs can achieve a high-computational throughput on neural network calculations because of two reasons:

(a) the result transferred from one MAC unit to the succeeding MAC unit serves as the input for this unit, and

(b) no memory access is required during the entire matrix multiplication process.

After completion of the computation, the results are loaded by the TPU in the outfeed queue. They are read from the outfeed queue and stored in the host's memory. Thus, TPUs are the proper hardware devices, which are carefully designed for performing large matrix operations that are commonly encountered in ML algorithms. So, TPUs can train the ML models efficiently, which otherwise take weeks or months for training.

(iii) Capability to process large ML models: HBM on the TPUs permits their application to larger models and batch sizes. Multiple-layer neural networks with a phenomenally large number of parameters benefit greatly from the parallel processing and systolic array architecture of TPUs. This is especially observed in models like transformers. The transformers are neural network architectures that transform or alter an input sequence into an output sequence. They can bring about this transformation by learning the background of a content and pursuing relationships between components of sequences, e.g., given the input sequence, 'What is the color of the grass?', the transformer model generates the output sequence, 'Grass is green in color'. It does so by using an internal mathematical representation to identify the appropriateness, pertinence and relationship among the words 'color', 'grass', and 'green'. Transformer models are extensively used in speech recognition and machine translation.

(iv) Energy efficiency: TPUs are designed to furnish high performance while minimizing power usage. Energy efficiency is measured in performance per watt, e.g., TPUs render the service of up to 4 teraflops of performance while consuming only 2 watts of power ~2 teraflops per watt. This achievement makes them the favored choices of data centers for large-scale AI applications at low operational costs (Rao 2024).

(v) Good scalability in AI environment with Google's software support and infrastructure: A software stack that includes TensorFlow is provided. It allows developers to leverage the capabilities of TPUs for AI.

7.5 Driving factors for motivation for TPU

Let us look into the annals of TPU development to find out how the quest for TPU began? The necessity for TPU arose from the rapidly growing computation requirements for neural networks.

A fundamental beginning step towards ML is the idea of perceptron. The perceptron is an ML method which is considered as the ancestor of DNNs (Kim and Kim 2021). It is an algorithm or mathematical model of the biological neuron. It represents the smallest component of a neural network. As it contains a single neuron, it is the simplest ML architecture (Surdeanu and Valenzuela-Escárcega 2024).

This artificial neural network is used for the supervised training of binary classifiers. Binary classification is the task of categorization of a set of elements into one of two groups (Mishra *et al* 2022). A binary classifier is a function capable of deciding whether or not an input represented by a dataset belongs to a particular class of numbers. As an example, the health parameters of a person are used to classify whether the person is suffering from a particular disease or not. Accordingly, the person is placed in either disease-afflicted or non-ailing group.

7.5.1 Components of a perceptron

Recalling our elementary discussion on the perceptron in sections 1.5.2 and 1.5.3, a perceptron consists of the following principal components (figure 7.2):

(i) Input layer: It consists of a series of input features. These features are supplied in the form of real numerical values, each of which represents an attribute of the input data.

(ii) Weights: Each input feature is associated with a weight. The weight of a given feature expresses the relative significance of that feature in determining the output of the perceptron with respect to other features. Hence, it is an indicator of the strength of connection between the input and the output.

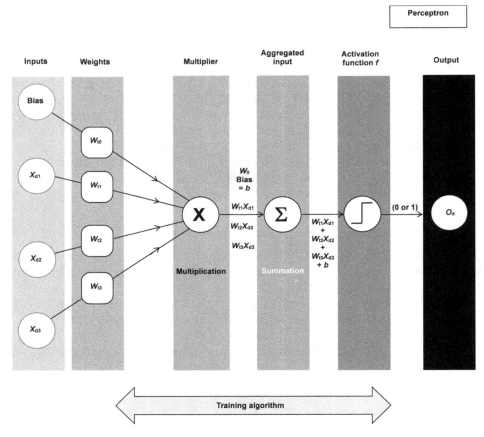

Figure 7.2. A single-layer perceptron consisting of the input layer with inputs X_{d1}, X_{d2}, X_{3d3} and bias; the weights layer with weights W_{t0}, W_{t1}, W_{t2}, and W_{t3}; the multiplier layer with outputs $W_{t1}X_{d1}$, $W_{t2}X_{d2}$, $W_{t3}X_{d3}$ and W_{00} Bias $= b$; the aggregated input layer obtained by applying the summation function with output $W_{t1}X_{d1} + W_{t2}X_{d2} + W_{t3}X_{d3} + b$; the activation function f layer giving the output 0 or 1; and the output layer showing the output O_d.

(iii) Bias term: It is a term included for modeling complex data. To illustrate, mention may be made of the intercept term in a linear equation denoting the intercept of the line with the coordinate axis. It gives the flexibility to shift the data curve upwards or downwards.

(iv) Summation function: It is a function which combines the input values with the corresponding weights to yield their weighted sum.

(v) Activation function: It is a function, such as the Heaviside step function, sigmoid function, ReLU function or some variant of ReLU such as the leaky rectified linear unit (LReLU), parametric rectified linear unit (PReLU), self-normalizing linear unit, and so forth (Xiangyang *et al* 2023). It is used to decide whether the neuron will fire or not depending on the weighted sum of the input values. It compares the weighted sum with a threshold to determine whether the output is 0 or 1.

(vi) Output: The output Y of the perceptron is a binary number, 0 or 1. It indicates whether or not the input data belongs to a class or category.

(vii) Training algorithm: A supervised learning algorithm, e.g., the perceptron learning algorithm or backpropagation is employed. In this algorithm, the weights and biases of the perceptron are updated to minimize the difference representing the error, between the predicted and true outputs.

7.5.2 Calculation steps performed in the working of the perceptron

The steps are enlisted below:

(i) Multiplication of weight W_{ti} with data X_{di}: Each value of input data X_{di} is multiplied with its respective coefficient or weight W_{ti} to represent the signal strength.

(ii) Addition of $(W_{ti}X_{di})$ products calculated in step (i): All the $(W_{ti} \ X_{di})$ products are added together to calculate:

$$\text{Weighted sum of products} = W_{t1}X_{d1} + W_{t2}X_{d2} + W_{t3}X_{d3} + \ldots + W_{tn}X_{dn} = \sum_i W_{ti}X_{di}. \quad (7.1)$$

(iii) Result offsetting for activation function shifting by addition of a bias term $W_0 = b$ to the weighted sum $\sum_i W_{ti}X_{di}$, as determined in step (ii): This is done to get the

$$\text{Weighted sum of products with bias term} = \sum_i W_{ti}X_{di} + b. \quad (7.2)$$

(iv) Applying the selected activation function to the weighted sum with bias term, as found in step (iii): This is done to obtain the output in binary form.

7.6 The structural units and instruction set of TPU

The TPU construction encapsulates the essence of perceptron's activity to work as a neural network inference accelerator. As shown in figure 7.3, the main components of a TPU are as follows.

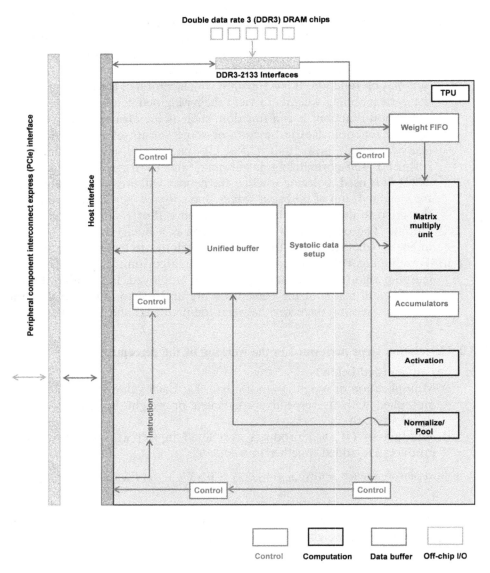

Figure 7.3. Parts and layout of a TPU. At the center lie the unified buffer and the systolic dataset up. On the right-hand side can be seen, from the top downwards, the weight FIFO, the matrix multiply unit, the accumulators, the activation unit and the normalize/pool unit. The weights are fetched into the weight first in, first out (FIFO) from the dynamic random-access memory (DRAM) chips communicating with the host CPU interface. Two-way communication linkage is done between the host interface and the unified buffer. The systolic setup is located adjacent to the unified buffer. The data from the systolic setup and the weights from the weight FIFO are multiplied in the MXU. The instructions move from the host interface along the control loop. The output of the normalize/pool unit is fed back to the unified buffer. The host interface is connected through a peripheral component interconnect (PCI) bus with off-chip devices. The control section is shown in red color, the computing section in yellow color, the data buffer in blue color and off-chip input/output in green color.

7.6.1 External (off-chip) components

The off-chip components are:

    (i) PCIe bus: The 'PCIe' stands for peripheral component interconnect express. It is a serial expansion bus standard. It provides a high-speed data transfer interface between the computer and its peripheral components. Serial data communication occurs in packets through pairs of transmit and receive signals. These pairs called lanes enable 250 MB s^{-1} bandwidth per direction per lane.

    (ii) DRAM chips: Dynamic random-access memory is a type of semiconductor memory. It is manufactured in the form of an integrated circuit chip. In this chip, each bit of data is stored in a memory cell consisting of a capacitor and a transistor. The capacitor–transistor combination is based on the metal-oxide semiconductor (MOS) technology.

7.6.2 Internal (on-chip) components

The on-chip components are:

    (i) Unified buffer: It is the static random-access memory (SRAM) that works as registers.

    (ii) Systolic data setup: The systolic array is a triumphant architecture for DNN hardware accelerators (Xu *et al* 2023). A systolic array is a group of processing elements, known as cells. Its beauty lies in the execution of an algorithm by rhythmical computing (Reinders 2011). In this type of computing, the data is transmitted from one cell to another cell by local communication only. A balance is maintained between computation and communication of data. Such a computing style is essential to reduce the energy consumption that takes place each time that the memory is accessed to fetch data during a calculation.

In a normal processor, a lot of energy is wasted in data movement because data processing is performed at a location far away from the memory unit. The systolic data setup obviates this problem by increasing the number of computations accomplished with each memory access. The number of computations is increased by replacement of a single computation with multiple computations (figure 7.4).

The increase in the number of computations in a systolic array is achieved by using each item of the input data a number of times. So, a larger throughput is attainable with a smaller memory bandwidth. Both the inputs to a matrix multiplier are used several times in the process of production of the output. Hence, each input value is read from the memory only once. Notwithstanding, it is used many times in different operations before storage back to the register. The spatially adjacent processing elements of the array are connected by shorter wires. The short wires reduce the incurred energy loss.

The data flow pattern described is referred to as 'systolic' for a special reason. The data moves through the processor in the form of waves. These waves remind us about the manner of blood circulation in the body by the pumping action of the

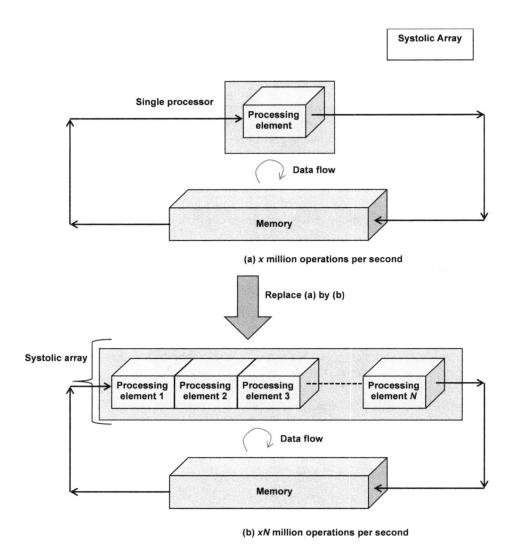

Figure 7.4. Systolic array concept for reducing power consumption: (a) normal processor consisting of a single processing element and (b) a systolic array-based processor comprising a sequence of processing elements from processing element 1 to processing element N. In (a) one data item is fetched from the memory to the processing element. Then calculation is done by the processing element. In the ensuing step, the result is stored in the memory. Then the next data item is brought from the memory to the processing element. Like the first data item, calculation is done on this data item. Subsequently, the result for this data element is stored in the memory, as done for the previous data item. This process is repeatedly followed. The difficulty with this repetitive sequential process is that a lot of time is wasted. Invaluable time is lost in fetching data from the memory and transferring the results back to memory. In (b), the calculation is planned in such a way that the result of execution of computation on one data item becomes the input data item for the next processing element. So, this processing element need not fetch the input data item for its calculation. Nor does it need to send its output to memory for storage because its output is the input data item for the ensuing processing element. By such planning, a large number of computations are executed. They are implemented without fetching data from memory or sending the results of intermediate steps to memory. The resulting saving in time

increases the computation speed of a multiple processing element systolic array by a large factor compared to a normal processor. The advantage gained becomes gigantic when the number of processing elements in the systolic array is large. The number of operations per second becomes xN million for x million operations per second on a normal processor. A high multiplication factor N is chosen. By placing participating processing elements in close vicinity, the calculation is speeded up still further because the inter-processing element communication time is shortened.

heart. Systole is a phase of the cardiac cycle. During systole, the ventricles of the heart contract. This contraction pumps out the blood from the heart into the circulatory system of the body. The network of TPU MAC units rhythmically compute and transmit data through the system like the heart pump. Every MAC sucks data in and drains data out after performing a short calculation. A continuous flow of data in the network is thus ensured.

A TPU matrix multiplier unit containing $256 \times 256 = 65\,536$ ALUs finishes $65\,536$ multiply-and-add operations on 8-bit integers in every cycle. Running at 700 MHz frequency, it can complete $65\,536 \times 700 \times 10^6 = 458\,752 \times 10^8 = 45.9 \times 10^{12}$ multiply-and-add operations per second $= 2 \times 45.9 \times 10^{12} = 91.8$ tera operations per second (TOPS). The intermediate results involved in this lengthy operation are transmitted directly between neighboring ALUs without resorting to any memory access.

For comparison, a reduced instruction set computer (RISC) CPU without vector extensions performs just a few operations per cycle, a CPU with vector extensions a few tens of operations per cycle and a GPU a few hundreds of thousands, e.g., 128 K operations per cycle. Consequently, a high performance per watt ratio is available with TPU, typically 83 times better than a CPU and 29 times superior to a GPU. However, note that the systolic array is optimized in terms of power and area efficiency for matrix multiplications. For optimization, a trade-off has been made by restricting the numbers of registers.

The control and flexibility of operation are lost in exchange for the higher efficiency derived. This makes the design of TPU suitable for the targeted application only. It must be reiterated that the advantages of its design are not extended to any general-purpose computation.

(iii) Weight memory and FIFO: This is a buffer (temporary storage area) for weights. In this storage, the data written first comes out first (FIFO).

(iv) MXU: This is made up of several tightly coupled data processing units. These data processing units are known as MAC units (figure 7.5(a)). They are hardwired units performing multiplication, addition and accumulation functions. To carry out these functions, they contain a multiplier, an adder and an accumulator as the three basic components.

From the operational viewpoint, a MAC unit reads the input values A, B from the memory cells. Then it multiplies them together to get the product $P = A \times B$. The result of the product $P = A \times B$ is stored in the the accumulator register. It may be noted that the output of the accumulator is fed back to the adder. Hence, in each clock cycle, the multiplier output is incremented by the value that was stored in the

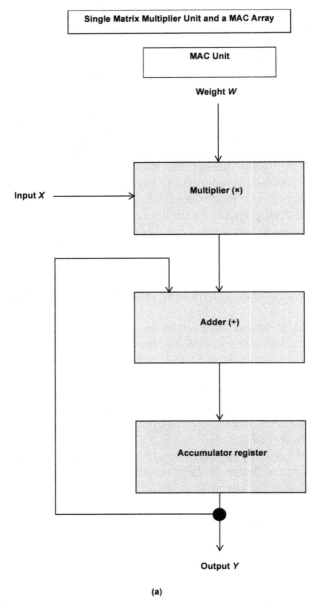

(a)

Figure 7.5. The MXU construction and its inclusion in a systolic array. (a) The MAC unit as a combination of a multiplier, an adder and an accumulator. (b) The MAC units inside a systolic array of computation cells. The array is utilized for performing large-scale matrix multiplications. Hence, it is known as the MXU. The MXU for 3 × 3 matrices is shown in the diagram. It consists of three rows and three columns of MAC units. In each unit, the weights are supplied form a horizontal bar above the unit. The input data are supplied from a vertical bar on the left-hand side of the unit. The weight and data values are multiplied together. Then the product is passed via an adder unit to the accumulator. The output of the accumulator is fed back to the adder unit for addition to the succeeding product from the multiplier. The weights shown in the first horizontal bar are W_{r11},

W_{t21} and W_{t31}; the weights shown in the second horizontal bar are W_{t12}, W_{t22} and W_{t32}; while those shown in the third horizontal bar are W_{t13}, W_{t23} and W_{t33}. For the first row of weights, the input data loaded in the vertical bars are X_{d11}, X_{d12}, X_{d13}. It is loaded in the order from right to left starting from rightmost vertical bar. For the second row of weights, the input data loaded in the vertical bars are X_{d21}, X_{d22}, X_{d23}. These data are loaded in the order from right to left, starting from the next vertical bar towards the left from the beginning vertical bar for the previous row. So, they are shifted leftward by one place with respect to the previous row. For the third row of weights, the input data loaded in the vertical bars are X_{d31}, X_{d32}, X_{d33}. They are loaded in the order from right to left starting from the next vertical bar towards the left from the inaugurating vertical bar for the preceding row, i.e., shifted leftward by one place with respect to the previous row. The outputs are shown by vertical arrows below the accumulators of multiplier units for the third row of weights.

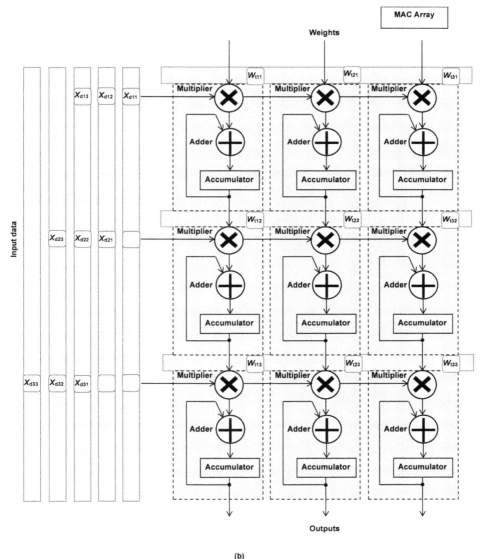

Figure 7.5. (Continued.)

accumulator register during the previous cycle. Thus, the MAC block implements the function

$$P = A \times B + P \tag{7.3}$$

To start a new accumulation, a reset button is pressed. Figure 7.5(b) shows an array of MAC units.

Figure 7.6 shows the MXU for 3×3 matrices. It comprises nine MAC units (ankur6ue 2018). The weight matrix $[W_t]$ is given by

$$[W_t] = \begin{bmatrix} W_{t11} & W_{t21} & W_{t31} \\ W_{t12} & W_{t22} & W_{t32} \\ W_{t13} & W_{t23} & W_{t33} \end{bmatrix} \tag{7.4}$$

and the activation matrix $[X_d]$ is expressed as

$$[X_d] = \begin{bmatrix} X_{d11} & X_{d21} & X_{d31} \\ X_{d12} & X_{d22} & X_{d32} \\ X_{d13} & X_{d23} & X_{d33} \end{bmatrix} \tag{7.5}$$

Let us apply the rules of matrix multiplication. It is noted that the number of columns in the weight matrix = 3 and number of rows in activation matrix = 3. The number of columns in the first matrix equals the number of rows in the second matrix. Since this equality condition is satisfied, the elements of the individual row of the activation matrix are multiplied by the elements of all columns in the weight matrix. Then the obtained products are added together. Finally, the sum of the products is arranged in the corresponding columns to get

$$[X_d] \times [W_t] \begin{bmatrix} X_{d11} & X_{d21} & X_{d31} \\ X_{d12} & X_{d22} & X_{d32} \\ X_{d13} & X_{d23} & X_{d33} \end{bmatrix} \times \begin{bmatrix} W_{t11} & W_{t21} & W_{t31} \\ W_{t12} & W_{t22} & W_{t32} \\ W_{t13} & W_{t23} & W_{t33} \end{bmatrix}$$
$$= \begin{bmatrix} W_{t11}X_{d11} + W_{t12}X_{d21} + W_{t13}X_{d31} & W_{t21}X_{d11} + W_{t22}X_{d21} + W_{t23}X_{d31} & W_{t31}X_{d11} + W_{t32}X_{d21} + W_{t33}X_{d31} \\ W_{t11}X_{d12} + W_{t12}X_{d22} + W_{t13}X_{d32} & W_{t21}X_{d12} + W_{t22}X_{d22} + W_{t23}X_{d32} & W_{t31}X_{d12} + W_{t32}X_{d22} + W_{t33}X_{d32} \\ W_{t11}X_{d13} + W_{t12}X_{d33} + W_{t13}X_{d33} & W_{t21}X_{d13} + W_{t22}X_{d23} + W_{t23}X_{d33} & W_{t31}X_{d13} + W_{t32}X_{d23} + W_{t33}X_{d33} \end{bmatrix} \tag{7.6}$$

Let us examine the implementation of matrix multiplication by the processor. A weight-stationary architecture is used for calculation of product of the weight and activation matrices. In the weight-stationary architecture, the weights W_{t11}, W_{t21}, W_{t31}; W_{t12}, W_{t22}, W_{t32}; W_{t13}, W_{t23}, W_{t33} are loaded into the MAC array. Coincidentally, the column entries of activation matrix are fed from the activation storage buffer from the left side. Consider the first column:

$$\begin{bmatrix} X_{d11} \\ X_{d12} \\ X_{d13} \end{bmatrix} \tag{7.7}$$

The entries are arranged in the order X_{d13}, X_{d12}, X_{d11}. Hence, X_{d11} will enter the MAC unit first followed by X_{d12} and then X_{d13}. The feeding of column entries of the activation matrix is offset in time. The entries of first column entry start from

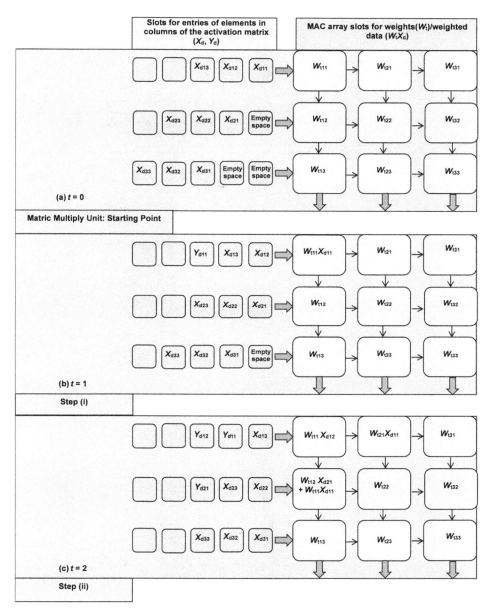

Figure 7.6. Matrix multiplication procedure on a weight-stationary two-dimensional systolic array: (a) $t = 0$, (b) $t = 1$, (c) $t = 2$, (d) $t = 3$, (e) $t = 4$, (f) $t = 5$, (g) $t = 6$, (h) $t = 7$. The weights are pre-loaded in the MAC array. A column of matrix $[X_d]$ (equation (7.5)) is fed from horizontal side to multiply with the matrix $[W_t]$. The same process is repeated for the succeeding columns of matrix $[X_d]$. The second column of $[X_d]$ is delayed by one clock cycle after the first column. Similarly, the third column of $[X_d]$ is delayed by two clock cycles after the first column. Parts (a)–(h) show the results of the computation at the beginning $t = 0$, and after $t = 1, 2, 3,$ ..., 8 cycles. The blank spaces left after evacuation of slots for entries of elements in columns of the activation matrix are occupied by elements of the next incoming activation matrix $[Y_d]$ given by equation (7.8).

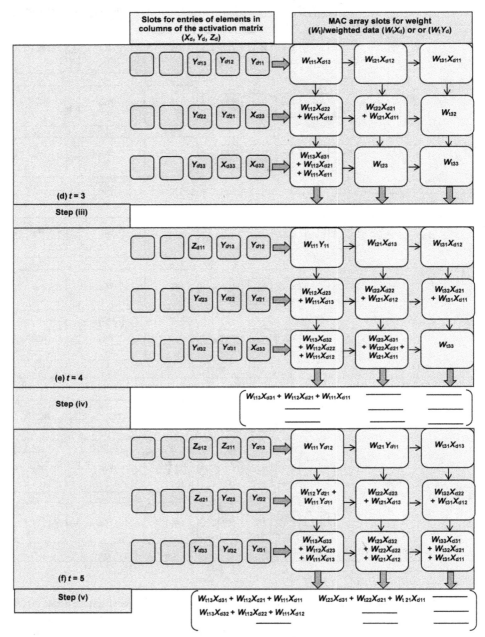

Figure 7.6. (Continued.)

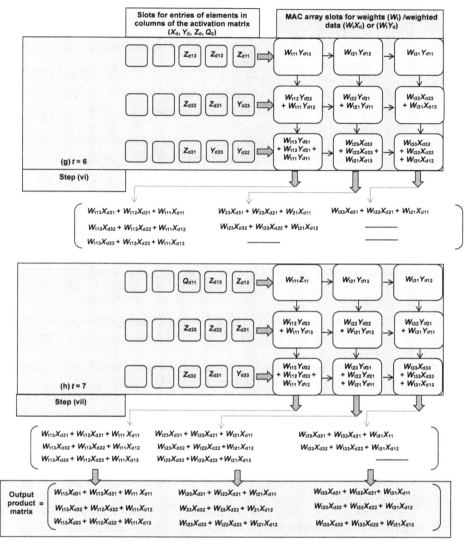

Figure 7.6. (Continued.)

extreme left as X_{d13}, X_{d12}, X_{d11}. The entries of second column start after leaving a gap of one space from left by inserting—in that space as X_{d23}, X_{d22}, X_{d21}, −. So, it is delayed by one clock cycle. The entries of third column start after leaving a gap of two spaces from left by inserting −, − in those two spaces as X_{d33}, X_{d32}, X_{d31}, − −. Hence, a delay of two clock cycles is introduced.

The blank spaces left after evacuation of slots for entries of elements in columns of the activation matrix are occupied by the next incoming activation matrix for computation such as

$$[Y_d] = \begin{bmatrix} Y_{d11} & Y_{d21} & Y_{d31} \\ Y_{d12} & Y_{d22} & Y_{d32} \\ Y_{d13} & Y_{d23} & Y_{d33} \end{bmatrix} \tag{7.8}$$

and the blank spaces in the MAC array slots for weights/weighted sums are filled by the results of computation of product of matrices Y_d and W_t.

The activations move horizontally from left to right. The psums (partial sums) formed move vertically from top towards the bottom. The formation of products and partial sums during the different stages of computations are shown in the diagram at $t = 0, 1, 2, 3, 4, 5, 6, 7, 8$. The final results of the matrix product are fed to the activation unit.

To examine the implementation of these steps by the processor:

(a) $t = 0$: All the activations are outside the MAC array slots. All the weights are inside the MAC array slots in their respective positions.

Activations in the first row slots for the first column of the activation matrix (equation (7.5)) are from left to right: $X_{d13}, X_{d12}, X_{d11}$

Activations in the second row slots for the second column of the activation matrix (equation (7.5)) are from left to right: $X_{d23}, X_{d22}, X_{d21}$, Empty space

Activations in the third row slots for the third column of the activation matrix (equation (7.5)) are from left to right: $X_{d33}, X_{d32}, X_{d31}$, Empty space, Empty space.

(b) $t = 1$: Activations in the first row slots for the first column of the activation matrix (equations (7.5) and (7.8)) are from left to right: $Y_{d11}, X_{d13}, X_{d12}$. The activation X_{d11} has moved inside the MAC array to the first row, first column, forming the product $W_{t11}X_{d11}$ in the first row, first column of MAC array slots.

Status: $W_{t11}X_{d11}$ in the first row, first column of MAC array slots.

Activations in the second row for second column of the activation matrix (equation (7.5)) are from left to right: $X_{d23}, X_{d22}, X_{d21}$. The activations have advanced rightwards by one space.

Activations in the third row for the third column of the activation matrix (equation (7.5)) are from left to right: $X_{d33}, X_{d32}, X_{d31}$, Empty space. The activations have advanced rightwards by one space.

(c) $t = 2$: Activations in the first row for the first column of the activation matrix (equation (7.8)) are from left to right: $Y_{d12}, Y_{d11}, X_{d13}$.

The activation X_{d11} has advanced rightward by one space into the MAC array forming the product $W_{t21}X_{d11}$ in the first row, second column of MAC array slots.

The activation X_{d12} has moved inside the MAC array to the first row, first column, forming the product $W_{t11}X_{d12}$ in the first row, first column of MAC array slots.

Status: $W_{t11}X_{d12}$ in the first row, first column slot and $W_{t21}X_{d11}$ in the first row, second column slot of MAC array.

Activations in the second row for the second column of the activation matrix (equations (7.5) and (7.8)) are from left to right: Y_{d21}, X_{d23}, X_{d22}.

The activation X_{d21} has moved inside the MAC array to the second row, first column of MAC array slots, forming the product $W_{t12}X_{d21}$i in the second row, first column of MAC array slots. The $W_{t11}X_{d11}$ product has come down from the first row, first column to the second row, first column of MAC array slots so that the sum $W_{t12}X_{d21} + W_{t11}X_{d11}$ is formed in the second row, first column, of MAC array slots.

Status: $W_{t12}X_{d21} + W_{t11}X_{d11}$ in the second row, first column of MAC array slots.

Activations in the third row for the third column of the activation matrix (equation (7.5)) are from left to right: X_{d33}, X_{d32}, X_{d31}. The activations have advanced rightwards by one space.

(d) $t = 3$: Activations in the first row for the first column of the activation matrix (equation (7.8)) are from left to right: Y_{d13}, Y_{d12}, Y_{d11}.

The activation X_{d11} has advanced rightward by one space forming the product $W_{t31}X_{d11}$ in the first row, third column of MAC array slots.

The activation X_{d12} has advanced rightward by one space forming the product $W_{t21}X_{d12}$ in the first row, second column of MAC array slots.

The activation X_{d13} has moved inside to the first row, first column of the MAC array forming the product $W_{t11}X_{d13}$ in the first row, first column of MAC array slots.

Status: $W_{t11}X_{d13}$ in the first row, first column of MAC array slots; $W_{t21}X_{d12}$ in the first row, second column of MAC array slots; $W_{t31}X_{d11}$ in the first row, third column of MAC array slots.

Activations in the second row for the second column of the activation matrix (equations (7.5) and (7.8)) are from left to right: Y_{d22}, Y_{d21}, X_{d23}.

The activation X_{d21} has moved to the second row, second column of MAC array slots forming the product $W_{t22}X_{d21}$ in the second row, second column of MAC array slots. At the same time $W_{t21}X_{d11}$ has come down from the first row, second column to the second row, second column of MAC array slots to be added to $W_{t22}X_{d21}$ forming the sum: $W_{t22}X_{d21} + W_{t21}X_{d11}$ in the second row, second column of MAC array slots.

The activation X_{d22} has moved inside the MAC array to the second row, first column of MAC array slots forming the product $W_{t12}X_{d22}$ in the second row, first column of MAC array slots. The $W_{t11}X_{d12}$ has come down from the first row, first column to the second row, first column of MAC array slots. Hence, the sum $W_{t12}X_{d22} + W_{t11}X_{d12}$ is formed in the second row, first column of MAC array slots.

Status: $W_{t12}X_{d22} + W_{t11}X_{d12}$ in the second row, first column; $W_{t22}X_{d21} + W_{t21}X_{d11}$ in the second row, second column of MAC array slots.

Activations in the third row for the third column of the activation matrix (equations (7.5) and (7.8)) are from left to right: Y_{d33}, X_{d33}, X_{d32}.

The activation X_{d31} has moved inside the MAC array to the third row, first column of MAC array slots forming the product $W_{t13}X_{d31}$. To this is added $W_{t12}X_{d21} + W_{t11}X_{d11}$ which has come down from the second row, first column to the third row, first column forming the sum: $W_{t13}X_{d31} + W_{t12}X_{d21} + W_{t11}X_{d11}$ in the third row, first column of MAC array slots.

Status: $W_{13}X_{d31} + W_{t12}X_{d21} + W_{t11}X_{d11}$ in the third row, first column of MAC array slots.

A similar sequence of operations up to (h) gives the complete matrix multiplication result as output.

(e) Accumulators and activations: Hardwired activation functions are included.

(f) Control: In TPU, the control logic is minimal occupying merely 2% of the die while the same uses much larger areas in CPU and GPU to cater to a broad range of applications.

7.6.3 CISC instruction set

The TPU design employs CISC style (Gilreath and Laplante 2003). The CISC design focuses on high-level instructions for running complex tasks. Notable among these tasks are the multiple multiply-and-add steps. This is a deviation from the standard practice followed with CPUs. The CPUs are usually based on the RISC design style. The RISC style concentrates on the simpler load, add, multiply and store instructions used in the majority of applications. In TPU, the instructions are not fetched from memory. Instead, they are sent to the TPU by the host computer over the PCIe interface.

7.7 TPU operation

7.7.1 Block diagram of the workflow of an inference task

The block diagram of the workflow of an inference task performed by a perceptron is depicted in figure 7.7. The inference process makes headway through the sequence of steps enumerated below:

Step 1: Fetching input data and weights: This is subdivided into two parts:
(a) the input data is placed from the host memory into the unified buffer of TPU, and
(b) the weights are transferred from the host memory into the weight memory FIFO of TPU.

Step 2: Performing large-scale matrix multiplication: This involves passage of the input and weights through the systolic array.

Step 3: Output storage: The result of matrix multiplication is stored in the accumulator.

Step 4: Applying activation function on the output: The activation function is applied on the product of matrices kept in the accumulator.

Step 5: Storing the post-activation function result: This result is stored in the unified buffer.

Step 6: Transferring the data to host memory: The result kept in the unified buffer is shifted to the host memory.

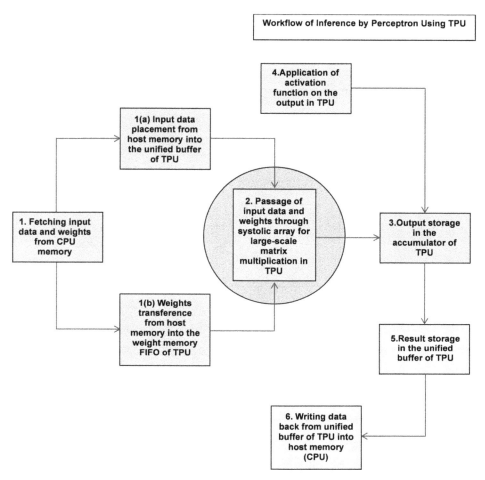

Figure 7.7. Plan for inference execution on a perceptron. The main steps involved include bringing the input data and weights from CPU and placing them into designated places in the TPU. The input data are placed in the unified buffer of TPU while the weights are kept in the weight buffer of TPU. After the input data and weights have reached their respective places, they are passed through the systolic array. The input data and weight matrices are multiplied with each other. The multiplication result is stored in the accumulator. The activation function is applied on the result. The outcome obtained after activation is stored in the unified buffer. From here, it is written to the host memory.

7.7.2 Inference implementation by TPU

The implementation methodology of the inference process by a TPU is shown in figure 7.8. The sequential steps involved are:

    (i) Input data and weights are loaded in their respective places:

        (a) Input data (X_{d1}, X_{d2}, bias) are fetched from the CPU memory and loaded in the unified buffer of TPU.

        (b) Similarly, the weights (W_{t0}, W_{t1}, W_{t2}) are brought form the CPU memory and kept in the weight FIFO of TPU.

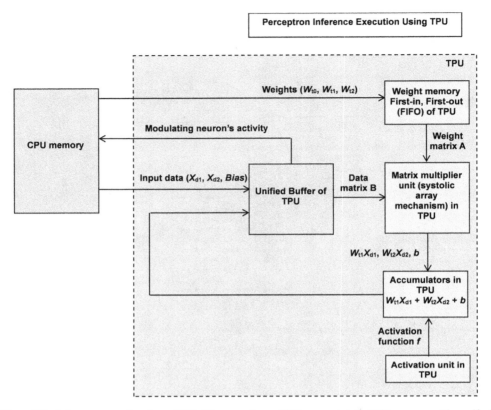

Figure 7.8. Inference execution by a TPU. Diagram shows TPU connected to CPU memory. The unified buffer of TPU, its weight memory FIFO, matrix multiplier unit, accumulators and activation unit are shown. The inference process begins with procurement of weights (W_{t0}, W_{t1}, W_{t2}) from the CPU memory to weight memory FIFO of TPU. At one and the same time, the input data (X_{d1}, X_{d2}, bias) are transferred from the CPU memory to the unified buffer of TPU. Inside the TPU, the weight matrix A from weight memory FIFO of TPU and the data matrix B from the data matrix of TPU are fed to its matrix multiplier unit in a systolic array mechanism. The products $W_{t1} X_{d1}$, $W_{t2}X_{d2}$, and the bias b come out from the matrix multiplier unit, and move to the accumulator. In the accumulator, the summation ($W_{t1} X_{d1} + W_{t2}X_{d2} + b$) is carried out. The activation function f from the activation unit is applied on the sum obtained. The outcome of activation is transferred to the unified buffer of TPU. A corrective signal is sent from the unified buffer of TPU to the CPU memory for modifying the activity of the neuron.

(ii) The weight matrix A and the data matrix B are loaded into the matrix multiplier unit of TPU following the systolic array mechanism.

(iii) The products $W_{t1}X_{d1}$, $W_{t2}X_{d2}$ along with the bias b are added together in the accumulator where the resulting sum $W_{t1}X_{d1} + W_{t2}X_{d2} + b$ is stored.

(iv) The activation function is applied on the result $W_{t1}X_{d1} + W_{t2}X_{d2} + b$ in the accumulator.

(v) The result with the activation function applied goes back to the unified buffer.

(vi) Finally, the output is sent from the unified buffer to the host CPU to modulate the activity of the neuron.

7.8 Summary and the way forward

7.8.1 Highlights of chapter 7 at a glance

(i) The terms 'Tensor', 'TensorFlow' and 'Tensor Processing Unit (TPU)' were introduced.

(ii) Chronological history and features of TPU were described and its specialties for AI work were enumerated (Jouppi 2016, Sato and Young 2017, Jouppi *et al* 2017, 2020, 2021, 2023, ankur6ue 2018, Cloud TPU 2024, Vahdat 2024).

(iii) TPUs are integrated into Google's cloud computing platform, Google Cloud to allow a seamless scaling of ML workloads. To handle larger datasets or more complex models, the services of additional TPUs are readily availed. In order to derive the full advantage of TPUs, Google provides a software stack including TensorFlow. The TensorFlow is an open-source ML framework that is optimized for this purpose.

(iv) The driving factors behind the motivation for TPU become evident by understanding the components of a perceptron. The calculation steps performed in its working must be clearly understood.

(v) The structural units of TPU and its instruction set were described. Both off-chip and on-chip components and the CISC instruction set were mentioned. The TPU operation was explained by drawing the block diagram of the workflow of an inference task and discussion of inference implementation by TPU.

(vi) The central hub of all activities in a TPU is the MXU. The MXU performs a major chunk of the computations required for ML. The ability of MXU to perform a large number of multiply–accumulate operations in parallel, is the primary cause of the high throughput of TPUs.

(vii) Various aspects of TPU architecture and its operational procedures were described. The systolic array architecture of TPU is a collection of processing elements called cells. These cells accelerate matrix computation in an elegant style. The style entails implementing an algorithm rhythmically to transfer data from one cell to the ensuing cell with only local communication. The time-consuming fetching-from-memory procedures are avoided throughout.

(viii) A TPU has a unified buffer that serves as a high-speed on-chip memory. It also has an activation unit that applies non-linear functions to the data.

(ix) The TPU is an energy-efficient ASIC providing several teraflops output with a power consumption of a few watts.

(x) The TPU is the successful culmination of efforts to create a faster AI chip by re-visualization of the fundamentals of AI computing. A recognition of limitations of conventional hardware to deal with the demands of new AI models is made. It is appreciated that these models involve a high volume of low-precision calculations without compromising accuracy.

The combined efforts paved the way to a variety of applications within the Google ecosystem.

(xi) Additionally, the requirements of localized, on-device processing outside the centralized data center facilities must be fulfilled. Therefore, a smaller size variant of TPU, the edge TPU with reduced power consumption was launched.

(xii) Among the unique features of TPU, mention may be made of:
  (a) its MXU to handle large matrix operations,
  (b) the high bandwidth memory for faster data access, and
  (c) the deployment of TPUs in pods and clusters networked together to work as a single interconnected unit. The clustering allows acceleration of the processing of large training datasets.

(xiii) The TPU pods are designed on a modular principle. So, the number of TPUs can be increased to meet the rising demands. The data flow inside pods is optimized in such a way that each TPU can acquire the needed data without incurring any delay.

(xiv) It is generally known that modern processors are designed for an intermixed functioning. Nevertheless, the CPUs have largely the characteristics of a scalar architecture. The GPUs have dominantly vector architecture properties. The TPUs display mostly matrix architecture features. It must be recapped that CPUs can do vector processing and GPUs too have tensor cores. Figure 7.9 brings out the distinguishing features of CPU, GPU and TPU on a common platform.

(xv) For clarification, the scalar architecture yields optimal performance on the type of workloads in which one instruction operates on one word of data in every cycle. The vector architecture is optimal for workloads in which one instruction operates on many words of data in every cycle. The matrix architecture is good for workloads in which one instruction multiplies matrices in every cycle.

(xvi) Let us look at the number of threads on which the aforesaid devices operate. It can be stated that CPUs work on a few threads only. The GPUs operate on thread blocks. The TPUs execute matrix operations on thread blocks.

(xvii) Keywords for this chapter are: TPU, TensorFlow, perceptron, systolic array, matrix multiply unit.

7.8.2 Getting ready to begin chapter 8

In the next chapter, we study the neural processing unit (NPU). In contrast to TPU which is based on a systolic array architecture, the NPU uses the traditional von Neumann architecture. TPUs have a lower latency and power consumption. NPUs show higher latency and power consumption. But NPUs have a higher peak performance than TPUs.

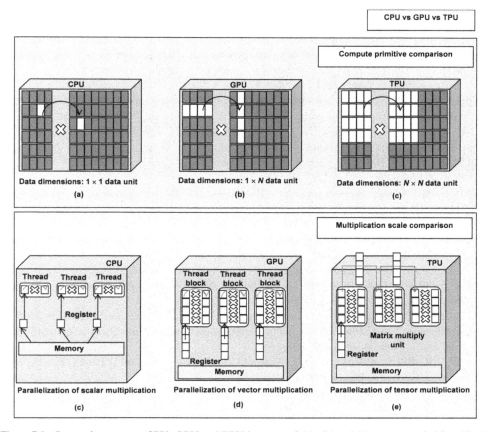

Figure 7.9. Comparison among CPU, GPU and TPU in terms of: (a), (b) and (c) compute primitive; (d), (e) and (f) multiplication scale.

References

Ankur6ue 2018 Understanding matrix multiplication on a weight-stationary systolic architecture, computer architecture https://telesens.co/2018/07/30/systolic-architectures/#:~:text=timing% 20hardware%20etc.),Matrix%20Multiplication%20on%20a%20Weight%20Stationary%202D% 20Systolic,MXU%20on%20a%20Google%20TPU)&text=(N%3D256)%20grid%20of,from%20the %20activation%20storage%20buffer

Cloud TPU 2024 https://cloud.google.com/tpu/docs/intro-to-tpu

Gilreath W F and Laplante P A 2003 Chapter 5: CISC, RISC, OISC: a tale of three architectures *Computer Architecture: A Minimalist Perspective* The Springer International Series in Engineering and Computer Science vol 730 (Boston, MA: Springer) pp 33–9

Jouppi N 2016 Google supercharges machine learning tasks with TPU custom chip *AI Mach. Learn.* May 19, https://cloud.google.com/blog/products/ai-machine-learning/google-super-charges-machine-learning-tasks-with-custom-chip

Jouppi N P *et al* 2017 In-datacenter performance analysis of a tensor processing unit *Proc. of the 44th Annual Int. Symp. on Computer Architecture (ISCA'17)(June 24–28, 2017) (Toronto, ON, Canada)* (New York: Association for Computing Machinery) pp 1–12

Jouppi N P, Yoon D H, Kurian G, Li S, Patil N, Laudon J, Young C and Patterson D 2020 A domain specific supercomputer for training deep neural networks *Commun. ACM* **63** 67–78

Jouppi N P *et al* 2021 Ten lessons from three generations shaped Google's TPUv4i: industrial product *2021 ACM/IEEE 48th Annual Int. Symp. on Computer Architecture (ISCA)(14–18 June) (Valencia, Spain)* pp 1 14

Jouppi N *et al* 2023 TPU v4: An optically reconfigurable supercomputer for machine learning with hardware support for embeddings *ISCA'23: Proc. of the 50th Annual Int. Symp. on Computer Architecture(June 17–21, 2023) (Orlando, FL, USA)* (New York: Association for Computing Machinery) Article No.: 82 pp 1–14

Kim E-H and Kim H-S 2021 Perceptron: basic principles of deep neural networks *Cardiovasc. Prev. Pharmacother.* **3** 64–72

Kundu K 2018 Google Announces Edge TPU, Cloud IoT Edge at Cloud Next 2018, July 26 https://beebom.com/google-announces-edge-tpu-cloud-iot-edge-at-cloud-next-2018/

Mishra S, Pandey M and Rautaray S S 2022 A review on machine learning algorithms for binary classification of heart disease patients *Proc. of the Int. Conf. on Innovative Computing & Communication (ICICC-2022)(February 19–20, 2022) (Delhi, India)* (Delhi: Shaheed Sukhdev College of Business Studies, University of Delhi) pp 1–6

Ramsundar B and Zadeh R B 2018 *TensorFlow for Deep Learning: From Linear Regression to Reinforcement Learning* (Sebastopol, CA: O'Reilly Media) p 256

Rao R 2024 TPU vs GPU in AI: a comprehensive guide to their roles and impact on artificial intelligence https://wevolver.com/article/tpu-vs-gpu-in-ai-a-comprehensive-guide-to-their-roles-and-impact-on-artificial-intelligence

Reinders J R 2011 Systolic arrays *Encyclopedia of Parallel Computing* ed D Padua (Boston, MA: Springer Science + Business Media) pp 2002–11

Sato K and Young C 2017 An in-depth look at Google's first Tensor Processing Unit (TPU) *AI Mach. Learn.* May 13, https://cloud.google.com/blog/products/ai-machine-learning/an-in-depth-look-at-googles-first-tensor-processing-unit-tpu

Surdeanu M and Valenzuela-Escárcega M A 2024 The perceptron *Deep Learning for Natural Language Processing: A Gentle Introduction* (Cambridge: Cambridge University Press) ch 2 pp 8–29

Vahdat A and Lohmeyer M 2023a Expanding our AI-optimized infrastructure portfolio: introducing cloud TPU v5e and announcing A3 GA *AI Mach. Learn.* August 30, https://cloud.google.com/blog/products/compute/announcing-cloud-tpu-v5e-and-a3-gpus-in-ga

Vahdat A and Lohmeyer M 2023b Enabling next-generation AI workloads: announcing TPU v5p and AI hypercomputer *AI Mach. Learn.* December 7, https://cloud.google.com/blog/products/ai-machine-learning/introducing-cloud-tpu-v5p-and-ai-hypercomputer

Vahdat A 2024 Announcing Trillium, the sixth generation of Google Cloud TPU, May 15, https://cloud.google.com/blog/products/compute/introducing-trillium-6th-gen-tpus

Xiangyang L, Xing Q, Han Z and Feng C 2023 A novel activation function of deep neural network *Hindawi Sci. Program.* **2023** 1–12

Xu R, Ma S, Guo Y and Li D 2023 A survey of design and optimization for systolic array-based DNN accelerators *ACM Comput. Surv.* **56** 20:1–20:37

IOP Publishing

AI-Processor Electronics
Basic technology of artificial intelligence
Vinod Kumar Khanna

Chapter 8

Neural processing unit

The use of approximation as a means of computation efficiency enhancement is emphasized. The procedure of conversion of general-purpose code into neural representation is explained. The parts of a neural processing unit (NPU) are mentioned and the constitution of its processing engine (PE) is elaborated. The NPU operation involves code generation, PE scheduling and functioning. The performance capability of an NPU is quantified. An analogy is drawn between it and a hardwired multilayered perceptron neural network. The ApproXimate Net (AXNet) accelerator is described and the multi-task learning idea is outlined followed by a discussion of the architecture and data flow of the AXNet during job execution. A reconfigurable accelerator, the reconfigurable neuromorphic computing accelerator (RENO) utilizes the memresistor-based resistive random-access memory (RAM). It works by a hybrid method of data representation. The RENO architecture and performance are elaborated. Components of the DianNao neural network accelerator, its operation by loop tiling or loop blocking and other hardware accelerators in the DianNao family are described. The main modules of the Ascend AI Processor, its Da Vinci artificial intelligence (AI) core architecture, software stack, neural network software flow, data flow and performance are briefly discussed. Examples of commercial chips with NPU integrated alongside central processing unit (CPU) and graphical processing unit (GPU) are given.

8.1 Introduction

This chapter surveys the main developments that have taken place in NPUs to acquire an overall view of the progress made in totality. The NPU is an AI accelerator which is built from the ground-up and tailored to speed up training and inference of deep neural network models. It excels at tasks like matrix multi-plications and activation functions, which are crucial for deep learning. However, its use is not restricted to these computing sectors only. It can also be used for general-

doi:10.1088/978-0-7503-6259-7ch8

purpose computations which are tolerant to approximations. So, we begin with a generalized description covering a broader perspective of the big picture of NPU.

8.2 Approximation as a means of computation efficiency enhancement

Survival during the difficult times in life is possible only by a flexible attitude. Flexibility is the foothold for resilience, and resilience is the way to happiness. The trees teach us a lot of lessons. During strong winds, the sturdy and brittle oak breaks. But the flexible willow tree bends and survives the storm. In the same way, we should have a bendable attitude towards calculations in which high accuracy is not required. In these calculations, we can adopt flexibility versus rigidity, and accept less accurate intermediate results as long as they do not affect our end results.

Let us explain in more detail. Conventional trends in computing concentrate on increasing speed and lowering energy requirement, which are mainly achieved by reduction of transistor dimensions to maintain accuracy. 'Approximate computation approach' is a huge departure from these trends. In this approach, the accuracy constraints are relaxed in favor of power consumed and promptness gained. Errors are permitted within reasonable limits to get better energy and speed solutions. In other words, a trade-off is made between computational accuracy and other parameters for realization of general-purpose processors. Applications that can benefit from approximation techniques are endowed with the pliable characteristics of tolerance to small errors in their outputs without any significant impact on the quality of result. Some of these applications are cited below (Esmaeilzadeh *et al* 2012, 2013, 2015):

  (i) Applications in which the input real-world data is frequently noise-afflicted, e.g. sensor signal processing, image and voice recognition.
 (ii) Applications in which the output signal to be perceived by human users can ignore errors which humans are unable to notice, e.g. audio synthesis, image reconstruction, etc.
(iii) Applications in which multiple answers to a question are acceptable, e.g., web search, machine learning, etc.
(iv) Applications in which data convergence can be stopped early and heuristic techniques give meaningful answers, e.g., iterative procedures in data analytics.

In these applications, a small sacrifice in accuracy of computing is unlikely to be seriously damaging. So, the flexibilities offered by a higher degree of permissible error can be gainfully utilized.

8.3 Conversion of general-purpose code into neural representation

Specific regions of approximate general-purpose code called the imperative source code are converted into neural networks by Parrot transformation (figure 8.1). This exposes the parallelism of operations. The simpler operations involved in neural computing are efficiently accelerated yielding noteworthy performance and energy

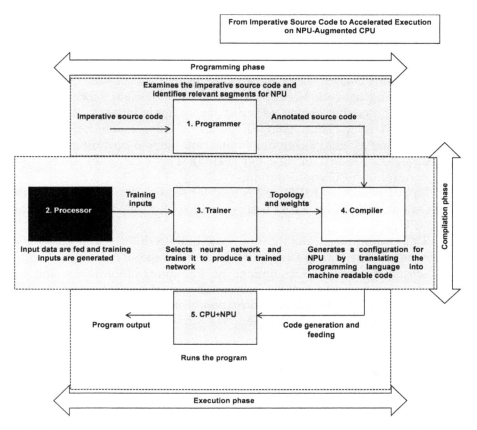

Figure 8.1. The Parrot transformation consisting of the programming phase, the compilation phase and the execution phase. In the programming phase, the programmer writes the annotated source code from the imperative source code. In the compilation phase, the processor produces training inputs from input data, the trainer selects a neural network and trains it, and the compiler uses the topology and weights to generate a configuration for NPU. In the execution phase, the code is generated and fed to the CPU and NPU combination to run the program, thus obtaining the output.

improvements. The Parrot transformation is an algorithm consisting of three main phases:

(i) Programming phase: The programmer identifies certain regions of the code that lend themselves easily to approximation. Hence, an annotated source code is generated from the imperative source code.

(ii) Compilation phase: The Parrot transformation is applied for three jobs: for observing the annotated source code, for training the neural networks and for producing binary code.

(iii) Execution phase: The transformed program is started on the main CPU core to configure the NPU. The services of NPU are availed to perform a neural network evaluation in place of the regular code region. Thus, NPU works as a tightly-coupled accelerator integrated with the main CPU.

8.4 Parts of NPU

Figure 8.2 shows the main parts of NPU:

(i) Processing engines: The NPU contains eight identical PEs. Each neuron in the neural network is assigned to one of these eight processing engines. Every processing engine performs the computations for all the neurons assigned to it. Thus, each PE performs the computation of a neuron consisting of multiplication, addition and sigmoid operations.

(ii) Configuration first in, first out (FIFO): This is used for setting up and retrieval of NPU configuration.

(iii) Input buffer: This FIFO unit is used for sending the inputs to the NPU.

(iv) Output buffer: This FIFO unit is used for retrieving the outputs from the NPU.

(v) Scaling unit: This scales the inputs and outputs of the neural network when required. The scaling factors are defined during NPU configuration.

(vi) Scheduling buffer: The timings of PE computations, bus and queue accesses are scheduled in accordance with the topology of the neural network. The scheduling information is a segment of the configuration information of the NPU. The bus scheduling data are stored in the scheduling buffer.

8.5 Processing engine

Its main components are (figure 8.3):

(i) Input FIFO: On receiving any input from the bus by a PE, its value is stored in the input buffer.

(ii) Weight buffer: After configuring the neuron weights for each PE, they are stored in the weight buffer.

(iii) Controller: The compiler-controlled schedule ensures the arrival of inputs to the PE in the same order as the corresponding weights appear in the weight buffer.

(iv) Multiply–add unit: This unit executes the operation expressed by the equation

$$y_i = \sum_i x_i \times w_i \tag{8.1}$$

where x_i is the input to a neuron and w_i is the associated weight.

(v) Accumulator registers: They store the y_i values.

(vi) Sigmoid unit: It executes the function

$$y = \text{sigmoid}(y_i) = \text{sigmoid}\left(\sum_i x_i \times w_i\right) \tag{8.2}$$

The sigmoid function, also called the logistic function is a function with an S-shaped characteristic curve. This function is mathematically expressed by the equation

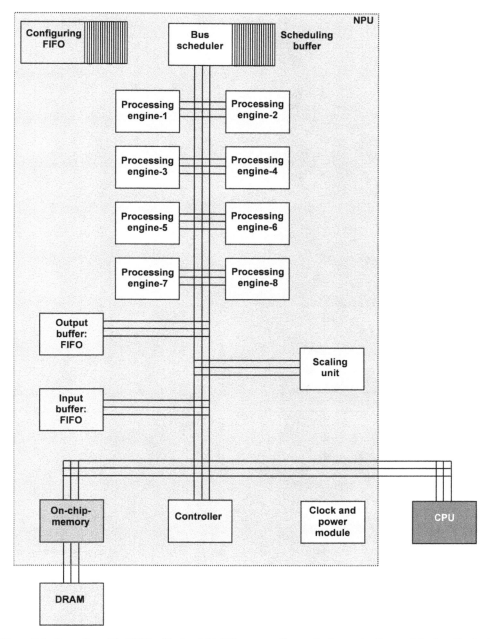

Figure 8.2. Reconfigurable NPU with its eight PEs, one scaling unit, the controller module, input/output buffers, the bus scheduler and scheduling buffer, and the on-chip memory. Also shown are the clock and power modules connected to CPU and the off-chip dynamic RAM (DRAM).

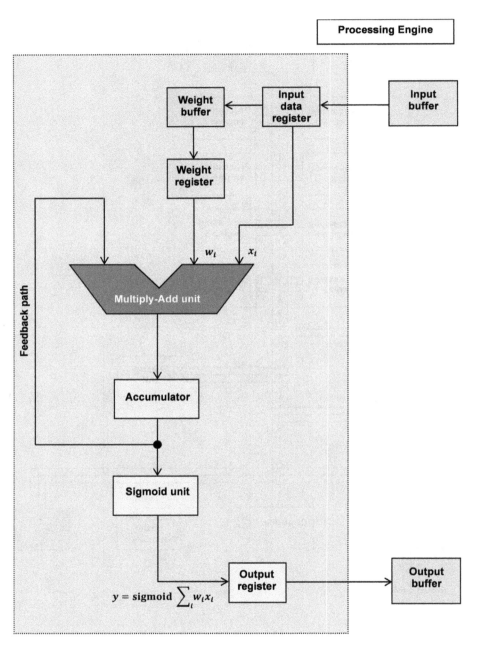

Figure 8.3. The organizational structure of the PE consisting of the input buffer, input data register and weight buffer. The x_i value from the input data register and the w_i value from the weight register are supplied to the multiply–add unit. The output of the multiply–add unit is sent to the accumulator. The output of the accumulator is fed back to the multiply–add unit. It is also applied to the sigmoid unit. This unit performs the sigmoid operation on the weighted sum $\Sigma x_i w_i$. The result is temporarily stored in the output register and conveyed to the output buffer.

$$y = \frac{1}{1 + \exp(-x)} = \frac{\exp(x)}{\exp(x) + 1} \tag{8.3}$$

It maps any real value between the limits 0 and 1. If x approaches $+\infty$, y becomes 1 and when x tends to $-\infty$, y reaches 0.

(vii) Output register file: The results are stored in the output register file.

8.6 NPU operation

8.6.1 Code generation

The code generation is done subsequent to the training phase of the neural network. During code generation, the compiler produces an instrumented binary code. This binary code runs on the CPU and invokes the NPU in place of calling the original function. When first loaded, the program performs NPU configuration. This is done by supplying the topology parameters and synaptic weights through the configuration interface to the compiler. The compiler substitutes the calls to the original function with special instructions that transmit the inputs to the NPU and gather the outputs from the NPU. The configuration and input–output communications are done using four extensions of instruction set architecture (ISA) for accessing the queues at the CPU–NPU interface, including those:

(i) for sending the configuration,
(ii) for retrieving the configuration,
(iii) for sending the inputs, and
(iv) for retrieving the outputs of the neural network.

These instructions are used to:

(i) enqueue the value of the specified register into the configuration FIFO,
(ii) dequeue a configuration value from the configuration FIFO to the specified register,
(iii) enqueue the value of the specified register into the input FIFO, and
(iv) dequeue the head of the output FIFO to the specified register.

For setting up the NPU, the program executes a series of instructions to enqueue the value of the specified register into the configuration FIFO for sending configuration parameters (number of inputs/outputs, network topology and synaptic weights) to the NPU. The operating system uses the instructions for dequeuing a configuration value from the configuration FIFO to the specified register for saving the NPU configuration during context switches.

For invoking the NPU, the program executes the instruction for enqueuing the value of the specified register into the input FIFO repeatedly for sending inputs to the configured neural network. No sooner than all the inputs are enqueued, the NPU begins computing, and delivers the results to the output FIFO. For retrieval of the output values, the program repeatedly executes the instruction for dequeuing the head of the output FIFO to the specified register.

Code generation involves the production of an NPU configuration by the compiler for implementation of the neural network for every entrant function. An order is allocated to the inputs of the neural network by the static NPU scheduling algorithm. This order decides the sequence of instructions for enqueuing the value of the specified register into the input FIFO, that the CPU will launch to the NPU during each invocation and the order of multiply–add operations among the PEs of the NPU. Then the scheduler:

    (i) designates each neuron to one PE,
    (ii) gives an order to the multiply–add operations unit taking into account the order assigned to the inputs of the layer,
    (iii) places an order to the outputs of the layer, and
    (iv) generates a bus schedule for the order of operations.

The order assigned for the final layer of the neural network determines the order of retrieval of the output of NPU by the program using the instruction to dequeue the head of the output FIFO to the specified register.

8.6.2 PE scheduling

The PEs are scheduled statically. In static scheduling, processes are mapped to PEs before their execution. In each piece of data, there is a source unit and a destination unit used for scheduling the bus for conveying a value from a PE or the input FIFO to a group of target PEs or the output FIFO. The source is the input FIFO or the recognizer of a PE accompanied by an index into its output register file. The destination is the output FIFO or a bit field representing the terminus PEs.

8.6.3 PE functioning

This works as follows: The value of the input received by a PE from the bus is stored in its input FIFO. After the neuron weights for each PE have been configured, they are loaded into the weights buffer. It is the responsibility of the compiler-directed scheduler to guarantee that the order of arrival of inputs is strictly the same as the order of appearance of correlative weights in the buffer. Hence, multiply–add operations are executed by the PE in the same order in which the inputs enter the FIFO of the PE. When the multiply–add operations of a neuron are done, each entry in the weight buffer is increased by one. After completion of these operations, the sigmoid function is applied by looking at a table. The result obtained by sigmoid function application is written in the output register file. The data stored in the weight buffer per neuron suggests which output register should be used.

8.7 NPU performance

NPU acceleration testing for diverse applications showed a speedup gain of 2.3 times with threefold energy saving incurring a maximum loss in quality ~9.6% in the NPU mode for the evaluated application compared to the regular mode (Esmaeilzadeh *et al* 2012).

8.8 NPU as a hardwired multilayered perceptron neural network

The NPU performs the task of a hardwired multilayered perceptron (Chen *et al* 2020); sections 1.5.2 and 1.5.3. It offers an interesting opportunity of accelerating some portions of a computer program. It can accelerate any segment of a program with well-defined inputs and outputs if it is frequently executed and can be approximated. For executing the program on the NPU, the programmer manually annotates the segment satisfying the aforesaid criteria. The compiler translates the program into NPU instructions. When the program is run, the marked parts of the computation task are offloaded from the CPU to the NPU. The CPU and NPU work in coordination to complete the program execution. Remember 'make hay while the sun shines' (NPU's role) by gainfully availing the advantage of a favorable situation whenever there is a chance to do so; otherwise 'slow but steady wins the race' (CPU's role) because the patience, persistence, and determination represented by gradual and stable progress are the virtues that help overcoming any obstacles to make the dream come true. Noteworthy examples of such program segments are the fast Fourier transform (FFT) and Sobel edge detection. The FFT is an algorithm that is used in AI to speed up the training of convolution neural networks (CNNs). The Sobel operator is used in computer vision. It consists of a pair of 3×3 convolution kernels performing a 2D spatial gradient measurement on an image to reveal the regions of high spatial frequency that correspond to edges.

8.9 AXNet accelerator

Undoubtedly, a precise number of representations are necessary during the training phase of deep neural networks. Nonetheless, a smaller bit-precision is adequate for inference to achieve lower power consumption. This has promoted the study of energy-efficient architectures employing various software–hardware co-optimization methods for inference (Lee 2021).

The energy-efficient neural network-based approximate computing for energy-resilient applications focuses on the acceleration in software-level. As neural networks are able to fit any continuous function, universal adoption of this method is possible for different tasks. Moreover, a neural network is easily deserialized and deployed in the cloud and on the edge, promising a high speedup. However, for satisfactory quality control, a single neural network looks unsafe and unreliable. The approximation quality is enhanced by deploying two neural networks designated as an approximator and a predictor. The approximator neural network provides the rough estimated results. The predictor neural network forecasts whether the approximator is successful in safely approximating the output when given the input data. An efficient cooperation between the two neural networks is essential to achieve the best combination of speed and approximation accuracy. The hinderance to this cooperation is that they have to accomplish different tasks, although sharing the same training data. One neural network is engaged in the prediction task. The other neural network is engrossed in the regression task.

A new neural network structure, AXNet works on the multi-task learning approach. It is able to merge the two neural networks to get an all-inclusive end-to-end trainable neural network (Peng *et al* 2018). The AXNet augments the percentage of safe-to-approximate samples known as the invocation. It also mitigates the approximation error. At the same time, the training effort is drastically reduced. A 50.7% higher invocation together with appreciable reduction of training time is demonstrated when making comparison with the present neural network-based approximate computing technique.

8.9.1 Multi-task learning

A vast proportion of learning algorithms are trained for solving one particular task. This method is commonly called single-task learning. It aims at the optimization of a solo metric. While paying attention towards one task, other vital information, which is not included in the problem, may be lost. Consequently, the performance of the model reaches a saturation limit. It is often advantageous to jointly optimize a model for several related tasks. Multi-task learning is a single shared machine learning model. It is able to do multiple different, notwithstanding interrelated tasks. In that way, it offers the advantages of high data efficiency, rapid model convergence, and lessened model overfitting. These advantages come naturally because of the shared representations.

An example of multi-task learning from human experience is provided by a person learning balancing on two wheels. The person finds it easy to learn riding a motorcycle. A person expert in motorcycle driving finds it easier to learn driving a car with manual transmission gears. At the base of the vehicle learning lies the ability to balance well on a two-wheeled vehicle in the first case, and the use of gears in the second case. It is an inductive transfer of knowledge. Taking cue from this learning, the approximator and predictor neural networks are not trained separately and in succession. Instead, they are subjected to parallel training and simultaneously too. A shared representation is used for this purpose. The common weight sharing method fails. However, a fusion of neurons between the approximator and predictor neural networks is able to meet the goal.

8.9.2 AXNet architecture

The AXNet architecture shown in figure 8.4 follows the NPU architecture described in the preceding section. The principal components of the AXNet are the processing engines (figure 8.5). A group of similar processing engines interlinked by a bus constitutes a tile. The input and output buffers of the tile are connected internally. Besides the tile, another component of AXNet is the controller with a bus scheduler. An on-chip memory is also built. Parallelism is established at the neuron level. Every PE calculates the output of a single neuron, in the predictor neural network or approximator neural network, by applying the equation

$$a^{(i+1)} = f\left(W^{i(T)}a^{(i)} \odot c^{(i)} + b^{(i)}\right) \qquad (8.3)$$

where f is the activation function of the hidden layer i, the symbol $\odot$ denotes the Hadamard product, a^i is the activation vector and c^i the corresponding control vector, $W^{(i)}$ refers to the ith vector of weight matrix, and $b^{(i)}$ is the ith bias vector.

The input/output vector is temporarily stored in the input/output buffer of the PE. The input/output buffer of the PE interfaces with the on-chip memory. Interfacing of

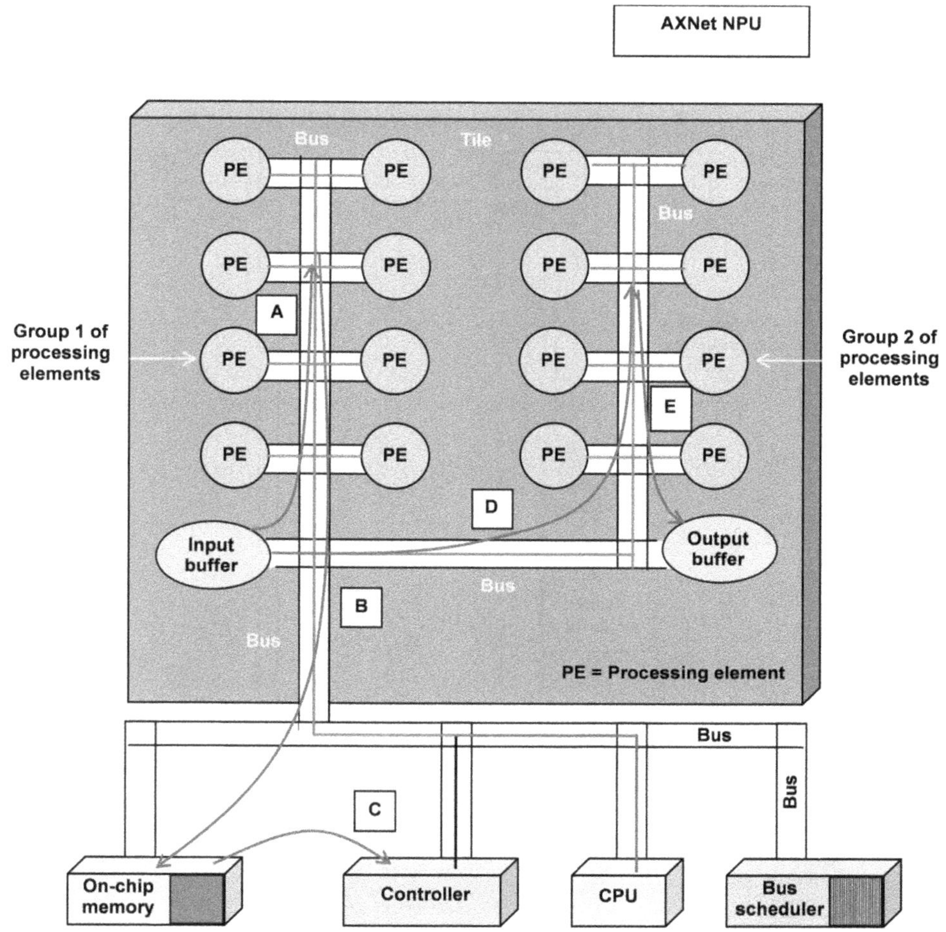

Figure 8.4. AXNet architecture. A tile contains two groups of processing elements. Each group consists of eight processing elements. In each group, the left and right processing elements are connected to each other by a horizontal bus. A vertical bus passes over these horizontal lines for a group to join them. The vertical buses for the two groups are connected by a horizontal bus. On one side of this horizontal bus, the input buffer is connected. On its opposite side, the output buffer is joined. This horizontal bus is connected by a vertical bus to a horizontal bus outside the tile. To this bus outside the tile, the on-chip memory, controller, CPU and bus scheduling buffer are joined by vertical buses.

the on-chip memory is done with the DRAM, input/output buffers in the tile and the CPU via the bus. Other items it can store are the weight matrix, the input samples and output results of AXNet, as well as the intermediate results that are transferred between two adjacent layers in the neural network. Depending on the result of prediction $P(x)$, the controller sends the invoke signal through the control bus shown as a blue line to PEs or CPU. The bus conflict is prevented from taking place by the bus scheduler during the simultaneous transference of data between the tiles, the CPU and the on-chip-memory.

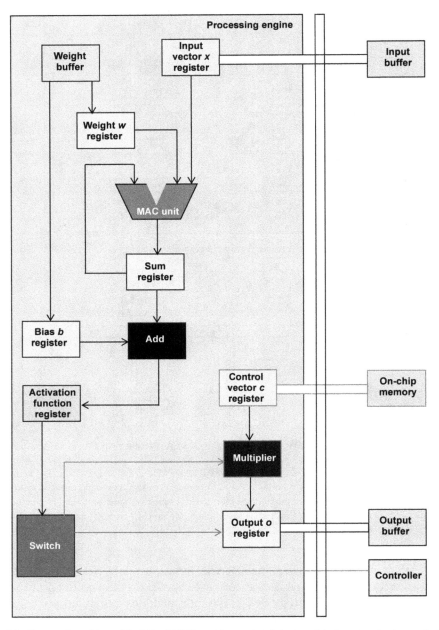

Figure 8.5. Parts and working of the PE of AXNet with associated devices. The input vector is supplied from the input buffer into the input vector *x* register. The weight buffer loads the data into the weight *w* register and bias *b* register. The multiply-and-accumulate (MAC) unit finds the dot product of the input vector *x* and the weight vector *w*. The product is kept in the sum register. The output of the sum register is fed back to input.

Bias is added to the output from the sum register. The output of the add unit is sent to the activation function register whose output goes to a switch unit. The switch enables the blue path to place the result of activation into the output o register and output buffer. In case an invoke signal is received from the controller, as shown by the brown line for computation of approximation subnet, the switch actuates the data flow on brown line. The product of the data in the control vector c register received from the on-chip memory and the activation value in the switch coming from the activation function register is calculated by the multiplier. Its output is sent to the output o register connected to the output buffer.

8.9.3 Dataflow of the AXNet during job execution

As soon as the input sample is received in the on-chip memory, the NPU schedules the computation for prediction subnet in the first three stages, and thereafter for approximation subnet in the last two stages. Effectively, an aggregate of five stages of computing, from stage A to stage E, is done, as summarized below:

Stage A: The input data x is fetched from the on-chip memory to the input buffer through the data bus. In each PE, the weight w vectors are brought from the on-chip memory to the weight buffer. After reception of the input x and weight w vectors, each PE carries out forward propagation of the prediction subnet. The prediction result $P(x)$ and control vectors $c^{(i)}$ are produced as the outcomes.

Stage B: The prediction result $P(x)$ and control vector $c^{(i)}$ of the prediction subnet are dispatched to the on-chip memory. Stated explicitly, $P(x)$ is placed in a specified address (the yellow region inside the on-chip memory).

Stage C: The controller receives the prediction result $P(x)$ from the on-chip memory. Looking at the value of $P(x)$, the controller invokes either the CPU or the approximation subnet through the control bus (blue line).

Stage D: If the approximation subnet is invoked, each PE procures input data and control vectors from the buffers. Forward propagation of the approximation subnet is performed.

Stage E: The result of the approximation is forwarded to the on-chip memory on the data bus.

In case the derived AXNet for an application is small, the NPU can statically allocate the computing/storage resource for the whole AXNet. In this situation, the weight vector can stay in the weight buffer of each PE all the time. Or else, the NPU can dynamically schedule the computation of AXNet layer by layer. In that circumstance, the input/output buffer of each PE and the on-chip memory will provisionally lodge the intermediate results between adjacent layers.

Modification of a general PE is done for computing the Hadamard product induced by the fusion of two neural networks. The internal structure of the modified PE is shown in figure 8.5. The computation flow consists of the steps enlisted below:

(i) loading of the input vector by the PE from input buffer into the x register,

(ii) loading of the weight data by the PE from the weight buffer into the w register and b register,

(iii) calculation of the dot product of the input x vector and the weight w vector by the multiply–add unit,

(iv) storage of the resultant product in the temporary register,

(v) sending the result to the activate unit after adding the bias b, and

(vi) implementation of the rectified linear unit (ReLU) function by the activate unit.

Though NPU performs different computations in the prediction subnet and approximation subnet, the PE has the same structure. A switch unit after activate unit sets up the path to be followed. At first, the switch unit enables the blue dotted path which directly pushes the activation result into the output register. When receiving an invoke signal in stage C, indicating the computation of approximation subnet, the switch unit activates the approximation subnet units (in orange lines). Inside the PE, the Hadamard product reduces to a standard multiplication operation between one activation value, say $a_j^{(i)}$, and the corresponding element in control vector $c_j^{(i)}$. Therefore, a multiplier can carry out the above computation. The output of $c_j^{(i)} \times a_j^{(i)}$ is stored in o register and waits for transference to the output buffer. Experimental investigations reveal that a superior invocation result and a gain of energy efficiency is obtained over the existing neural approximate computing frameworks (Peng *et al* 2018).

8.10 RENO: a reconfigurable accelerator

The RENO (figure 8.6) is a reconfigurable memristor-based neuromorphic computing accelerator (Liu *et al* 2015). It uses on-chip memristor-based crossbar (MBC) arrays to realize a perceptron network. Its objective is the acceleration of artificial neural network (ANN) computations to offer an economical and error-tolerant ANN computation platform supplementing the generalized computations of CPU.

8.10.1 Memresistor and ReRAM

Similar to the NPU designed for broad-purpose programs, RENO is an accelerator for neural networks using an identical concept for designing the processing element. The difference between RENO and NPU is that the basic computation unit of RENO is an MBC for performing matrix vector multiplications.

A memristor (named as a combination of memory and resistor) is a non-volatile electronic memory device. It consists of a metal-oxide layer sandwiched between two electrodes. Its resistance can be programmed by applying a current or voltage of appropriate pulse width or magnitude. The programmed value of resistance can be stored even with power off for enabling a memory function. The memresistor forms the basis of a resistive random-access memory (ReRAM). Each ReRAM cell is a memresistor cell with programmable resistance. The data storage in a cell is done by a representative resistance level. A low value of resistance stands for bit 1, while a high resistance value signifies bit 0. For reading the stored data, a small sensing voltage is applied across the device. Then the amplitude of the resulting current varies according to the resistance of the device. The current amplitude is used for the read operation.

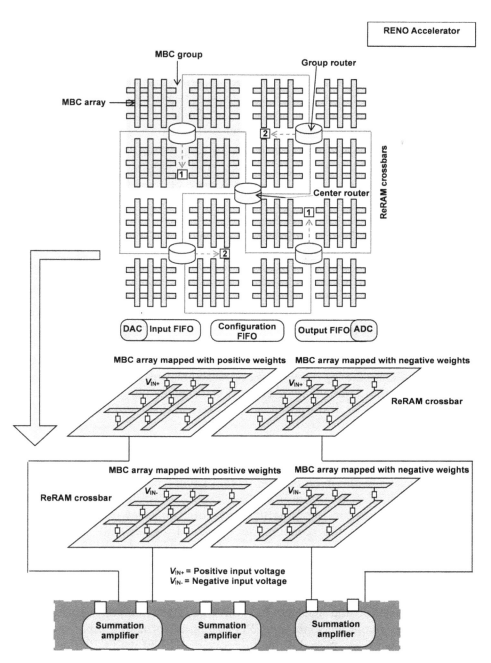

Figure 8.6. The RENO accelerator. Top: The diagram illustrates four MBC groups. Each group is constructed from four MBC arrays joined through a group router. An MBC array is segregated into four sub-crossbars to multiply the combinations of the signed signals and the signed synaptic weights. Each group router dialogs with the four local MBC arrays within the group as well as three other group routers and the central router. As the MBC arrays calculate in analog format, digital-to-analog/analog-to-digital (DA/AD) convertors are placed

at the interfaces of MBC arrays and routers. Furthermore, input, output and configuration FIFOs are located at the interfaces of RENO for receiving the commands and data, and conveying the results back to the processor in digital format. Bottom: MBC array connections are shown. The MBC arrays mapped with positive weights located on the left-hand side are connected to one summation amplifier while the MBC arrays mapped with negative weights appearing on the right-hand side are connected to another summation amplifier. Here, V_{IN+} and V_{IN-} are the positive and negative input voltages, respectively.

8.10.2 Hybrid method of data representation

Each PE of RENO consists of four ReRAM crossbars. Two crossbars correspond to the processing of positive and negative inputs. The remaining two crossbars pertain to positive and negative weights. Data transference between the PEs is coordinated by routers. The RENO design adopts a crossbreed method of data representation. This contrasts with several neuromorphic systems that perform digital computations with arithmetic-logic units or approximate analog calculations using analog-to digital and digital-to-analog interfaces.

In the RENO, computations within the MBC arrays and the signal communications among the MBC arrays are carried out in the analog mode, whereas the control information is predominantly in the form of digital signals. Hence, unlike complementary metal-oxide semiconductor (CMOS) routers which transfer digital results, the routers of RENO transfer the analog midway computation results from the one neuron to the succeeding neuron.

In RENO, the intermediate results are completely analog. They are managed by analog routers. Only the input and final output signals are digital. Hence, converters of both types, digital-to-analog converters (DACs) and analog-to-digital converters (ADCs) are essential for transferring data between RENO and the CPU.

8.10.3 RENO architecture

The role of RENO is to work as a complementary functional unit to CPU. Its main task is to accelerate the execution of ANN-relevant portions. In RENO, MBC arrays perform analog neuromorphic computations. There are four MBC groups. Each MBC group consists of four MBC arrays interconnected through a router. An MBC array multiplies the combination of signed signals and signed synaptic weights. A mixed-signal interconnection network (M-net) connects the MBC arrays and topologically reconfigures the RENO. In M-Net, the data is maintained in analog form. But the transference of control and routing information is done digitally. This is essential for simplification of the synchronization and communication between CPU and RENO. For receiving the commands and data, and transmitting the results back to the CPU in digital form, the input, output and configuration FIFOs are placed at the interface of RENO with CPU.

8.10.4 RENO performance

In comparison to a usual CPU, RENO reaches on average 177.67 × (27.2 ×) performance speedup and 184.71 × (25.18 ×) energy diminution over the simulated yardsticks processed by the MLP (AAM) (Liu *et al* 2015). Here, MLP and AAM are neural networks; MLP stands for the multilayer perceptron and AAM represents the active appearance model.

8.11 The DianNao family of energy-efficient hardware accelerators

The DianNao family is a series of hardware accelerators. It is especially designed for neural networks. During design, emphasis is laid on the influence of memory on the design, performance and energy aspects of the accelerator. Testing and evaluation studies have been performed on a number of representative neural network layers. The revelations of these studies show that a speedup of 450.65 × is attained over a GPU. The speedup is accompanied by reduction in energy by 150.31 × preponderantly for a 64-chip member of the DianNao family (Chen *et al* 2016).

8.11.1 The DianNao neural network accelerator

8.11.1.1 Components of DianNao
This consists of one computational block known as the neural functional unit, one control processor, and a three-part on-chip memory (figure 8.7):

(i) On-chip storage memory: The on-chip storage components include three buffers for the three types of data. The three data types are concerned with the input neurons, the output neurons and the synapses. They have different characteristics, e.g., read width and reuse distance.

    (a) NBin: This is an input buffer for input neurons.

    (b) NBout: This is an output buffer for output neurons.

    (c) SB: This is a buffer for synaptic weights.

All these are modified buffers of scratchpads. A scratchpad in a dedicated accelerator realizes efficient storage, combined with systematic and easy exploitation of locality by manually adapting algorithms. The subdivision of storage helps in customization of the static random-access memory (SRAM) to the proper read/write width, and in circumventing conflicts that are likely to take place in a cache. Spatial locality of data is availed with three DMAs (direct memory accesses). One DMA is used for each buffer resulting in two load DMAs for inputs and one store DMA for outputs.

(ii) Neural functional unit (NFU): This is the computational block. It looks after the synapses as well as neuron computations. Depending on the layer type, computations at the NFU are broken into either two or three stages. For the classifier and convolutional layers, these stages are:

    (a) multiplication of synapses with inputs,

    (b) summation of all multiplications, and

    (c) application of the sigmoid function.

The last stage may differ in characteristics using either sigmoid or another nonlinear function. The sigmoid function for the classifier and convolutional layers is implemented by piecewise linear interpolation. For pooling layers, averaging or maximization operations are undertaken. The adders having multiple inputs are actually adder trees. Shifters and max operators are used.

(iii) CP: This is the control logic or control processor.

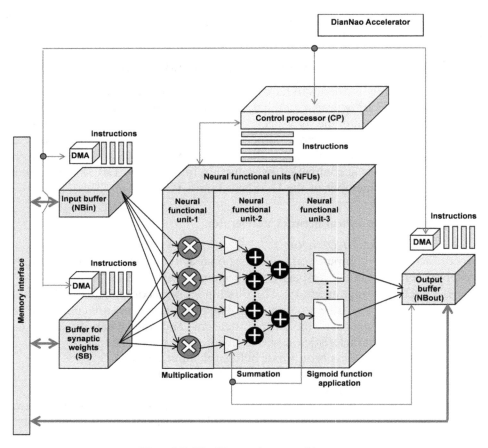

Figure 8.7. DianNao accelerator architecture.

8.11.1.2 Operation of DianNao by loop tiling or loop blocking

Loop tiling is used. The loop tiling helps in reducing memory accesses and thereby accommodating large neural networks. It is a technique of optimizing the computer program to improve the locality of reference. Locality of reference is the tendency of the program for accessing instructions under temporal locality or spatial locality. Temporal locality gives a high probability of accessing an already-accessed memory location in the near future. Spatial locality provides a high probability of accessing a neighboring location to an already-accessed location.

Loop tiling works by modification of the memory access pattern of the loop by decomposing the data processing into smaller segments. These segments are called tiles or blocks; hence the name 'loop tiling'. Being smaller than original work sets, they easily fit into data caches. Loop tiling functions by reusing part of the data which is previously available in the data cache before its expulsion from the cache. Tiling decreases the memory bandwidth requirement.

8.11.2 Other hardware accelerators in the DianNao family

These include Dadiannao: a machine learning (ML) supercomputer, Shidiannao: a low-power CNN accelerator, and Pudiannao: a polyvalent ML accelerator (Chen *et al* 2016).

8.12 The Ascend AI processor

8.12.1 Main modules of an Ascend AI processor

The Ascend AI processor is a neural processing unit of Huawei, a leading information and communication technology company. It comprises four main modules (Liao *et al* 2021, Huawei Technologies 2023):

- (i) control CPU module,
- (ii) AI computing engine module = AI core (Da Vinci architecture) + AI CPU; the Da Vinci architecture will be explained in section 8.12.2,
- (iii) multi-level caches module, and
- (iv) digital vision pre-processing (DVPP) module.

The modules are illustrated in figure 8.8.

The AI core is the computing core of the processor which is assigned large computing tasks, e.g., the matrix correlation calculation of neural network.

The AI CPU has the responsibility of complex computing and executive control, performing the general calculation of control operator, scalar and vector.

The DVPP dedicated hardware module performs data pre-processing. It is usually activated to pre-process the image and video data, and for providing data formats to AI core to fulfil the computing demands in specific conditions.

8.12.2 Da Vinci architecture of AI core

This consists of three parts:

- (i) Computing unit: This houses a cube unit, a vector unit and a scalar unit:
    - (a) Cube unit: The cube unit and accumulator execute matrix correlation operations.
    - (b) Vector unit: This is used for the calculations between a vector and a scalar, as well as dual vectors.
    - (c) Scalar unit: This is a μCPU. It controls the entire AI core operation: the cycle control and branch decision of the complete program, provision of data address and related parameter calculation for matrix and vector, and basic arithmetic operations.
- (ii) Memory system: This includes the AI core's on-chip memory unit, and the compliant data path. The memory unit is composed of:
    - (a) Memory control unit: This provides direct access to lower-level caches besides AI core via bus interface. It also directs access to the memory through the double data rate synchronous dynamic random-access memory (SDRAM, DDR) or high bandwidth memory (HBM).

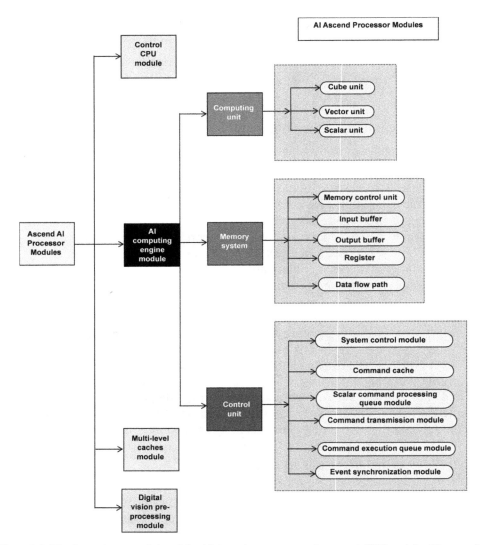

Figure 8.8. The four primary modules of the AI Ascend processor are the control CPU module, AI computing engine module, multi-level caches module, and digital vision pre-processing module. The AI computing engine module is subdivided into three parts: computing unit made of cube, vector and scalar units; the memory system consisting of the memory control unit, input and output buffers, register and data flow path; and the control unit comprising the system control module, command cache, scalar command processing queue module, command transmission module, command execution queue module and event synchronization module.

    (b) Input buffer. This is used to temporarily store the data that is frequently used. In this way, it avoids external reading to AI core via the bus interface each time.

    (c) Output buffer: This is used to store the intermediate results of calculation of each layer in the neural network. Storage of the intermediate results helps in obtaining data conveniently when entering the next layer.

(d) Register: This is mainly used by scalar units.

(e) Data flow path: This is the flow path of data in AI core during completion of a computing task by the AI core.

(iii) Control unit: This provides command control for the whole calculation process for the operation of the entire AI core. The control unit is made up of:

(a) System control module: This controls the execution process of the task block.

(b) Command cache: Subsequent commands are prefetched in advance. Also, multiple commands are read into the cache at one time to improve the efficiency.

(c) Scalar command processing queue module: Here matrix computing command, vector computing command and memory conversion command are decoded. They are imported into the scalar queue to realize address decoding and computing control.

(d) Command transmission module: This reads the command address and decoding parameters configured in the scalar command queue. According to their command types, it guides them to the corresponding command execution queue. However, the scalar command resides in the scalar command processing queue for the ensuing execution.

(e) Command execution queue module: Here different commands enter different queues: matrix queue, vector queue and memory conversion queue. They are executed in the order of their entry.

(f) Event synchronization module: This controls the executing state of each command pipeline at any time. It adjusts the data dependence and synchronization between command pipelines by analyzing the dependencies of different pipelines.

8.12.3 Software stack of the Ascend AI processor

It is divided into four levels and one auxiliary tool chain (figure 8.9):

(i) L3 application enabling layer: This provides different processing algorithms for specific applications, apart from offering the engine of computing and processing for various fields. It includes:

(a) Computer vision engine: This provides the encapsulation of video or image processing algorithms, which is specialized to deal with algorithms and applications in the field of computer vision.

(b) Language engine: This provides the encapsulation of basic processing algorithms for voice, text and other data for language processing according to specific application scenarios.

(c) General business execution engine: This provides general neural network reasoning ability.

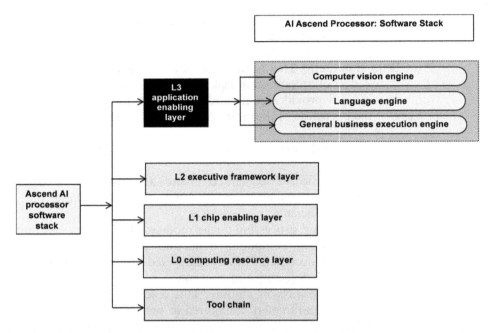

Figure 8.9. Software layers of the Ascend AI processor. The layers include from bottom upwards: the tool chain, the L0 computing resource layer, L1 chip enabling layer, L2 executive framework layer and L3 application enabling layer. The topmost L3 application enabling layer comprises the computer vision engine, the language engine and the general business execution engine.

  (ii) L2 Executive framework layer: This is the encapsulation of framework calling ability and offline model generation capability. It contains the framework manager and the process choreographer.
  (iii) L1 chip enabling layer: This is the bridge between offline model and Ascend AI processor.
  (iv) L0 computing resource layer: This is the hardware computing power foundation of the processor. It provides computing resources and performs specific computing tasks. It is composed of the operating system, the AI CPU, the AI Core and the DVPP dedicated hardware modules.
  (v) Tool chain: This is a platform offering a set of tools for the convenience of programmers. Support is provided for the development, debugging, network transplantation, optimization and analysis of custom operators.

8.12.4 Neural network software flow of Ascend AI processor

It is used to generate, load and execute an offline model of neural network application. It contains functional modules such as:
  (i) Matrix module: This is used for the landing and implementation of a neural network on the processor.
  (ii) DVPP module: This is used for data processing and modification prior to input to meet the needs of the computing format.

(iii) Tensor boost engine: This is a storehouse of neural network operators.
(iv) Framework: This is used to build the original neural network model into the form supported by the processor.
 (v) Runtime: This provides various resource management channels for task distribution and allocation of neural network.
(vi) Task scheduler: This provides specific target tasks for the processor as a task driver of hardware execution.

8.12.5 Data flow of Ascend AI processor

For the reasoning application of face recognition, the data is collected and processed by the camera. Then the data is reasoned, and finally the face recognition results are delivered as the output. The stages involved in this process are:

 (i) Data collection and processing by the camera: The compressed video stream is passed from the camera, and the data is stored in double data rate synchronous dynamic random-access memory (DDR-SDRAM) written as DDR in abridged form, through the PCIe channel. The DVPP reads the compressed video stream into the cache. After pre-processing, the DVPP writes the decompressed frame into the DDR memory.
 (ii) Reasoning on data: The task scheduler sends commands to DMA to preload AI resources from DDR to on-chip buffer. It configures the AI core to execute tasks. When the AI core works, it reads feature graph and weight, and writes the result into DDR or on-chip buffer.
(iii) Results of face recognition as output: After the AI core completes the processing, it sends a signal to the task scheduler. Then the task scheduler checks the result and assigns another task if necessary, and returns to sending commands to the DMA, as in step (ii). When the last AI task is completed, the task scheduler reports the result to the host.

8.12.6 Performance of the Ascend 910 processor

The Ascend 910 made in a semiconductor foundry on 7 nm process node delivers 256 TeraFLOPS for half-precision floating point operations. It delivers 512 TeraFLOPS for integer precision calculations. Maximum power consumption is 310 W (Xu 2019, Feldgoise and Dohmen 2024).

8.13 Further NPU chip examples

Many of these chip examples consist of systems-on-chip (SoCs) with the NPU integrated alongside CPU and GPU:

 (i) Intel® Core™ Ultra Processors: The Intel® Core™ Ultra processors provide three dedicated engines (CPU, GPU, and NPU) for high-performance low-power processing, including graphics and AI.
 (ii) Intel® AI Boost/Intel® NPU: It is an integrated AI engine enabling low-power AI acceleration and CPU/GPU off-load.
(iii) Intel® Gaussian & Neural Accelerator processes AI speech and audio applications (Intel® Core™ Ultra Processor (Series 1) Product Brief).

(iv) Intel® AI Boost (NPU) is a power-efficient AI accelerator integrated into Intel® Core™ Ultra processors (Intel product brief) in combination with built-in Intel® Arc™ GPU for compute-intensive AI operations.

(v) The Intel AI Boost has a scalable architecture of multiple tiles. These are neural compute engines containing hardware acceleration blocks for matrix multiplication, convolution, and similar operations. The neural compute engines also have streaming hybrid architecture vector engines (SHAVE). These are intended for high performance parallel computing (Trivedi *et al* 2024).

(vi) Apple's Neural Engine (ANE): This is a group of specialized cores functioning as an NPU dedicated to AI/ML, as part of SoC designs from Apple fabricated by TSMC, e.g., the Apple M1 is an ARM-based SOC introduced by Apple in 2020. The ARM (Advanced RISC Machines or Acorn RISC Machine) is a family of RISC instruction set architectures (ISAs) for computer processors. The. Apple M1 SoC of 5 nm technology node features four high-performance CPU cores, four high-efficiency CPU cores, seven or eight GPU cores and 16 neural engine cores (Newsroom 2020).

(vii) Qualcomm's Snapdragon X Elite Processors: Snapdragon® X Elite has a 4 nm SoC architecture with low-power consumption. It features a 12-core CPU optimizing demanding workloads, an integrated GPU delivering graphics for entertainment, an AI Engine with an integrated NPU for creativity, video conferencing and security. It can run generative AI large language models. The Micro NPU inside the ultra-low-power Qualcomm® Sensing Hub bestows improved security, login experience, and privacy. It includes waking-up-the-device ability during sleep mode. (Snapdragon® X Elite DCN 87-71417-1 Rev F).

(viii) NVIDIA Deep Learning Accelerator (NVDLA): This is an open-source hardware neural network AI accelerator made by NVIDIA. It is written in Verilog. It is configurable and scalable to meet different architectural needs (NVDLA 2024).

(ix) Hanguang 800: It is a high-performance AI inference chip made by Alibaba (Alibaba Cloud 2019).

(x) Exynos 9820, Exynos 9825, Exynos 980, Exynos 99: These are SoC devices embedding an NPU, made by Samsung Electronics (Samsung NPU 2024).

(xi) AWS Inferentia Accelerators: These are designed by Amazon Web Services (AWS), Inc. The aim is to achieve high performance at the lowest cost in Amazon Elastic Compute Cloud2 (Amazon EC2) for deep learning and generative AI inference applications.

The first-generation Amazon Web Services (AWS) Inferentia accelerator powers Amazon EC2 Inf1 instances. The first-generation Inferentia accelerator has four first-generation NeuronCores. It has up to 16 Inferentia accelerators per EC2 Inf1 instance.

AWS Inferentia2 accelerator supplies up to 4× higher throughput. Its latency is up to 10× lower than Inferentia. Inferentia2-based Amazon EC2 Inf2 instances are optimized for complex large language and other models. The Inferentia2 accelerator has two second-generation NeuronCores. It has up to 12 Inferentia2 accelerators per EC2 Inf2 instance. The Inferentia2 accelerator supports up to 190 tera floating operations per second (TFLOPS) of FP16 performance (AWS Inferentia 2024).

8.14 Summary and the way forward

8.14.1 Highlights of chapter 8 at a glance

(i) The use of approximation as a means of computation efficiency enhancement was emphasized. The procedure of conversion of general-purpose code into neural representation was explained (Esmaeilzadeh *et al* 2012).

(ii) The parts of an NPU were described and the constitution of its PE was elaborated.

(iii) The NPU operation involves code generation, PE scheduling and functioning.

(iv) The performance capability of the NPU was quantified and an analogy was drawn between it and a hardwired multilayered perceptron neural network.

(v) The AXNet accelerator was expounded. The multi-task learning idea was outlined followed by discussion of the AXNet architecture and data flow of the AXNet during job execution (Peng *et al* 2018).

(vi) A reconfigurable accelerator, the RENO utilizes the memresistor-based resistive RAM (Liu *et al* 2015). It works by a hybrid method of data representation. RENO architecture and its performance were elaborated.

(vii) Components of the DianNao neural network accelerator and operation of DianNao by loop tiling or loop blocking were deliberated (Chen *et al* 2016). Other hardware accelerators in the DianNao family were mentioned.

(viii) The main modules of the Ascend AI Processor, its Da Vinci AI core architecture, software stack, neural network software flow, data flow and performance were briefly discussed (Liao *et al* 2021, Huawei Technologies 2023).

(ix) Examples of commercial chips with NPU integrated alongside CPU and GPU were given.

(x) Overall, the neural processing unit processes large amounts of data in parallel. It performs trillions of operations per second. It consumes less power and surpasses CPU or GPU in AI tasks. So, CPU and GPU are set free to look after other tasks. The NPU provides expeditious, high-bandwidth AI in real time. Real-time performance is suitable for using voice commands, and creating images quickly. ML and NPU form a powerful problem-solving team (Microsoft 2024).

(xi) In contrast to the powerful parallel processing capability of GPU and the versatility of GPU cores (figure 8.10(a)), the NPU (figure 8.10(b)) is designed with a primary emphasis on the type of workloads encountered

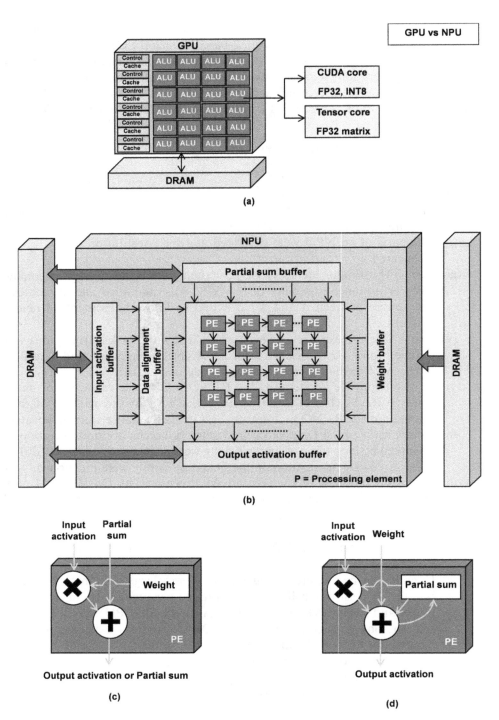

Figure 8.10. GPU and NPU comparison: (a) the GPU and its cores, (b) the NPU; (c) and (d) two types of NPU processing elements.

in deep neural network tasks. These workloads rely heavily on matrix multiplications and additions.

The GPU sometimes becomes too heavy-duty processor for these loads. So, the NPU is built with a different types of processing elements keeping these dissimilar end goals in mind (figure 8.10(c) and (d)). Each processing element of NPU holds a weight in its register, multiplies it with the input activation data, and adds the result to the partial sum input (psum) to generate final output activation data. The NPU avoids off-chip memory access by adopting the strategy of data-reuse pattern. So, the neural network tasks are handled more efficiently with NPUs at a comparatively lower power consumption than the GPU (Ishida *et al* 2020).

(xii) Keywords for this chapter are: NPU, AXNet, RENO, DianNao, Ascend, Core Ultra and X Elite processors, ANE.

8.14.2 Getting ready to begin chapter 9

In the next chapter, we continue our pursuit of processors dedicated to selected applications. In this expedition, we come to a class of processors aimed at visual applications such as image processing and computer vision in which convolutional neural networks are widely used. So, we address processors for CNNs and then move to vision processing units.

References

Cloud A 2019 Community: Alibaba unveils AI chip to enhance cloud computing power https://alibabacloud.com/blog/alibaba-unveils-ai-chip-to-enhance-cloud-computing-power_595409

AWS Inferentia 2024 Why Inferentia? 2024, Amazon Web Services, Inc., https://aws.amazon.com/machine-learning/inferentia/

Chen Y, Chen T, Xu Z, Sun N and Temam O 2016 DianNao family: energy-efficient hardware accelerators for machine learning *Commun. ACM* **59** 105–12

Chen Y, Xie Y, Song L, Chen F and Tang T 2020 A survey of accelerator architectures for deep neural networks *Engineering* **6** 264–74

Esmaeilzadeh H, Sampson A, Ceze L and Burger D 2012 Neural acceleration for general-purpose approximate programs *45th Annual IEEE/ACM Int. Symp. on Microarchitecture (1–5 December) (Vancouver, BC, Canada)* pp 449–60

2013 *IEEE Micro* **33** 16–27

2015 *Commun. ACM* **58** 105–15

Feldgoise J and Dohmen H 2024 *Data Snapshot, Pushing the Limits: Huawei's AI Chip Tests U.S. Export Controls* (CSET: Center for Security and Energy Technology) https://cset.georgetown.edu/publication/pushing-the-limits-huaweis-ai-chip-tests-u-s-export-controls/

Huawei Technologies Co., Ltd 2023 Huawei Atlas AI computing solution *Artificial Intelligence Technology* (Singapore: Springer) ch 6 pp 163–219

Intel® Core™ Ultra Processor (Series 1) *Product Brief, Features at a Glance* Intel Corporation https://intel.com/content/www/us/en/products/docs/processors/core-ultra/core-ultra-series-1-product-brief.html

Intel Product Brief: Intel® Core™ *Ultra Processors PS Series Product Brief Elevate Edge Innovation with Power-efficient AI, Graphics, and Versatility Combined* Intel Corporation https://intel.com/content/www/us/en/products/docs/embedded-processors/core-ultra/ps-product-brief.html

Ishida K *et al* 2020 SuperNPU: an extremely fast neural processing unit using superconducting logic devices *2020 53rd Annual IEEE/ACM Int. Symp. on Microarchitecture (MICRO)(17–21 October) (Athens, Greece)* pp 58–72

Lee K J 2021 Architecture of neural processing unit for deep neural networks *Advances in Computers* **vol 122** ed S Kim and G C Deka (Amsterdam: Elsevier) ch 7 pp 217–45

Liao H, Tu J, Xia J, Liu H, Zhou X, Yuan H and Hu Y 2021 Ascend: a scalable and unified architecture for ubiquitous deep neural network computing: industry track paper *2021 IEEE Int. Symp. on High-Performance Computer Architecture (HPCA)(27 February–3 March, 2021) (Seoul, Korea (South))* pp 789–801

Liu X *et al* 2015 RENO: a high-efficient reconfigurable neuromorphic computing accelerator design *2015 52nd ACM/EDAC/IEEE Design Automation Conf. (DAC)(8–12 June) (San Francisco, CA, USA)* pp 1–6

Microsoft 2024 All about neural processing units (NPUs) Microsoft 2024 https://support.microsoft.com/en-us/windows/all-about-neural-processing-units-npus-e77a5637–7705-4915-96c8-0c6a975f9db4

Newsroom 2020 Press release, Apple unleashes M1, 2024 Apple Inc. https://apple.com/newsroom/2020/11/apple-unleashes-m1/

NVDLA 2024 NVDLA open source project documentation, https://nvdla.org

Peng Z, Chen X, Xu C, Jing N, Liang X, Lu C and Jiang L 2018 AXNet: ApproXimate computing using an end-to-end trainable neural network arXiv:1807.10458 [cs.LG] 8

Samsung NPU 2024 NPU (Neural processing unit) 2024 Samsung https://semiconductor.samsung.com/support/tools-resources/dictionary/the-neural-processing-unit-npu-a-brainy-next-generation-semiconductor/#:~:text=Samsung%20Electronics%20launched%20the%20premium,previous%20model%20(the%209810), https://blog.longterm.io/samsung_npu.html

Snapdragon® X Elite DCN 87–71417-1 Rev F Performance reborn, https://docs.qualcomm.com/bundle/publicresource/87–71417-1_REV_F_Snapdragon_X_Elite_Product_Brief.pdf

Trivedi R, Ambati M and Nagi J S 2024 Scaling Intel neural processing unit (NPU) in AI client ecosystem, with DirectML on Windows MCDM (Microsoft compute driver model) architecture, https://community.intel.com/t5/Blogs/Tech-Innovation/Artificial-Intelligence-AI/Scaling-Intel-Neural-Processing-Unit-NPU-in-AI-Client-Ecosystem/post/1574577

Xu E 2019 Huawei launches Ascend 910, the world's most powerful AI processor, and MindSpore, an All-scenario AI computing framework, Futurewei News, August 23 https://edge-ai-vision.com/2019/08/huawei-launches-ascend-910-the-worlds-most-powerful-ai-processor-and-mindspore-an-all-scenario-ai-computing-framework/

IOP Publishing

AI-Processor Electronics
Basic technology of artificial intelligence
Vinod Kumar Khanna

Chapter 9

Convolution neural network processor, and the vision processing unit

The convolution operation, the convolution neural network (CNN), the training and inference phases of a CNN and the constituent layers of a CNN are explained. The constituent parts of the CNN accelerator are described and the achieved performance is affirmed. Differences between a vision processing unit (VPU) and graphical processing unit (GPU) are elaborated. VPUs are extensively used in healthcare, agricultural and automotive sectors, retail, security, safety, and industrial robotics. Deficiencies of traditional units in fulfilling the vision processing needs are mentioned. A single-unit design for vision tasks based on the hybrid chip architecture is discussed and the VPU workflow and experiments are summarized. Performance capabilities of commercial VPUs are mentioned.

9.1 Introduction

This chapter describes the salient features and working of the CNN processor, and the VPU. The VPU is an ideal tool for machine vision. Its keen watchfulness renders real-time, lightning-fast, correct analysis similar to what an Eagle eye can provide.

9.2 Convolution neural network

CNN is a deep neural network (DNN) widely used in performing image processing and classification tasks for computer vision (Krichen 2023). By a 'deep neural network' is meant a neural network made up of several layers.

9.2.1 Convolution operation

Convolution is a mathematical operation that is performed on two functions f and g. As a consequence of this operation, a third function $f(t) \otimes g(t)$ is produced. The third function, $(f \otimes g)(t)$ resulting from convolution of the first two functions as well as the process applied for obtaining it are referred to as convolution. It is an integral

doi:10.1088/978-0-7503-6259-7ch9   9-1   © IOP Publishing Ltd 2024. All rights,

expressing the degree of overlapping of a function g as it is shifted over the function f to blend the two functions together. The integral is written as

$$[f \otimes g](t) = \int_0^t f(t - \tau)g(\tau)d\tau \qquad (9.1)$$

involving the steps of folding, shifting, multiplication and addition.

9.2.2 CNN training and inference phases

A CNN used for image recognition passes through two phases:
(i) Training phase: The CNN is fed with a large dataset of labeled images of the objects of interest. Let us consider the transit of a single image through the CNN. The image is fed to the input layer of the CNN. The input layer conveys the image to the first hidden layer. Here, a set of filters are applied to the image for extraction of specific features, e.g., edges and textures, from it. The output of the first hidden layer is transported to the second hidden layer. The second hidden layer applies more filters on the image to extract higher-level features from it. The process is repeated over several layers till the final layer is reached. The set of features extracted are passed to the output layer of the CNN. The output layer produces a probability distribution over the classes of objects examined. From this distribution, the network makes a decision in favor of the class with the highest probability.

An algorithm called backpropagation is applied for supervised learning by the neural network using the gradient descent. It is an optimization algorithm in machine learning. In this algorithm, the gradient of an error function is calculated with respect to the weights of the neural network. Accordingly, the weights are adjusted for minimization of error between the predicted and actual outputs. Thus, training is done by repeatedly refining the model.

(ii) Inference phase: The trained model is used to make predictions in real time when applied on a brand-new set of images of objects that are submitted for examination. Making use of the intelligence gathered during training, the CNN gives the output generally in the form of a class of objects or a numerical score signifying the probability value.

In many CNN applications, the training is done off-line while the inference is done on-line on computer terminals. So, the inference phase lays more crucial speed and power demands on the processor.

9.3 Constituent layers of a CNN

The CNN is composed of a multiplicity of layers (figure 9.1):
(i) Input layer: This layer contains an image or a sequence of images.
(ii) Convolution and activation layers.
(iia) Convolution layer: This is the core layer of the CNN. It is engaged in the task of feature extraction from the image. As its name

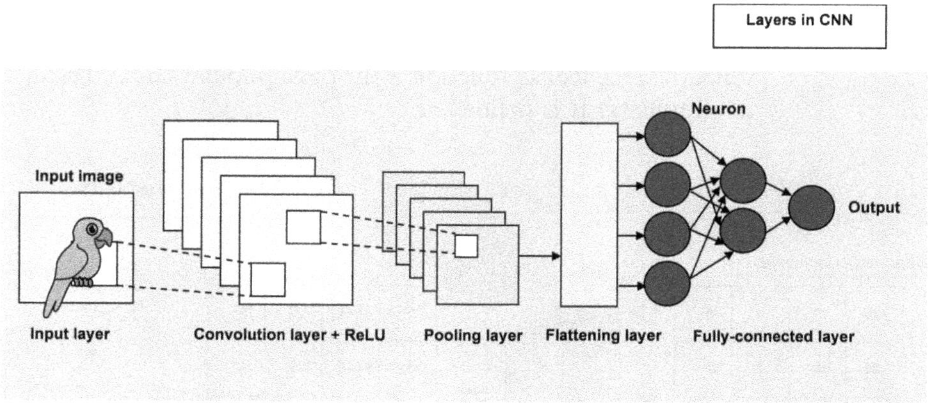

Figure 9.1. The CNN architecture. Its constituent layers are displayed. The first layer is the input layer. The convolution layer and the activation function layer are shown as merged into a second single layer. The third layer is the pooling layer. The fourth layer is the flattening layer and the fifth layer is the fully-connected layer. The input layer represents the pixel matrix of the image. The convolution layer is used for extracting various features from the image, e.g., lines, edges and corners. The ReLU or rectifier activation function layer introduces non-linearity in the deep learning model and interprets the positive part of its argument. The pooling layer makes the computation fast, lowers its cost and prevents overfitting. It reduces the sizes of feature maps without loss of any vital information to provide a compressed data representation. The flattening layer makes the multi-dimensional input one-dimensional. The fully-connected layer combines the information transmitted in the preceding layers to capture the global patterns and relationships in the input data. The output is given by the classifier as a category or probability.

implies, it performs a convolution operation on the received image. The convolution is done using a set of filters or kernels which are smaller in size than the image (figure 9.2). They have small widths and heights but the same depth as the input volume. Actually, the filters used are small matrices. They slide over the whole input volume. During this sliding, a dot product is computed between the kernel weights and a small region of the input volume. A 2D output is obtained for each filter. All the results are stacked together. From these results, a set of feature maps is obtained as the output of this layer. These feature maps are two-dimensional arrays representing the various features of the image. They reflect the extent of matching of local regions of an image with the applied filters.

(iib) Activation layer: This layer introduces non-linearity in the network model. The non-linearity is introduced by applying a nonlinear activation function to the output of the preceding layer. Non-linearity is imperative because many real-world phenomena exhibit complex nonlinear behavior. So, a linear model will not capture these phenomena. The activation functions that have been most widely used for this purpose are the rectified linear unit (ReLU), the sigmoid and the hyperbolic tangent functions. Amongst these functions, the ReLU function is the most popular choice because of its simplicity. It is defined as

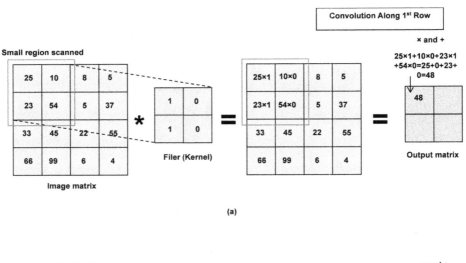

(a)

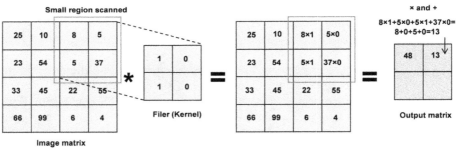

(b)

Figure 9.2. Mapping the kernel on the input image matrix: (a) and (b) along the first row, (c) and (d) along the second row. (a), (b), (c) and (d): All contain the image matrix, the output matrix, the small region scanned and the filter (kernel). (a) The kernel slides over the four elements on the top-left corner of the image matrix and performs the multiplications, 25×1, 10×0, 23×1 and 54×0. Then it calculates the sum $25 \times 1 + 10 \times 0 + 23 \times 1 + 54 \times 0 = 25 + 0 + 23 + 0 = 48$. The sum thus determined is placed in top-left side corner of the output matrix. (b) The kernel slides over the four elements on the top-right corner of the image matrix and performs the multiplications, 8×1, 5×0, 5×1 and 37×0. As before, it calculates the sum $8 \times 1 + 5 \times 0 + 5 \times 1 + 37 \times 0 = 8 + 0 + 5 + 0 = 13$, which is placed in top-right side corner of the output matrix. (c) The kernel slides over the four elements on the bottom-left corner of the image matrix, performing the multiplications, 33×1, 45×0, 66×1 and 99×0 and calculating the sum $33 \times 1 + 45 \times 0 + 66 \times 1 + 99 \times 0 = 33 + 0 + 66 + 0 = 99$, which is placed in bottom-left side corner of the output matrix. (d) The kernel slides over the four elements on the bottom-right corner of the image matrix and performs the multiplications, 22×1, 55×0, 6×1 and 4×0, whereafter it calculates the sum $22 \times 1 + 55 \times 0 + 6 \times 1 + 4 \times 0 = 22 + 0 + 6 + 0 = 28$, which is placed in bottom-right side corner of the output matrix.

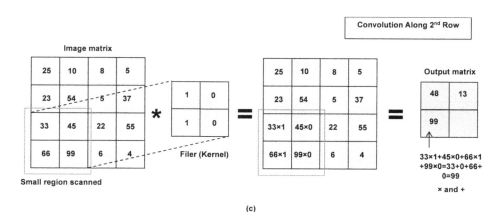

(c)

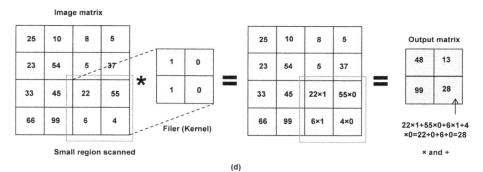

(d)

Figure 9.2. (Continued.)

$$f(x) = \max(0,\ x) \tag{9.2}$$

and returns the maximum of 0 and x. The output of this function is the input value if the input value is positive. The output is zero if the input value is negative.

(iii) Pooling layer: This layer is also termed a down-sampling or sub-sampling layer. It reduces the spatial dimensions of the feature maps by down-sampling them. Retention of the important data is a vital aspect of this process otherwise accuracy will be lost. The pooling layer works by dividing the input data into small non-overlapping regions known as pooling windows or receptive fields. After subdivision of input data, it performs an aggregation operation on each window. In the aggregation operation, either the maximum or average value of the matrix elements is taken (figure 9.3).

(iv) Flattening layer: Flattening is necessary because predictive algorithms experience difficulty in processing nested structures. They need input in a

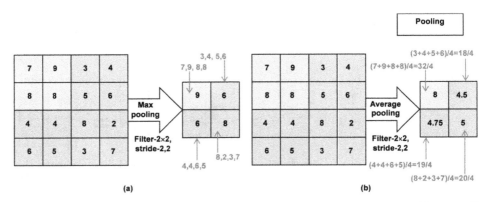

Figure 9.3. Types of pooling: (a) Max pooling and (b) average pooling. Pooling involves sliding a 2D filter over each channel of given feature map. The filter sliding is accompanied by summarizing the features falling within the region covered by the filter. Max pooling selects the maximum element from the region of the feature map covered by the filter. Based on the maximum elements thus collected from the regions of sliding, a down-sampled output feature map is generated containing the most prominent features of the input feature map. In average pooling, the attention is directed towards the average value, not the highest value of the elements in a particular region over which the filter moves. It calculates the average values of the elements present in the territory of the feature map confined within the filter. By calculating average values of the elements, it generates a down-sampled output feature map. The obtained feature map contains the average values of features of the input feature map. Apparently, max pooling extracts more pronounced features such as edges from the feature maps. Average pooling performs a smooth feature extraction.

flat tabular format. Flattening the data transforms it into a more easily digestible format. Learning patterns are more convenient from this format. The flattening layer acts as a bridge between the pooling and fully-connected layers. Its duty is to reduce the dimensionality of data and hence improve the efficiency of computation.

(v) Fully-connected layer: This layer is used at the end of the network to connect every neuron in a preceding layer to every neuron of the current layer, i.e., every input neuron to every output neuron. It produces the final output of the CNN giving the classification or regression result.

9.4 CNN accelerator

Low-power, more specialized but less computationally complex accelerators are demanded for the cloud and mobile applications because the machine learning algorithm is only operated in the feedforward or inference mode without the any training. The CNN accelerator shown in figure 9.4 is controlled by the instructions received from host processor (central processing unit, CPU) (Scanlan 2019, Kim *et al* 2020). The off-chip or external memory is a large dynamic random-access memory (DRAM).

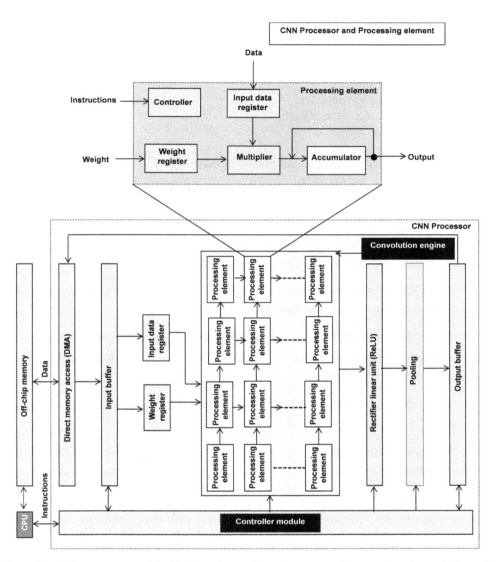

Figure 9.4. Main components of the CNN accelerator. Internal structure of a processing element is shown in the inset. Bottom: The CNN processor functions with the help of a convolution engine. This engine contains a large number of interconnected processing elements in a 2D format. The inputs to the convolution engine are provided from an input data register and a weight register. Both these registers are fed by the input buffer. The input buffer itself receives data by direct memory access from the off-chip memory. A controller module interfaced to a CPU is connected by two-way linkages to input and output buffers. It is connected by unidirectional flow to the convolution engine, a ReLU and a pooling unit. The CPU has bidirectional communication with the controller module as well as with the off-chip memory. Top: The main computational parts of the processing element are the multiplier and the accumulator. The multiplier receives data from the input data register and weights from the weight register. It operates under instructions received from the controller.

9.4.1 Parts of the CNN accelerator

The CNN accelerator consists of the following parts (figure 9.4):
  (i) The on-chip memory: It includes the:
      (a) input buffer,
      (b) the weight register,
      (c) the input data register, and
      (d) output buffer.

The on-chip memory is typically restricted in the range 139–344 kB because of the constraints imposed by die size. This small memory size is adequate to process neural networks. Nevertheless, large memory sizes in the range of MB magnitudes may be necessary for storage of the weight parameters of uncompressed network models. Around 1 mm^2 die area is consumed by 3.84 MB memory.

  (ii) The convolution engine: It is the computational core of the system. It comprises an array of processor elements (PEs) that perform convolutions or matrix multiplications. Apart from these operations, it looks after pooling and local response normalization as well as nonlinear functions such as the sigmoid or ReLu activation functions.

  (iii) PEs: Each PE (figure 9.4 (inset)) is constructed from multiply and accumulate (MAC) units. The PEs performing vector multiplication as the primitive operations are also built. The PEs allow the data input to be registered and transmitted to the adjoining PEs. In this way, they permit re-use of data. Data re-use decreases the requirements to read and write data from/to the higher memory levels. Identically, the accumulated values are conveyed to the neighboring PEs. Thus, the output values are allowed to be accumulated across the array of PEs working on individual calculations. From the output buffer, the output data values are delivered to the off-chip memory.

  (iv) The ReLU: This unit performs the ReLU function. As already mentioned in section 9.3, the ReLU is a simple function which does not involve heavy computation or complicated mathematics. Hence, the model requires a shorter time for training as well as running. A desirable feature of ReLU is the possibility of sparsity of network. The sparsity is achievable because ReLU yields zero output for all negative inputs. Hence, any negative input is converted to zero by ReLU meaning that the concerned neuron is not activated. Consequently, all the neurons are not activated simultaneously. Only a few neurons are activated. Needless to mention that the computation is comparatively straightforward when using ReLU than with other activation functions.

  A supervised learning algorithm used to train a neural network consists of calculation of the loss function C by moving backwards form the output to input nodes. It is called the backpropagation. The function C in this algorithm is the difference between the predicted and actual values. It quantifies the error. Hence, it measures the closeness with which the network models the training data. The network is trained by error correction by reducing the loss function to the least possible value. The loss function is decreased via adjustment of weights and biases of the network. In supervised learning, the loss functions are categorized into the two types, namely regression and classification loss functions.

To calculate the new weight $W_{\text{CORRECTED}}$ of a node, the product of the learning rate (η) and the gradient of the loss function ($\partial C/\partial W$) is subtracted from its previous weight W_{PREVIOUS} as represented by the equation

$$W_{\text{CORRECTED}} = W_{\text{PREVIOUS}} - \eta\left(\frac{\partial C}{\partial W}\right) \tag{9.3}$$

The chain rule of partial derivatives is applied to the gradient of the loss function. With the help of this rule, the gradient ($\partial C/\partial W$) is expressed as a product formed by the multiplication of the gradients of the activation functions of the various nodes in the neural network. The gradients are calculated with respect to the weights of the nodes. The calculated values of gradients of the activation functions of the nodes are used to determine the corrected weights of nodes in a neural network.

What happens when a sigmoid function (suppose) is used in place of the ReLU function? A major concern stemming from the use of the sigmoid function deserves attention. It is a common knowledge that the partial derivative of the sigmoid function exhibits saturation at a maximum value of 0.25. When the sigmoid function is used with deep networks, a serious problem arises because the partial derivative of the loss function approaches a value close to zero. The disappearance of the partial derivative leads to the vanishing gradient problem. However, the same does not happen for shallow networks.

As the derivative vanishes, the weights of the network remain unchanged. We know that the updating of weights and biases during backpropagation is the key mechanism through which a neural network learns. But with vanishing of the gradient, the updating of weights ceases. As the weights are not updated, the network cannot learn. Consequently, the performance of the network deteriorates.

This problem is not faced with ReLU as its slope does not plateau or saturate. It is zero for negative values and 1 for positive inputs. Hence, models using ReLU function exhibit fast convergence.

When does the ReLU function fail? It must not be forgotten that the derivative of the ReLU function is 0 for inputs < 0. Then the partial derivative of the loss function has a value of 0 because the gradient of 0 is 0. When the partial derivative is 0, the node becomes a dead node since the corrected value of the weight is the same as its previous value. It is not activated and is therefore caught in a perpetually inactive state. This problem is avoided by using a leaky ReLU function, which is an improved version of the ReLU function in which the activation function is defined as an extremely small linear component of input x for negative values of x, such as

$$f(x) = \max(0.01x, \, x) \qquad (9.4)$$

instead of the usual zero value. Then the gradient is a non-zero value so that dead neurons are no longer encountered, thereby eliminating the dying ReLU problem.

(v) The pooling unit: Here pooling is done by aggregating and summarizing information dispersed in patches of feature maps. Elaborating further on pooling described in section 9.3, the pooling layer is a crucial component of CNN because the spatial dimensional reduction of feature maps lowers the computational cost and avoids overfitting. The classification output is independent of the object's position because of identification of the same features irrespective of the position of the object. This kind of independence provides translational invariance in feature maps. Furthermore, robustness with respect to scale or rotation is obtained during capturing of the principal features of an image done by avoiding subtle variations or noise in input data.

The major disadvantage of pooling is the loss of details of information by over-smoothing of feature maps. The resultant decrease of feature resolution may sometimes critically influence the classification or regression task. Moreover, introduction of a hyperparameter, e.g., the size of pooling region, to optimize performance requires skills in building models. A hyperparameter is a parameter that is manually set before training a model to define any configurable part of the learning process.

9.4.2 Performance of CNN accelerator

Adoption of various refurbishments such as dataflow optimization, reduced precision, model compression and sparsity techniques have enabled DCNN (deep convolution neural network) accelerators to attain $4\times-20\times$ superior efficiency over GPUs at 16-bit precision with a power consumption <3.3 W (Scanlan 2019).

9.5 Eyeriss CNN accelerator

Eyeriss is an accelerator for DCNNs. It is implemented in 65-nm complementary metal-oxide semiconductor (CMOS) technology with a chip size 4 mm × 4 mm, logic gate count 1176k, and on-chip SRAM 181.5 Kbytes. It uses a processing dataflow called row stationary on a spatial architecture with 168 processing elements (Chen *et al* 2017). The row stationary dataflow performs a reconfiguration of the spatial architecture to map the computation of a given CNN shape. This reconfiguration enables optimization of operation to achieve the best energy efficiency by maximum re-use of data locally to diminish expensive data movement, e.g., DRAM accesses.

Application of run-length compression and PE data gating yields further energy efficiency improvement. Run-length encoding is a lossless compression technique in which a repeatedly occurring symbol is replaced by one copy of the symbol together with a count of the number of times it has been found. Data gating is a dynamic power reduction technique. In this technique, power is saved by reducing the superfluous transitioning in a circuit by applying different gating mechanisms to stop unnecessary toggling in the circuit (Kumar 2006).

The performance of the Eyeriss chip is benchmarked against two widely used state-of-the-art CNNs:

(i) AlexNet (a CNN with eight layers, five convolutional layers, two fully-connected hidden layers, and one fully-connected output layer, named after one its inventors Alex Krizhevsky (Krizhevsky *et al* 2017). For AlexNet, Eyeriss is able to process the convolutional layers at 35 frames/s (a processing throughput of 23.1 giga multiply–accumulate operations per second, GMACS) and 0.0029 DRAM access/MAC at 278 mW (batch size $N = 4$).

(ii) VGG-16 (Visual geometry group-16, a CNN consisting of 16 weight layers, 13 convolutional layers and three fully-connected layers). For VGG-16, Eyeriss processes 0.7 frames s^{-1} and 0.0035 DRAM access/MAC at 236 mW ($N = 3$).

9.6 Vision processing unit

9.6.1 Difference between VPU and GPU

The vision processing unit is a chip used to accelerate computer vision jobs. It has a different objective from the GPU. The GPU is a specialized device for rasterization and texture mapping for 3D graphics. It can carry out video encoding and decoding for quick, concurrent rendering of high-resolution images and video.

Conversely, the VPU is a component which is geared towards a fundamentally different functionality. It is focused on the acceleration of the tasks of analyzing and interpreting visual data. It is specifically designed to perform different types of complex calculations. These are the calculations involved in tasks such as object detection and image recognition. The prime motives behind these tasks are management and processing of images, videos, etc, efficiently in real time. Object tracking, facial recognition and scene understanding have enabled AI equipment to quickly make decisions by visual inspection of the scenario.

The VPU runs a machine vision algorithm such as a CNN, or computer vision algorithm, e.g., scale-invariant feature transform. These algorithms are used for detection, description and matching of local features in images. The VPU is able to achieve a balance of power efficiency and computing performance. The balance is obtained by combining highly parallel programmable calculations with workload-specific hardware acceleration in an architecture that minimizes data movement.

9.6.2 Typical applications

VPUs are invaluable aids to humans, machines and robots in deciphering the information content and interpretation of an image or video. They are like intelligent cameras which make them capable of working faster and in a better way with static and animated scenes. They are empowered by capabilities to comprehend and respond adeptly and promptly to visual stimuli. Their power efficiency is ideally suitable for portable mobile devices and embedded systems. So, they can deal with intricate visual analysis without excessively loading the battery.

9.6.2.1 Healthcare
VPUs are used for x-ray and MRI image analysis for faster and accurate detection and diagnosis of diseases, and fractures/tumors/wounds, etc. They can also monitor patient movements and behavioral patterns for improved patient care and safety.

9.6.2.2 Agricultural
VPU-enabled machine vision systems help to reduce manual tasks and the cost of crop production. They upgrade the productivity and competitiveness of farmers to ensure agricultural supplies.

A method for detection of curved and straight crop rows in images captured in maize fields is developed providing accuracies $\sim$86.3%–92.8% (García-Santillán *et al* 2017). It comprises linked phases of image segmentation, and identification of starting points for determining the beginning of the crop rows and for detection of their bending or flexure. The aim is to identify the crop rows for accurate autonomous guidance and for carrying out site-specific treatments, including weed removal. Weeds are the plants outside the crop rows. The performance of the method is compared against existing strategies, achieving accuracies $\sim$86.3%–92.8%.

9.6.2.3 Automotive
VPUs are integral parts of advanced driver-assistance systems for pedestrian and lane detection. They assist in traffic sign recognition by processing images from cameras.

9.6.2.4 Retail, security and safety
VPUs help by analyzing customer behaviors in knowing about customer preferences and marketing patterns. Such an analysis leads to personalized shopping and enhancement of customer experiences. Smart surveillance systems equipped with VPUs are able to analyze video footage of shops quickly. Thus, thefts are prevented and proper security of stores is ensured.

9.6.2.5 Industrial robotics
VPUs impart autonomous navigation capabilities to robots. These capabilities help the robots in moving and performing tasks independently without human aid. They help robots in object detection and identification, and precisely handling the objects to carry out complex assignments.

9.6.3 Inadequacy of traditional units for vision processing needs

Let us examine the needs of vision processing and the customer expectations on the capabilities of processors. Machine vision tasks are composed of the following segments (figure 9.5):

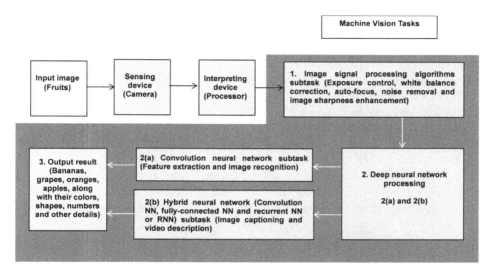

Figure 9.5. Machine vision tasks. The image (here assorted fruits) is acquired by an image sensor such as a camera and interpreted by a processor such as the VPU. The work of the processor is divided into two major activities: image signal processing and DNN processing. The DNN processing is subdivided into CNN subtask and hybrid NN subtask. The CNNs look into feature extraction and perform image recognition. The hybrid NNs provide image captions and describe the video content. The output is a declaration of all the information about the image.

(i) Image signal processing task;
(ii) DNN task;
     (a) CNN subtask, and
     (b) Hybrid neural network subtask.

The hybrid DNNs are combinations of convolutional neural networks (CovNets), fully-connected neural networks (FCNNs) and recurrent neural networks (RNNs). They are used for image captioning and video description.

Many DNN processing units (DNPUs) can neither handle the image signal processing (ISP) algorithms nor the CNNs and CNN–RNN frameworks used in machine vision. They struggle under the burden of the large volumes of data in images/videos. On the other side, ALU-array-based VPUs cannot process the DNNs. By confronting our problems, we acquire perseverance to face difficult situations.

9.6.4 Single unit design for vision tasks

'When one door closes, another door opens' (Alexander Graham Bell). Evidently, a heterogeneous VPU architecture is necessary for implementation of the ISP task, and execution of the CNNs and hybrid DNN subtasks on a single unit. Furthermore, the VPU must work on a sharing scheme for multiplexing the hardware resources for the various subtasks (Liu *et al* 2022).

9.6.5 Hybrid chip architecture

The VPU shown in figure 9.6 consists of a hybrid processing element (PE) array. The array comprises four PE blocks PEB0, PEB1, PEB 2 and PEB3. Each PE block is equipped with horizontal and vertical buffers, one ALU, a 1D row processor made of several multiply–accumulate (MAC) units and enhanced ALUs, coupled with input buffer and row buffer. An image sensor interface is provided.

The hybrid array is connected to the weight buffer, Hauffman decoder/encoder and global buffer. Finite-state controller and instruction buffer are also included. The instruction buffer and Hauffmann decoder/encoder are linked with external memory. The Huffman encoder/decoder module exchanges data between the VPU and external memory. Two-symbol Hauffman coding is used for compression of data and weights. Hauffman coding is a prefix-free data encoding method in which no codeword is used as a prefix to another codeword making the compressed data easy to decode. Decoding is the conversion of the encoded data to the original sequence of characters.

The finite-state controller decodes the instructions to produce the control signal for all modules of the VPU. A finite-state machine is a machine with a finite number of states, in which only one single state is active at a time, and which transitions from one state to another to perform different activities.

9.6.6 VPU workflow

The ALU and the MAC arrays carry out ISP tasks. They also perform operations on the convolutional layers, and also the non-MAC operations in the CNNs. The row processor acts on the FC layers and the RNNs. In the beginning, the ALU array executes image signal processing on the red, green and blue RGB-RAW image pixels. The intention is to convert them into fine-tuned RGB images. Then the PE array processes the RGB image as three input channels for the convolutional layers and pooling/activating layers. The idea is to generate the feature vectors of the image. The row processor analyzes the feature vectors for the fully-connected layers or RNNs. The results are communicated to the external memory to undergo further processing.

9.6.7 Experiments

The VPU is built on a field-programmable gate array (FPGA) for experimentation on vision operations. The vision tasks are done efficiently with an average perform-ance of 2.26×10^{10} operations per second/W (GOPS/W) (Liu *et al* 2022).

9.7 Commercial VPU performance capabilities

Intel® Movidius™ Myriad™ X vision processing unit (lithography 16 nm, processor base frequency 700 MHz, package size 14 mm × 14 mm × 0.9 mm) (Intel 01) contains a specially built hardware accelerator. This accelerator is the neural compute engine. It is designed to spectacularly increase performance of DNNs without losing the low power characteristics (Intel 02). The noteworthy feature of

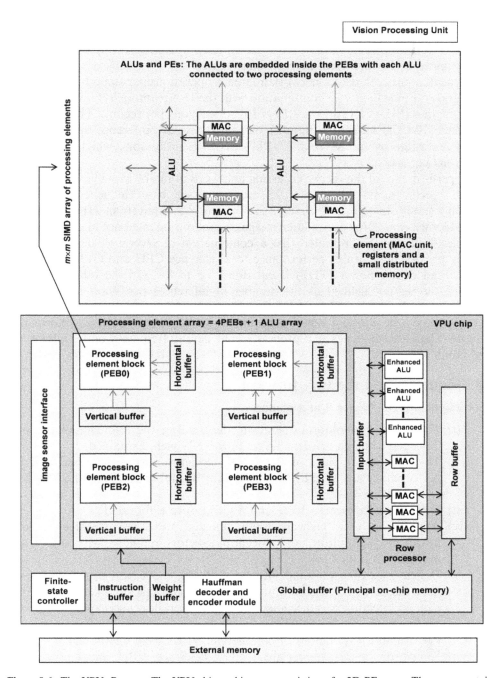

Figure 9.6. The VPU: Bottom: The VPU chip architecture consisting of a 2D PE array. The array contains four PE blocks. These blocks have horizontal/vertical buffers, and an ALU array with ALUs embedded in the processing element blocks. Also, it has several buffers: a row processor (RP) containing MACs and enhanced ALUs, a row buffer, a global buffer, an input buffer, a weight buffer, and an instruction buffer. The other components of the VPU chip are an image sensor interface, a finite-state controller and a Hauffman decoder/encoder. Top: Connection of ALUs and PEs (MAC unit + registers + small memory) inside the VPU.

the chip is an array of multiply–accumulate blocks. The neural compute engine speedily executes the calculations required for deep inference. It does so by directly interfacing with the intelligent memory network.

The peak number of neural network inference operations per second attainable by the neural compute engine in combination with the 16 streaming hybrid architecture vector engines (SHAVE) cores is 9.16×10^{11} operations per second. This number is more than 10× the maximum number of neural network inference operations per second realizable by the Myriad 2 VPU's SHAVE processors, which is 8×10^{10} operations per second.

The performance of the vision processing unit 4 GB of Intel® Movidius™ Myriad™ 2, with specifications: lithography 28 nm, processor base frequency 933 MHz, marketing status discontinued, package size 9.5 mm × 8.0 mm (Intel 03) was explored (Rivas-Gomez *et al* 2018) over a large image dataset during inference in convolutional networks. Initial results indicated that a combination of several of these chips can possibly provide comparable performance to a reference CPU and GPU enactment. The thermal design power (TDP) is cut down up to 8×. On the other hand, the observed throughput defined as the number of inferences per Watt, is more than 3× higher than the CPU–GPU embodiment. Thus, the VPU is touted as a class of chips that seek to provide ultra-low power consumption without deprivation of performance.

Note: The base frequency of a processor is its default frequency under idle or low load condition.

9.8 Summary and the way forward

9.8.1 Highlights of chapter 9 at a glance

(i) The CNN, convolution operation, the training and inference phases of a CNN, and the constituent layers of a CNN were explained.

(ii) The constituent parts of the CNN accelerator were described and the performance achieved by the accelerator was affirmed (Scanlan 2019, Kim *et al* 2020).

(iii) Differences between a VPU and a GPU were elaborated.

(iv) Applications of VPUs were briefly itemized. VPUs are extensively used in healthcare, agricultural and automotive sectors, retail, security, safety, and industrial robotics.

(v) Deficiencies of traditional units in fulfilling the vision processing needs were mentioned. A single-unit design for vision tasks based on the hybrid chip architecture was discussed. The VPU workflow and experiments were summarized (Liu *et al* 2022).

(vi) Performance capabilities of commercial VPUs were mentioned.

(vii) Keywords for this chapter are: CNN accelerator, VPU, image processing, computer vision.

9.8.2 Getting ready to begin chapter 10

(i) From a broad perspective, it is reiterated that the main driving force behind the escalation of CNN processor and VPU demands is the increasing complexity and sophistication of computer vision algorithms and deep learning. These intricacies have led to a growing need for dedicated hardware solutions to efficiently process the immense amount of visual data generated.

(ii) The proliferation of smart sensors and IoT (Internet of Things) applications is raising the demand in edge computing setups. The automotive industry has significantly raised hopes and aspirations with advancements in driver-assistance systems (ADAS) and autonomous driving technologies. VPUs are indispensable components aiding vehicles in perception of their surroundings. With the enhanced perception, they can make intelligent decisions in real time to promote driving experience along with safety.

(iii) Target markets for CNN and VPUs include robotics, IoT, emerging digital cameras for virtual and augmented reality, machine vision integrated smartphones and mobile devices.

(iv) In view of the huge impact of these processors on vision tasks, sparse tensors-based processors for CNN models have garnered considerable attention for efficient processing of high-resolution shapes. This topic will be investigated in the next chapter.

References

Alexander Graham Bell ForbesQuotes https://forbes.com/quotes/253/#:~:text=When%20one%20door%20closes%20another,Alexander%20Graham%20Bell%20%2D%20Forbes%20Quotes

Chen Y-H, Krishna T, Emer J S and Sze V 2017 Eyeriss: an energy-efficient reconfigurable accelerator for deep convolutional neural networks *IEEE J. Solid-State Circuits* **52** 127–38

García-Santillán I D, Montalvo M, Guerrero J M and Pajares G 2017 Automatic detection of curved and straight crop rows from images in maize fields *Biosyst. Eng.* **156** 61–79

Intel 01 *Intel® Movidius™ Myriad™ X Vision Processing Unit 0GB* Intel Corporation, https://intel.com/content/www/us/en/products/sku/204770/intel-movidius-myriad-x-vision-processing-unit-0gb/specifications.html

Intel 02 *Intel® Movidius™ Myriad™ X VPU Product Brief* Intel Corporation, https://intel.com/content/www/us/en/products/docs/processors/movidius-vpu/myriad-x-product-brief.html

Intel 03 *Intel® Movidius™ Myriad™ 2 Vision Processing Unit 4GB* Intel Corporation https://intel.com/content/www/us/en/products/sku/122461/intel-movidius-myriad-2-vision-processing-unit-4gb/specifications.html

Kim Y, Kim H, Yadav N, Li S and Choi K K 2020 Low-power RTL code generation for advanced CNN algorithms toward object detection in autonomous vehicles *Electronics* **9** 1–14

Krichen M 2023 Convolutional neural networks: a survey *Computers* **12** 1–41

Krizhevsky A, Sutskever I and Hinton G E 2017 ImageNet classification with deep convolutional neural networks *Communications of the ACM* **60** 84–90
pp 1097–105 Original version of this paper was published in the *Proc. of the 25th Int. Conf. on Neural Information Processing Systems (Dec. 2012 Lake Tahoe, NV)*

Kumar A 2006 Dynamic power reduction using data gating *UT Electronic Theses and Dissertations* The University of Texas at Austin 2023 http://hdl.handle.net/2152/30338

Liu P, Yang Z, Kang L and Wang J 2022 A heterogeneous architecture for the vision processing unit with a hybrid deep neural network accelerator *Micromachines* **13** 1–27

Rivas-Gomez S, Peňa A J, Moloney D, Laure E and Markidis S 2018 Exploring the vision processing unit as co-processor for inference *IEEE Int. Parallel and Distributed Processing Symp. Workshops (IPDPSW) (21–25 May) (Vancouver, BC, Canada)* pp 589–98

Scanlan A G 2019 Low power and mobile hardware accelerators for deep convolutional neural networks *Integration* **65** 110–27

IOP Publishing

AI-Processor Electronics
Basic technology of artificial intelligence
Vinod Kumar Khanna

Chapter 10

Compressed and sparse neural network processors

The concept of sparsity in neural networks is introduced. Sparse and dense matrices are defined. The advantages of sparsification of a neural network are clarified. Pruning-induced sparsity is elaborated upon. Weight and activation sparsity; structured and unstructured sparsity are distinguished. Acceleration of 2:4 fine-grained sparsity pattern is deliberated. The workflow for network training is planned. The practicalization of 2:4 sparsity adoption by the Ampere graphical processing unit (GPU) architecture of NVIDIA is explained. An energy-efficient inference engine (EIE) is realized by deep compression. Its implementation methodology is presented highlighting the roles of different components, viz., the leading non-zero detection node, activation queue, the pointer read unit, the sparse matrix read unit, the arithmetic unit and the activation read/write unit. A sparse convolution neural network (SCNN) accelerator is considered explaining its novelties, integral parts and operation. The front-end and back-end operations of a sparse neural accelerator processor (SNAP), and the working of its associative index matching unit are summarized. Its processing element consists of a sequence decoder, a weight register file, an input activation register file, a compute path and an address path. The functioning of the controller, memory and computation modules of the SNAP system is discussed along with its operational efficiency.

10.1 Introduction

This chapter elucidates the sparsity concept in the context of neural networks, and explains the operation of a sparsity and data compression-enhanced neural network processor. Sparsity in natural data allows the use of only a few parameters for data representation. Compressibility allows a reduction in the number of values that are required to be stored for accurate representation of data. The two ideas of signal processing, sparsity and compressibility help in breaking new ground in computing at the grassroot level of data.

doi:10.1088/978-0-7503-6259-7ch10

10.2 Sparsity in neural networks

10.2.1 Sparsity: sparse and dense matrices

In general terms, 'sparsity' means to be in a scattered or scanty state giving the impression of a state lacking in density. A sparse population of people is interpreted as a small number of people inhabiting a large geographical area in a distantly spread-out distribution. Such a terrain is said to be sparsely populated.

In machine learning, the sparsity notion alludes to matrices. It refers to a matrix of numbers containing several zeros or values that do not appreciably influence a calculation. A sparse matrix is one in which the number of zero elements exceeds the number of non-zero elements. It is a matrix with lots of zero elements. The other side of the coin is that a dense matrix has a large number of non-zero elements.

10.2.2 Rationale behind weight sparsification of a neural network

Redundant weights in a dense neural network make the network training highly time-consuming. Their ramifications are raising of the carbon footprint and escalation of costs. But we are aware that the redundant weights tend to generate overlapped features. Therefore, they can be readily disregarded without degrading accuracy. At the same time, their removal decreases the number of floating-point operations. Hence, the same computation task is accomplished with a smaller-size model. On the whole, their omission yields faster inference with reduced energy consumption without sacrificing accuracy. Thus, it can be construed that there always exists a sparse network within a dense network. This subnetwork can be trained to give the same result as the dense network.

10.2.3 Pruning-induced sparsity in a neural network model

The size of a neural network model is considerably reduced by setting the values of certain weights to zero. Zeroing of some weights effectively removes connections between some neurons. So, for trimming a neural network in this manner, the weights in the given network are ranked in terms of their magnitudes. After the ranking, a threshold value is decided. A comparison of weights is made with the threshold value whereafter the weights whose magnitudes fall below the threshold value are discarded.

10.2.4 Types of sparsity

Various kinds of sparsity are (figure 10.1):
   (i) Unstructured sparsity (US): Here, specific elements within a data set have zero value without conformance to an organized arrangement.
   (ii) Coarse-grained structured sparsity (CSS): In CSS, large groups or blocks of elements of a dataset are zero. Sparsity can be enforced either along the input or output axis of the weight tensor. Separate shrinkage of the input axis is called channel pruning, and results in channel sparsity, whereas separate shrinkage of the output axis is known as filter pruning, resulting in

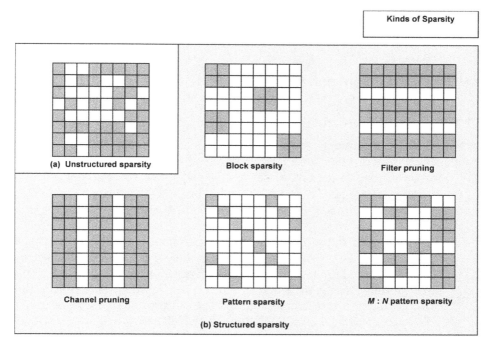

Figure 10.1. Sparsity: (a) unstructured and (b) structured. Five types of structured sparsity are shown, viz., block, filter, channel, pattern and *M:N* pattern.

       filter sparsity. Besides channel sparsity and filter sparsity, block sparsity can be created in a network. Pruning a square block along both the input and output axes causes block sparsity.

(iii) Fine-grained structured sparsity (FSS): In FSS, the ratio of zero elements to non-zero elements, remains constant across the weights. In pattern-based sparsity, predefined patterns are enforced in a model. The enforcement is brought about either by domain knowledge or algorithmic requirements. A familiar example is the $N{:}M$ sparsity obtained by retention of N non-zero weights for every M continuous weights.

10.3 Acceleration of 2:4 fine-grained sparsity pattern

10.3.1 The 2:4 sparsity pattern

A 2:4 sparsity pattern requires that every two of a group of four values must be zero. The resulting 50% sparsity provides three benefits, namely:

   (i) efficient accesses to memory,

  (ii) a compressed format with a low overhead, and

 (iii) double the mathematical throughput.

A dense matrix A with M rows and K columns becomes a sparse matrix B written as

$$B = M \times \frac{K}{2} \tag{10.1}$$

with M rows and $K/2$ columns (Mishra *et al* 2021).

10.3.2 Workflow for network training

The workflow for training a network for maintaining accuracy with a 2:4 sparsity compliance consists of three steps:

   (i) Basic training of the neural network is performed without any sparsity to produce a dense matrix $A = M \times K$ of trained weights shown in figure 10.2(a).

  (ii) Pruning of the model is carried out in the 2:4 pattern to produce a sparse matrix $B = M \times K/2$.

 (iii) Fine-tuning of the model involves making small adjustments so that the model works as best as possible, thus attaining the desired performance. This is achieved by retraining the model with weights initialized to those after pruning, using the same optimizer and schedule as during training without sparsity, and keeping the sparsity pattern calculated after pruning. The sparse matrix B after fine tuning the weights is shown in figure 10.2(b).

 (iv) The sparse matrix B is written in a compressed format in figure 10.2(c). The non-zero data values and non-zero 2-bit indices are shown separately.

10.3.3 Making sparsity adoption pragmatic by the NVIDIA Ampere GPU architecture

The sparsity support is extended by the NVIDIA Ampere GPU architecture in its specialized matrix-math units, the tensor cores (NVIDIA 2021). The sparse tensor core hardware selects only the elements from matrix C that correspond to the non-zero values in matrix B. Hence, the unnecessary multiplications by zeroes are left out. The tensor core allows the first argument to be stored in the sparse format (matrix B), and performs the operation

$$\text{Sparse matrix } B \times \text{Dense matrix } C = \text{Dense matrix } D \tag{10.2}$$

in which the second input matrix C and the output matrix D are both dense matrices (figure 10.2(d)). To clarify, the matrix B is a sparse matrix of size $M \times K/2$ while matrices C and D are two dense matrices of input and output activations, respectively.

10.4 Energy-efficient inference engine (EIE) by deep compression

Matrix vector multiplication is the part and parcel of neural network calculations. Whenever the matrix exceeds the capacity of the cache memory, the large matrix size necessitates main memory access. As we know, this memory access is a lethargic activity and so becomes problematic for the calculation.

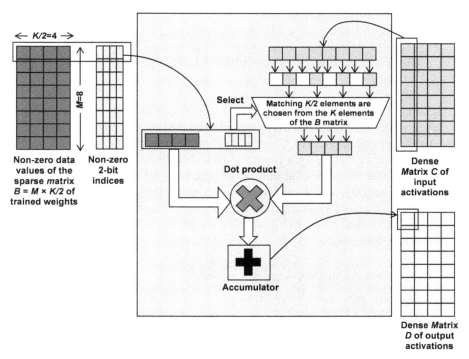

2:4 Sparsity Pattern Generation, and its Implementation by Tensor Core

Dense Matrix
A = M × K of
trained weights

(a) Basic training.

Sparse matrix B = M × K/2 of
trained weights with blank squares
showing zero values.

(b) After pruning and fine-tuning.

Only nonzero
data values of
the sparse
matrix
B = M × K/2 of
trained weights

Non-zero
2-bit
indices

(c) After compression.

Non-zero data
values of the
sparse matrix
B = M × K/2 of
trained weights

Non-zero
2-bit
indices

Select

Matching K/2 elements are
chosen from the K elements
of the B matrix

Dot product

Accumulator

Dense
Matrix C of
input
activations

Dense Matrix
D of output
activations

(d) The sparse operation of tensor core.

Figure 10.2. Inference acceleration with sparsity: (a) dense trained weights, (b) pruned and fine-tuned weights, (c) sparsed and compressed weights, and (d) sparse tensor core operation. In (a), the basic training is carried out with a dense matrix of trained weights. In (b), the weights are pruned by fine-grained structured sparsity with a 2-out-of-4 non-zero pattern. The pruning is followed by implementation of a recipe for fine-tuning the non-zero weights. In (c), the data footprint and bandwidth are slashed by compressing the weights. In (d), the sparse tensor core operation increases the output twofold by skipping the zeroes.

10.4.1 Deep compression

10.4.1.1 Process steps in deep compression

To get rid of the problem caused by the large matrix size, the neural network needs to be compressed by a technique known as deep compression. The deep compression technique is able to decrease the network size drastically without compromising accuracy (Han *et al* 2016). It is a three-step process as follows:

(i) Network pruning: After normal training, the small-weight connections, viz., those lower than a threshold are removed from the network, as discussed in section 10.2.3. Then the network is retrained to determine the final weights for the remnant sparse connections.

(ii) Quantization and weight sharing: Quantization decreases the number of bits needed for storage of weights. The weights are quantized into bins. All the weights in a bin have the same value. Hence, only a small index is stored as an indicator for each weight in a table of shared weights. The *k*-means clustering algorithm is used for identification of the shared weights for each layer of the network. This algorithm is a centroid-based unsupervised learning algorithm.

(iii) Hauffman coding: The Hauffman code is an optimal prefix code. It uses variable length codewords for encoding source symbols. Here, a screwed or biased distribution of values is utilized.

10.4.1.2 Energy savings

The significant advantage derived from compression is that the network model becomes so small that it can be easily accommodated on the on-chip static random-access memory (SRAM). Fetching of weights from SRAM instead of DRAM provides a saving of energy by a factor of 120. Additionally, an order of magnitude benefit is achieved through exploitation of sparsity. A further eight-fold improvement accrues from the weight sharing. Combining all these energy savings with a threefold factor through skipping zero activations from rectified linear unit (ReLU) leads to the realization of an EIE presented by Han *et al* (2016). We give details of EIE in ensuing sections.

10.4.2 Implementation methodology

10.4.2.1 Leading non-zero detection node

The EIE consists of a large number of processing elements (PEs). Each group of four PEs performs a local leading non-zero detection. The detection is done primarily for the selection of the first non-zero result (figure 10.3(a)). The root leading non-zero detection (LNZD) node is the central control unit (CCU). The CCU controls an array of PEs. It monitors the condition of every PE by communicating with the master, e.g., a central processing unit (CPU).

10.4.2.2 Activation queue

In each PE (figure 10.3(b)), the activation queue receives the non-zero elements of the input activation vector a_j and their index j, as broadcast by the CCU. For a PE with a full queue, the broadcast is disabled.

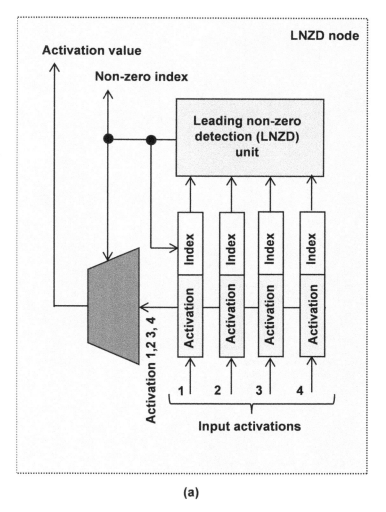

(a)

Figure 10.3. Architectures of: (a) the LNZD node and (b) the PE. In (a), the LNZD node receives the result of leading non-zero detection carried out on the input activations of a group of four processing elements. In (b), the activation value and the index from the LNZD node are fed to the activation queue. For any column of the matrix, acquisition of the start and end pointers of the arrays is made by using the index of the particular entry which is positioned at the head of the activation queue. Two SRAM banks are employed for the storage of pointers. Selection between the banks is done by using the least significant bit (LSB) of the address. The pointers help the sparse-matrix read unit to read the non-zero elements of the slice of a column of a particular PE from the sparse-matrix SRAM. On receipt of sparse matrix data from the sparse matrix read unit, the arithmetic unit performs the multiply–accumulate operation. The activation read/write unit contains two activation register files. These files lodge the source and destination activation values during a single cycle of fully-connected layer computing. Leading non zero detection logic is applied to choose the first non-zero result for availing the advantage of input vector sparsity.

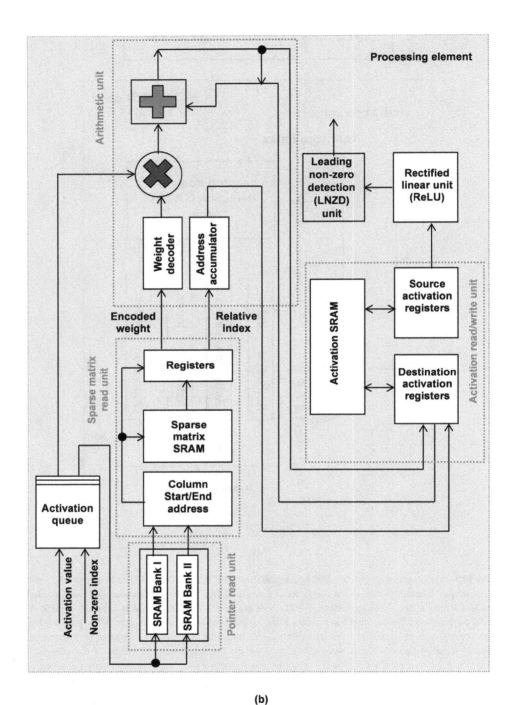

(b)

Figure 10.3. (Continued.)

10.4.2.3 Pointer read unit

The index j of the entry lies at the head of the activation queue. This index is used to look up for the beginning and end pointers (p_j, p_{j+1}) for the (v, x) arrays for column j. The (v, x) entries are used in the arithmetic-logic unit, as explained below. The pointers are placed in two SRAM banks. The LSB of the address allows choosing between the banks.

10.4.2.4 Sparse matrix read unit

The pointers (p_j, p_{j+1}) are used to read the non-zero elements of the slice of column $I_{i,j}$ of the PE from the sparse matrix SRAM. $I_{i,j}$ is the index to the shared weight $W_{i,j}$.

10.4.2.5 Arithmetic unit

On receipt of an entry (v, x) from the sparse matrix read unit, the arithmetic unit performs the multiply–accumulate operation. An accumulator array, constituted by the destination activation registers, is indexed by the x value. During this time, the activation value a_j at the head of the activation queue is multiplied by v to get the output activation vector b as

$$b_x = b_x + a_j \times v \tag{10.3}$$

10.4.2.6 Activation read/write unit

The two activation register files of this unit keep the source and destination activation values in a single round of fully-connected layer computation.

10.4.2.7 Performance parameters

In comparison to a GPU, the energy required for computing a fully-connected layer is decreased by 3400 times. A PE of area 0.64 mm^2 can execute 1.6 giga operations per second (GOPS) with 9 mW power dissipation. The energy and performance scale nearly linearly from 1 PE to 256 PEs. The EIE outperforms a CPU by a factor of 189, a GPU by 13 times and a mobile GPU by 307 times. The energy consumption is 2.4×10^4 times less than that for the CPU, 3400 times less than a GPU and 2700 times less than a mobile GPU (Han *et al* 2016).

10.5 Sparse convolution neural network (SCNN) accelerator

The SCNN accelerator described by Parashar *et al* (2017) speeds up the inference manifold in CNNs.

10.5.1 Novelties introduced

Several novel features are introduced in the SCNN architecture by taking advantage of:

(i) Sparsity: The SCNN exploits sparsity in both weights and activations. These are derived from zero-valued weights and zero-valued activations. The zero-valued weights arise from network pruning in the training phase.

The zero-valued activations result from ReLU operations during inference. Many operations are optimized by sparsity exploitation because multiplication by zero always results in zero. Moreover, addition of zero to a partial sum is irrelevant. Last but not the least, a data containing many zeros is expressible in a compressed format.

(ii) Data compression: Sparse weights and activations are encoded to reduce the data footprint. The favorable consequences of compression are prevention of memory overloading, and avoidance of large-size data motion through the memory hierarchy.

(iii) Disregarding meaningless computations: Zero weight and activation operands are not sent to the multiplier.

(iv) Using the PT-IS-CP-sparse dataflow: The dataflow represented by this shortened form is the planar tiled-input stationary-Cartesian product-sparse mechanism. It enables the utilization of a Cartesian product-based architecture yielding maximal reusage of weights and activations within a group of processing elements.

Provision of a dense data representation in a compressed encoded format is an added advantage. Only the densely encoded weights and activations are supplied to the multipliers. Hence, only non-zero weights and activations are accessed from the input storage for delivery to the multiplier array because they are multiplied like a Cartesian product. Activation vectors are reused against several weight vectors. For this reason, unwarranted data transfers are eliminated, and consequently less memory storage suffices.

10.5.2 SCNN parts

The SCNN (figure 10.4(a)) consists of an array of PEs, which are connected to their nearest neighbors for exchanging halo values during processing. Halos are the cross-tile dependencies at the edges of the smaller tiles into which the activation plane is partitioned for distribution across the PEs in the PT-IS-CP-sparse dataflow.

A layer sequencer drives the array. It looks after the movements of weights and activations. A DRAM controller is connected to it. The DRAM controller sends weights to the PEs. To-and fro- movements of activations take place between the DRAM controller and the PEs.

Principal components of a PE (figure 10.4(b)) are:

Input side: This is divided into two parts. It has a weight buffer FIFO, and an input activation RAM (IARAM).

Computation unit: This has four parts. It comprises a multiplier array, a scatter crossbar, a bank of accumulator buffers and a computation coordinator unit.

Output side: This consists of two parts: It includes a PPU and an output activation RAM (OARAM).

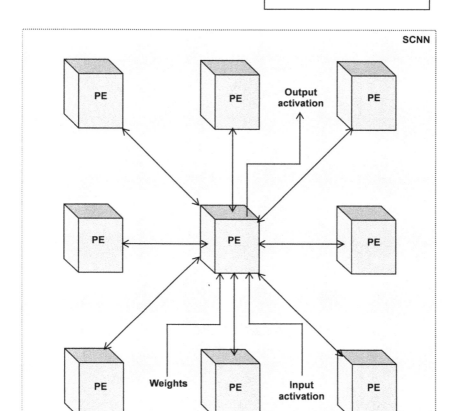

(a)

Figure 10.4. The ScNN: (a) Parts of ScNN and (b) parts of a processing element of SCNN. The complete SCNN shown in (a) is constructed of a multiplicity of PEs. The PEs are connected to their proximate neighbors by two-way linkages. Each PE has channels for receipt of input activations and weights. Each PE also has channels for supplying output activations. A layer sequencer drives the array of PEs. It plans the movements of weights and activations. It is connected to a dynamic RAM (DRAM) controller. The DRAM controller broadcasts weights to the PEs. It also regulates the flow of activations to the PEs and from PEs. A single PE of SCNN shown in (b) consists of a weight buffer first in, first out (FIFO) and input activation RAM as input section; and an output activation RAM and a post-processing unit (PPU) as the output section. Within its computation unit resides a multiplier array, a scatter crossbar, and a series of accumulator buffers along with a computation coordinator.

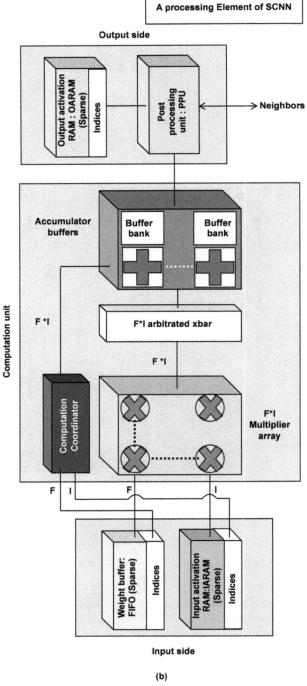

(b)

Figure 10.4. (Continued.)

10.5.3 Operation

(i) For processing any CNN layer, a portion of the input image is sent by the layer sequencer into the IARAM of each PE. The image conveyance is accompanied by the broadcasting of compressed-sparse weights into the weight buffer of each PE. As soon as a layer of the network is completed, a distribution of sparse-compressed output activations is made across the OARAMs of the PEs. When the output activation of a layer is the input activation of the succeeding layer, IARAM and OARAM are swapped between the two layers.

(ii) According to the order defined by the dataflow, the weight and input activations are acted upon followed by dumping an output channel group of partial sums in the accumulation buffers. A vector F of compressed weights and a vector I of input activations are fetched from their respective buffers. The fetching of F and I is followed by their distribution into $F \times I$ multiplier array to compute the Cartesian product of the vectors. Simultaneously, the indices from the sparse-compressed weights and activations are acted upon for calculating the coordinates in the dense output activation. The $F \times I$ products are sent to an array of accumulator banks. Indexing is done by the output coordinates.

(iii) After completion of the output-channel group, the PPU exchanges the partial sums with neighboring PEs. Then it applies ReLU, pooling and dropout functions. Finally, it compresses the output activations and writes them into the OARAM.

Performance improvement by a factor of 2.7 times and energy improvement by 2.3 times are achieved over contemporary dense CNN accelerators (Parashar *et al* 2017).

10.6 Sparse neural accelerator processor (SNAP)

10.6.1 Front-end and back-end operations

The SNAP is a sparse deep neural network processor of Zhang *et al* (2021), which works as follows:

(i) At the front end: The compressed, sparse weights and input activations are fetched. The valid weight and input activation pairs are discovered from the data and indices. The valid weight and input activation pairs are supplied to compute non-zero partial sums.

(ii) At the back end: The output activation partial sums are compressed and written back to the output activation buffer.

The objectives of the front-end operations are achieved by using an associative index matching unit. This unit encodes the addresses of the valid weight and input activation pairs. These pairs are discovered from compressed, unstructured, and sparse data arrays of weights and input activations. The discovery is based on parallel associative search. A sequence decoder in the PE decodes the addresses of the valid weight and input activation pairs. Then the weight and input activation pairs are dispatched for parallel computation.

Accordingly, we shall first describe the working of the associative index matching unit. Then we shall look into the processing element containing the sequence decoder, and afterwards we will consider the complete SNAP system.

10.6.2 Associative index matching unit

The associative index matching (AIM) unit (figure 10.5) works on the extraction of non-zero valid weight and input activation pairs of matching channel index. It consists of an array of $N \times N$ comparators. Each line of the array is connected to a priority encoder. The weight and input activation arrays of channel indices are received from a processing element. Then each weight channel index is compared with each input activation channel index. Thereafter, the match result of each line is encoded into a valid bit. This bit indicates the match and the match position in the index channel array. Immediately after the task is completed, a list of valid position pairs is served to a PE.

Note: Taking the image tensor as an example, the image tensor is an n-dimensional array of numbers containing the values of pixels of the image. The index of the tensor is a number that specifies the position of an element along a dimension of the tensor. A channel of an image tensor is the value used for encoding each pixel in the tensor, e.g., a channel in a CNN represents a feature map produced by applying filters to the input data through convolution.

10.6.3 Processing element

The PE (figure 10.6) is made of several parts, as described in the subsections below:

10.6.3.1 Sequence decoder
This part is used for conversion of a list of valid-position pairs into pairs of weight-input activation data. The sequence decoder itself consists of three parts (figure 10.7):

    (i) a valid register file,
    (ii) a position register file, and
    (iii) a three-way priority encoder.

It operates as under:
    (i) The valid position pairs from the AIM unit are fed to the valid register and position register files, respectively.
    (ii) The three-way priority encoder transforms three valid-position pairs having valid bit = 1 at a time into addresses of weight-input activation data. These addresses are supplied to the weight register file and the input activation register file, respectively.

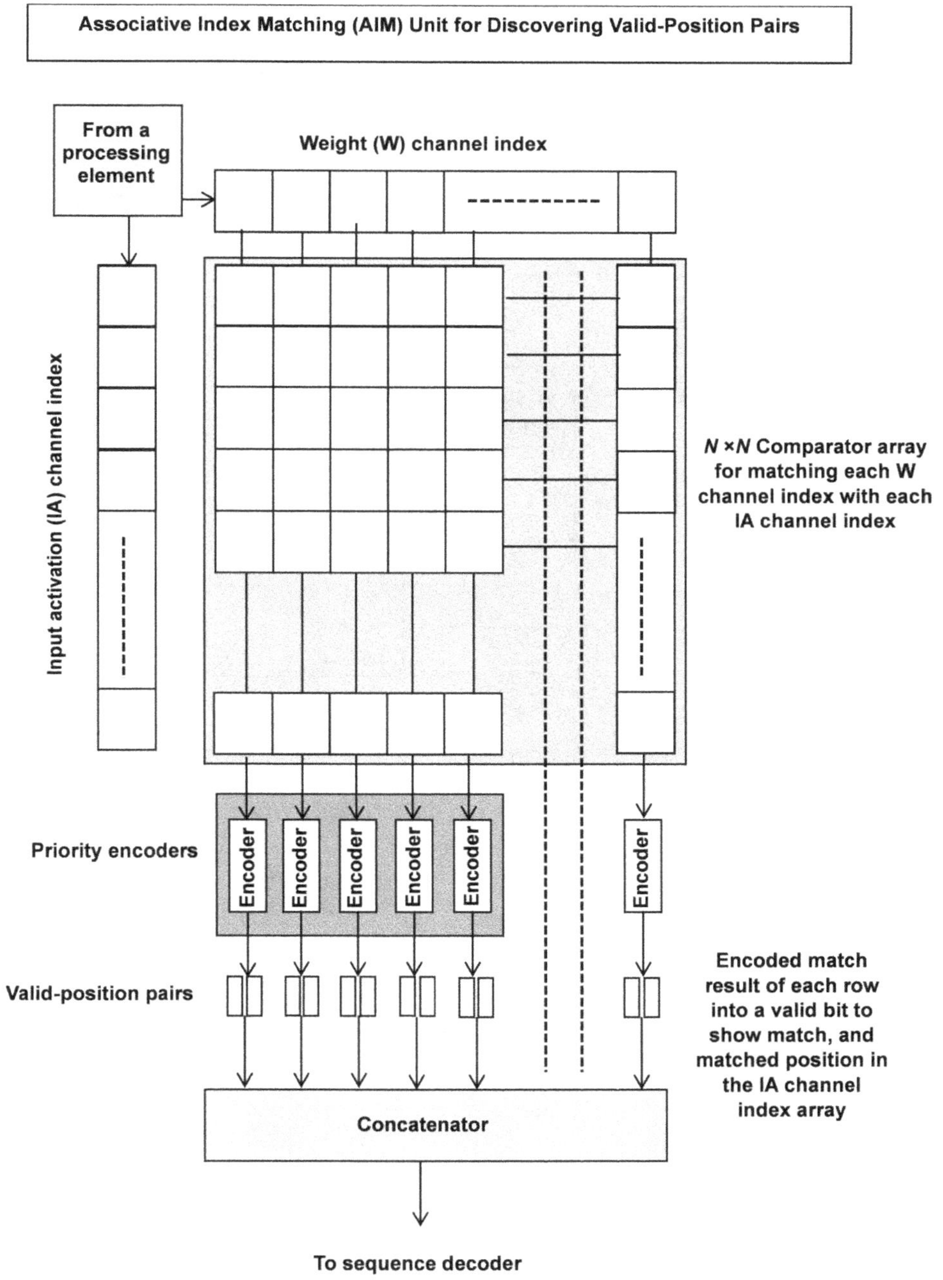

Figure 10.5. Associative index matching unit microarchitecture consisting of an $N \times N$ comparator array. Each row of the array is connected to a priority encoder. A PE delivers the weight and input activation channel index arrays to the AIM. After comparing each weight channel index with each input activation index, the match result of each line is encoded by the priority encoder into a valid bit. The process completes with the associative index matching unit providing a list of valid-position pairs to a PE.

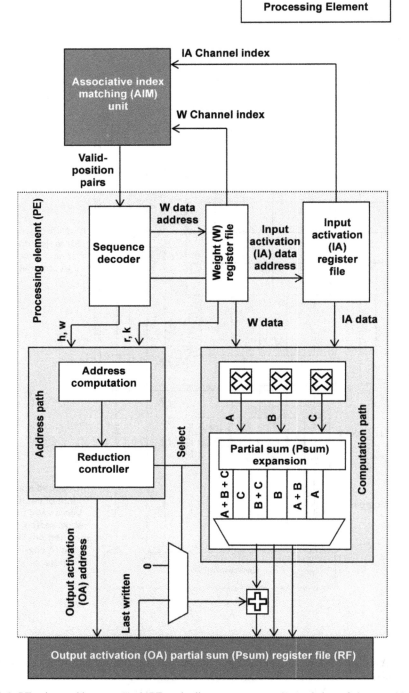

Figure 10.6. PE microarchitecture. Each PE embodies a compute path consisting of three multipliers and partial sum (psum) accumulators. It also contains an address path for computing the addresses. For supplying the inputs, the PE selects the psum reduction pattern, a sequence decoder, a weight register file and an input activation register file. For storing the outputs, the PE selects an output activation psum register file.

10-16

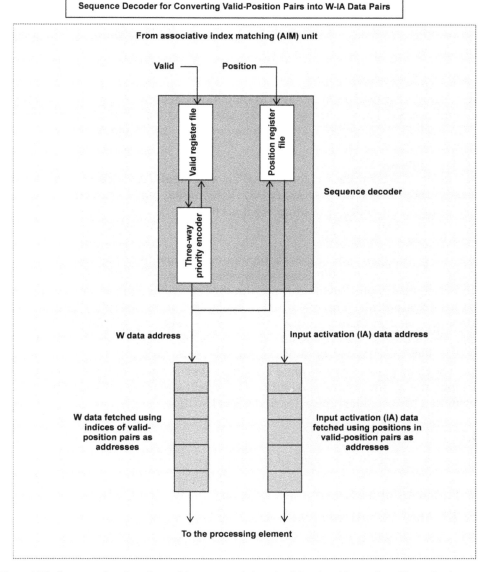

Figure 10.7. Sequence decoder microarchitecture consisting of valid and position register files and a three-way priority encoder.

(iii) Overwriting the valid bits to 0 invalidates the three valid-position pairs.

(iv) The PE completes the received list of valid-position pairs. After its completion, the processing element requests a new list of valid-position pairs from the associative index matching unit.

10.6.3.2 Weight register file
The addresses to fetch weight data are represented by the indices of the valid position-pairs. The indices go to the weight register file from where the weight channel index is sent to the AIM unit.

10.6.3.3 Input activation register file
The addresses for fetching input activation data are represented by the positions in the valid-position pairs. The positions are supplied to the input activation register file from where the input activation channel index is sent to the AIM unit.

10.6.3.4 Compute path
The weight data from the weight register file and input activation data from input activation register file are transmitted to the compute path. The compute path is made of three multipliers and three psum accumulators.

10.6.3.5 Address path
It has two components: the address computation unit and the reduction controller. It calculates the addresses. Subsequently, it chooses a psum reduction pattern along with the inputs supplied by a sequence decoder, a weight register file and an input activation register file. The intent is to obtain an output activation address. This address is stored in the output activation address psum register file. During every cycle, the sequence decoder sends three pairs of weight-input activation data together with their indices (h, w) for input activation data and (r, s, k) for weight data. The three pairs of weight-input activation data are guided to the compute path for generation of the psums A, B and C. The weight and input activation indices are transmitted to the address path for address calculations.

10.7 The SNAP system

This system (figure 10.8) is made of three modules.

10.7.1 Modules of the SNAP system

10.7.1.1 Module 1: the controller module
It has three components: the dataflow controller, the core controller and the dispatcher.

10.7.1.2 Module 2: the memory module
It contains a weight multi-banked buffer and an input activation/output activation unified multi-banked buffer. It also contains a data alignment unit and a compression unit. Both the buffers of the memory module receive inputs from a data alignment unit fed by the input/output interface.

The compressed weight and input activation data are obtained from the external memory and aligned in packets. The weight packets are stored in the weight buffer of each core following the configuration of the system. The input activation packets are placed inside the input activation buffer. The input activation/output activation

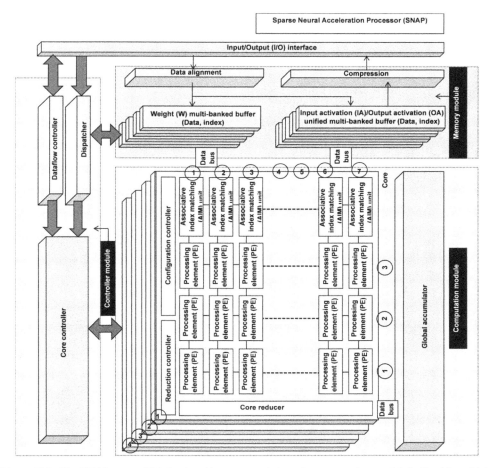

Figure 10.8. The SNAP system high-level architecture consisting of a computation module built with multiple cores, a controller module for configuring the compute cores and coordinating communication with the external interface, and a memory module composed of multi-banked input/output activation buffers that are shared between the compute cores and unshared weight buffers of each compute core.

unified multi-banked buffers send their outputs to a compression unit. This unit supplies the outputs to the input/output interface. Hence, the output activations undergo compression before they are written to external memory.

10.7.1.3 Module 3: the computation module

It has two components: a core and a global accumulator. A 2.3 mm^2 SNAP chip is fabricated in 16 nm complementary metal-oxide semiconductor (CMOS) technology. The chip has four cores containing 7 × 3 array of PEs. Inside each core, 7 AIM units are shared between 3 PEs in a line in a time-multiplexed scheme. Each PE contains 3 multipliers and 1 sequence decoder which total up to $4 \times 7 \times 3 \times 3 = 252$ multipliers; the SRAM is 280.6 kB. Each PE yields psums. These psums are accumulated by the core reducer. Before final writing back, a global accumulator further accumulates the received psums.

10.7.2 Efficiency of the system

The measured peak effective efficiency of the chip is 2.55 $TOPsW^{-1}$ on synthetic sparse workloads and 3.61 $TOPsW^{-1}$ on a pruned network; TOPs = trillion operations per second (Zhang *et al* 2021).

10.8 Summary and the way forward

10.8.1 Highlights of chapter 10 at a glance

(i) The concept of sparsity in neural networks was introduced. Sparse and dense matrices were defined. The advantages of sparsification of a neural network were clarified.

(ii) Pruning-induced sparsity was elaborated upon. Kinds of sparsity, viz., weight and activation sparsity, structured and unstructured sparsity were distinguished.

(iii) Acceleration of 2:4 fine-grained sparsity pattern was deliberated (Mishra *et al* 2021). The workflow for network training was planned. The practicalization of 2:4 sparsity adoption by the NVIDIA Ampere GPU architecture was explained (NVIDIA 2021).

(iv) Realization of an EIE by deep compression was studied (Han *et al* 2016). Its implementation methodology was explicated. The roles of different components were mentioned. The components are the leading non-zero detection node, activation queue, the pointer read unit, the sparse matrix read unit, the arithmetic unit and the activation read/write unit.

(v) An SCNN accelerator was considered (Parashar *et al* 2017). The novelties introduced, SCNN parts and operation were explained. The front-end and back-end operations of a SNAP were discussed (Zhang *et al* 2021). The working of its associative index matching unit was summarized. Its PE consists of a sequence decoder, a weight register file, an input activation register file, a compute path and an address path. The functioning of the controller, memory and computation modules was discoursed. Its operational efficiency was highlighted.

(vi) Sparsity-aided efficient hardware accelerators for CNNs inference task are especially beneficial for edge devices. These devices typically have limited memory and compute resources. They impose strict requirements on energy usage. The inference task is performed both in the cloud and at the edge devices. The operations are speeded up if there are lots of zeros in the matrices. Presence of zeroes implies that some multiplications may be neither computed nor stored.

But sometimes near-human or better accuracy is compulsory. Examples of such tasks are video processing, speech recognition and natural language processing. In such demanding circumstances, the training phase involves large datasets with several weight-update iterations. Therefore, it is likely to take several hours or days to complete. The lengthy process is inconvenient and needs to be shortened. To reduce the time, the neural network training may be performed in the cloud or on a large cluster of machines (Soltaniyeh *et al* 2022).

(vii) Keywords for this chapter are: 2:4 sparsity, EIE, SCNN, SNAP.

10.8.2 Getting ready to begin chapter 11

We devoted two successive chapters to CNN processors. At this juncture, it is opportune to divert our attention to another important subject, that of processing graph algorithms. The processing of graph algorithms is very useful in a wide range of applications. These are the applications where data is represented as a graph, and where discerning patterns and relationships in data are insightful.

References

Han S, Mao H and Dally W J 2016 Deep compression: compressing deep neural network with pruning, trained quantization and Huffman coding *4th Int. Conf. on Learning Representations, ICLR 2016(2–4 May) (San Juan, Puerto Rico)* pp 1–14

Han S, Liu X, Mao H, Pu J, Pedram A, Horowitz M A and Dally W J 2016 EIE: efficient inference engine on compressed deep neural network *Proc. ACM/IEEE 43rd Annual Int. Symp. on Computer Architecture (ISCA) (Seoul, South Korea, 18–22 June)* (IEEE Computer Society) pp 243–54

Mishra A, Latorre J A, Pool J, Stosic D, Stosic D, Venkatesh G, Yu C and Micikevicius P 2021 Accelerating sparse deep neural networks 1–18 arXiv: 2104.08378v1

NVIDIA 2021 *NVIDIA Ampere GA102 GPU Architecture* (Second generation RTX) pp 1–53

Parashar A, Rhu M, Mukkara A, Puglielli A, Venkatesan R, Khailany B, Emer J, Keckler S W and Dally W J 2017 SCNN: an accelerator for compressed-sparse convolutional neural networks *ACM/IEEE 44th Annual Int. Symp. on Computer Architecture (ISCA)(24–28 June) (Toronto, ON, Canada)* pp 27–40

Soltaniyeh M, Martin R P and Nagarakatte S 2022 An accelerator for sparse convolutional neural networks leveraging systolic general matrix-matrix multiplication *ACM Trans. Archit. Code Optim. (TACO)* **19** 1–26

Zhang J -F, Lee C-E, Liu C, Shao Y S, Keckler S W and Zhang Z 2021 SNAP: An efficient sparse neural acceleration processor for unstructured sparse deep neural network inference *IEEE J. Solid-State Circuits* **56** 636–47

Chapter 11

Graph analytics processor for graph algorithm computations

Graphs are valuable tools for real-world data modeling. Graph algorithms are a set of instructions that traverse the nodes of a graph to determine specific nodes or paths between two nodes. The design, analysis, and implementation of graph algorithms helps in leveraging relationships within a dataset to uncover hidden connections and reveal crucial factors for developing intelligent solutions. Problems faced by common processors in handling graph algorithms are elucidated. Multi-core and networked parallel processors are able to handle these algorithms with limited success. So, the graph processor evolved as a convenient solution. Its parallel architecture is emphasized and the advantages of implementation of graph algorithms as sparse matrix operations on a sparse matrix algebra instruction set are mentioned. An important module in the graph processor is the matrix reader and writer module. Matrix data formats and reader/writer design considerations are outlined. Other modules of the graph processor include the arithmetic-logic unit (ALU) module, the communication module, the node controller module and the systolic merge sorter module. A high-bandwidth, low-power communication network is used in this processor. The graph processor is prototyped on an field-programmable gate array (FPGA) at MIT Lincoln Laboratory. The GraphCore intelligence processing unit is described. The Intel PIUMA architecture is introduced.

11.1 Introduction

As we have seen in the foregoing chapters, neural network computations are facilitated by neural processing units. On the same lines. graph algorithm calculations are efficiently executed by graph processors. These are actually processors of graph algorithms that are specially designed by taking cognizance of the obstacles encountered when regular processors are deployed for this purpose, and making the coerced improvements to expunge their shortcomings. This chapter will unfold and demystify the complexities of graphical analysis by:

(i) defining graphs and graph algorithms,
(ii) highlighting the formidable problems faced when normal processors are used for computation of graph algorithms, and
(iii) suggesting admissible remedies for laying the groundwork on the way headed for the development of the graph processor.

11.2 Graph and graph algorithm

11.2.1 Graph

A graph is a non-linear data structure. It consists of vertices or nodes denoted by V, and edges in the form of lines or arcs. These edges hook up any two nodes in the graph, and are denoted by E (Deo 2016). Hence a graph G is represented by the symbol G (V, E). The graph evokes the relationship between different entities. The entities of interest for drawing graphs differ widely. They can be a person, place, thing or a property of a material. Graphical data are omnipresent such as in graphical analysis of the structure of world wide web (Boldi and Vigna 2004), graphic manipulation for assembly of genomic sequences (Zerbino and Birney 2008), drawing constitutional graphs in chemistry (Balaban 1985), etc. A close scrutiny of graphs helps in various ways, e.g.,

(i) In performing a social network analysis such as the relationship among a group of friends (figure 11.1).
(ii) In the inspection/visualization of a transportation system of roads linking different routes to a destination.
(iii) In knowing about the coordination, cooperation and camaraderie spirit in team sports, e.g., between hockey, soccer or cricket players on the field.
(iv) For examining the topology of computer networks regarding optimality of interconnections between the switches and routers for their better utilization, and so forth.

Drawing graphs, processing them and pondering methodically over the results furnishes treasured information and insights into the dynamics of real-life situations. The information received provides assistance in making necessary corrections and improvements for inching towards the goal. The adage, 'A picture is worth a thousand words' aptly expresses the capability of graphs because they can convey an idea or tell the inside story of an incident more effectively than a verbal description by providing more information in a terse manner similar to putting the ocean in a pot.

Graphs are placed into two classes: undirected and directed according as their edges are bi- or uni-directional (figure 11.2):

(i) Undirected graph: In an undirected graph, the edges do not have a specific direction. Their bidirectional nature implies non-existence of a parent-child relation between the entities.
(ii) Directed graph: In a directed graph, the edges have specific directions. So, unidirectionality of action is implied.

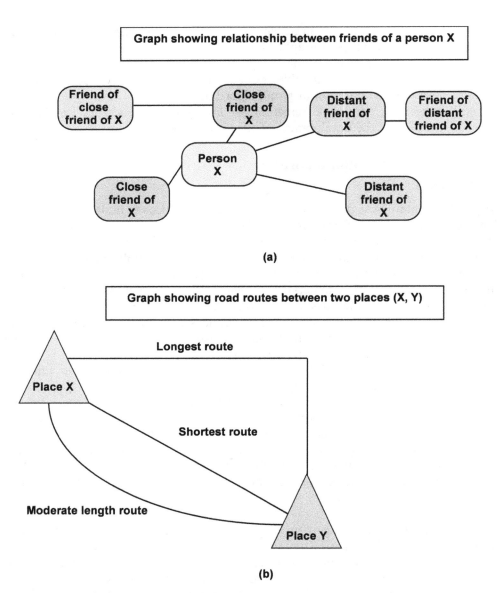

Figure 11.1. Graphical representation of relationships showing: (a) a social interaction between a group of friends, and (b) the road route connections between two places. In (a), a person *X* is related to close and distant friends as well as to the friend of close friend and the friend of distant friend. In (b), two places *X* and *Y* are connected by the shortest, longest and moderately long routes.

> *What information do the graphs convey?*
> *Graphs represent complex data in a simple way,*
> *Offering an eloquent visual display*
> *Of the scenario they portray,*

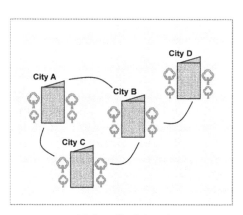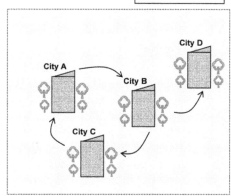

Types of Graphs

(a) An undirected graph

(b) A directed graph

Figure 11.2. Types of graphs: (a) undirected and (b) directed graphs connecting cities. The graph in (a) can be imagined as showing persons visiting different cities for the sake of tourism so that they stay temporarily for a few days at the cities visited, and return to their native abodes, while that in (b) shows persons permanently moving between cities on a transferable job basis so that they reside at their destination cities. Therefore, in (a), the movements in both directions are allowed and no arrow is marked, while in (b) movement takes place in one direction only, as indicated by the arrow.

And a concise approach to data documentation,
For easy data comprehension and visualization
Whereby patterns and relationships between entities are highlighted,
Any unusual deviations from standards are easily sighted.

11.2.2 Graph algorithm

A graph algorithm is a suite of commands or procedures used for analysis of a graph to solve real-life problems (Needham and Hodler 2019). A few examples of graph algorithms are:

(i) Bread-first search algorithm: This is a graph-traversing algorithm along the breadth direction. It is used to determine the shortest path between two nodes.

(ii) Depth-first search algorithm: This is a similar algorithm to the bread-first search algorithm. It differs from that algorithm in the aspect that here the search is carried out while moving downwards from the root node instead of travelling sideways.

(iii) Dijkstra's and A* algorithms: These are iterative algorithms. They are applied for finding the shortest path connecting a source node to every other node in the graph.

11.3 Overcoming the difficulties encountered in processing graph algorithms

11.3.1 Awkward predicaments of using common processors for handling graph algorithms

Common processors are able to handle graph algorithms satisfactorily and perform fairly well when the database is small or moderate in size. But for a large database, they do not provide adequate throughput and fail despondently. Lack of compatibility of the processor architecture with the computational flow of the graph is the reason for this failure. Most processors have a cache-based memory. The memory access patterns during graph processing are generally haphazard in nature. Therefore, high cache miss rates are observed with these processors. Hence, these processors are debilitated by cache-miss issues.

Let us explain the episode of cache miss occurrence in more detail. Cache miss is an event which is opposite of the cache hit. In a cache hit, data requested by an application from the cache is successfully retrieved from the cache. In a cache miss, the requested data is not currently found in the cache memory. So, the application makes a second attempt. This time it strives to locate the data in the slower secondary database. If this trial is able to trace out the data, the data found is copied to the cache in anticipation of a possible request for the same data in the near future.

Cache hit ratio is defined as the ratio between the number of cache hits and the total number of efforts made = (the number of cache hits + number of cache misses). It is a metric which indicates the extent or degree of fulfilment of content requests with respect to the total number of requests received, irrespective of positive or negative results obtained. The time delay caused by cache misses increases the latency of the system and proportionately degrades its performance.

11.3.2 Multi-core and networked parallel processors

Two approaches that have been ventured upon to deal with graph algorithms are:

(i) Using multi-core processors: The impediments faced with common processors are alleviated by using multi-core processors. A graph algorithm will obviously run faster on a multi-core processor than a single-core processor. But a multi-core processor too has a cache-based architecture, and is crippled by cache-related issues. So, the advantage gained by using it is restricted in extent offering only a trivial improvement over a normal processor. In contrast to the multi-core central processing units (CPUs), graphical processing units (GPUs) having more threads and a much broader memory bandwidth, appear more promising.

(ii) Using networked parallel processors: Graph processing is significantly speeded up by distribution of computation among several processors, as accomplished in a networked parallel processor. Parallelism, as exhaustively discoursed in chapter 4, is much like division of labor, the separation of a huge work into tasks given to different persons. Regrettably, the increase in speed of the processor attained by parallelism flattens off with a small number of processors. This happens because the non-localized database structure heavily loads the inter-processor communication requirements.

The parallelism and high memory access bandwidth of GPU are exploited for graph processing (Shi *et al* 2018). The main issues are examined and confronted, e.g., layout of data, pattern of accessing memory, mapping of the workload and programming GPU for these applications. A GPU excels in graph processing as long as the data can be correlated with the single instruction, multiple data (SIMD) model (Liu and Huang 2019). Nonetheless, graph algorithms, e.g., breadth-first search are unable to derive the full benefit of GPU capability. Irregularity in accessing memory and disorderly control flow are implicated for these failures. Furthermore, compared to the gigabyte/terabyte scale memory of server machines, GPU has a limited memory capacity making it useful for graph processing on small datasets only (Shao *et al* 2022). Memory oversubscription occurs when the memory demand exceeds the available memory on GPU. The unified virtual memory (UVM) technology solves the oversubscription problem by provision of a single virtual address space between the CPU and the GPU across which automatic data migration can take place via demand paging, a scheme for memory management. But it introduces overhead loading, e.g., address translation and page fault handling. CGgraph system vanquishes GPU-memory oversubscription by pulling out a subgraph to be loaded into GPU memory on one occasion (Cui *et al* 2014).

The aforementioned remarks suggest the need of a dedicated approach that is technologically put forward singularly for graph algorithms. Every cloud has a silver lining. Looking at the inadequacies of both the approaches of multi-core and networked processors, an advanced multiprocessor architecture is designed for optimal performance on large databases based on sparse matrix operations (Song *et al* 2010a, 2010b). Sparse matrices are those which contain a large number of zero elements, as discussed in section 10.2.1. We will try to systematically grasp and shed light on the elegant techniques used for graph processing by Song and co-workers in this chapter. For clarity, we shall move in a piecemeal manner following their footsteps.

11.4 Structural dissection, and understanding the distinctive features of graph processor architecture

11.4.1 Performing computing operations parallelly with a node processor architecture

The graph processor works with a parallel algorithm based on the distributed sparse matrix operations. By a distributed sparse structure is meant a configuration in which the various operations are carried out on sparse matrices in a distributed memory system. The distributed memory system is a multiprocessor system consisting of independent processors called nodes. Each node of this system has its own local memory. Data is exchanged between fellow processors via a network interface.

The distributed memory concept becomes readily comprehensible by describing the example of an office in which each officer is working in his/her own separate cabin and can communicate with other officers over the telephone. This is because in a distributed memory processor, each processor has its independent memory and can communicate with other processors over a network. The knowledge about distributed memory will be helpful in appreciating the architecture of the graph processor as we proceed further.

A common criterion to declare a matrix as sparse is that

$$\text{Number of non} - \text{zero elements} \approx \text{Number of rows or columns} \qquad (11.1)$$

For quantification of sparsity of a matrix, a score S is defined as

$$S = \frac{\text{Number of zero values inside the matrix}}{\text{Total number of elements constituting the matrix}} \qquad (11.2)$$

Because of sparsity, only a few non-zero values need to be stored in the memory, and the computation time is reduced by creating a data structure that navigates through the non-zero elements of the matrix only.

The full and sparse matrix representations of a graph are shown in figure 11.3. A graph $G(V, E)$ is considered having vertices V and edges E. The graph of figure 11.3 (a) is a directed graph. It has vertices $V = 0,1,2,3,4$, and edges E connecting these vertices. The edges are the arrows with their wedge-shaped heads pointing towards the allowed directions. The full matrix for the graph of figure 11.3(a) is given in figure 11.3(b). The different elements of this matrix denote the edges joining any two identified vertices, e.g., an element A_{ij} of this matrix represents the edge joining the vertex i to the vertex j. In figure 11. 3(b), the numbers for vertices 0, 1, 2, 3, 4 are marked in the leftmost column and topmost row. The full matrix for the graph is written by visual observation of the connections between the ith vertex numbered 0,1, 2, 3, 4 along the ordinate to jth vertex numbered 0, 1, 2, 3, 4 along the abscissa. The precept followed for writing the matrix is that a connection is written as '1' wherever we see an arrow joining the vertices from i to j. A connection is set as '0' whenever there is no arrow joining from i to j. Thus, a table is constructed in figure 11.3(b) showing the presence and absence of connections between any two

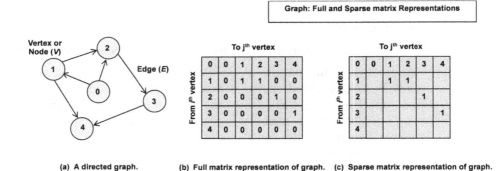

Graph: Full and Sparse matrix Representations

(a) A directed graph.

Full matrix representation (b) — To j^{th} vertex / From i^{th} vertex:

	0	1	2	3	4
0	0	1	1	0	0
1	0	0	0	1	0
2	0	0	0	0	1
3	0	0	0	0	0
4	0	0	0	0	0

Sparse matrix representation (c) — To j^{th} vertex / From i^{th} vertex:

	0	1	2	3	4
0		1	1		
1				1	
2					1
3					
4					

(b) Full matrix representation of graph. (c) Sparse matrix representation of graph.

Figure 11.3. Illustrating the matrix representations of a graph: (a) the directed graph, (b) full matrix for the graph and (c) the corresponding sparse matrix for it.

vertices in the directed graph of figure 11.3(a). This table is the full matrix representing the directed graph of figure 11.3(a).

The equivalent sparse matrix for this graph is constructed in a duplicated tabulated pattern simply by deleting all the zeroes in the full matrix of figure 11.3 (b). The resulting sparse matrix after dropping zeroes is given in figure 11.3(c). This sparse matrix is tantamount to the full matrix in figure 11.3(b). Hence, the sparse matrix is the matrix obtained by obliteration of all zero entries in the full matrix for the graph. Sparse matrix operations for graph algorithm implementation are applied after completing the following steps:

(i) writing the full matrix for the graph, and
(ii) transformation of the full matrix to the sparse matrix format.

Acceding to the sparse matrix approach confers two-fold advantages:

(i) An immense simplification of the instruction set for graph processing is manifested.
(ii) Sophistication in the architectural design of the multiprocessor for execution of graph algorithms is diminished.

These advantages will be elaborated further in section 11.4.2 below.

The recognizable element-level operators that are used in the matrix operations are the multiply and accumulate operators for matrix multiplication. As an alternative for these operators, arithmetic or logical operators are used for graph algorithm computations. These include the logical conjunction, logical inclusive disjunction, logical exclusive disjunction, minimum, maximum, etc. Thus, the graph algorithm acquires the demeanor of a chain of sparse matrix operations.

The aims and scope of the graph processor are now clearly spelt out. It is a computer architecture using graph algorithms and the related calculation subroutines for the analysis, management and manipulation of graph structures in large databases. To grease the wheels of graph processing, the graph algorithms will be converted to the sparse matrix format. Definitely, large matrices cannot be accommodated in the memory of a single processor. So, a single processor cannot provide the prerequisite large throughput, and we have to seek the services of multiple processor nodes, as we had indicated in the beginning of this section in the backdrop of the distributed memory idea. We shall distribute the massive data for these matrices over a large number of processors located at different nodes or intersections. These are the so-called node processors defined earlier (figure 11.4).

Taking recourse to multiple-node processor paradigm amounts to building an array of sparse matrix processors. Each sparse matrix processor is a self-contained, self-sufficient and self-governing computer comprising arithmetic and matrix calculation circuits that are supported by a private memory. Fast interconnections among the node processors are established through a high-speed broadband computer network. Hence, the interconnections are hooked up by coupling the node processors with the global communication network. The node processors are also fastened to the global control processor. The global control bus makes this connection. The interconnections thus secured permit the node processors to

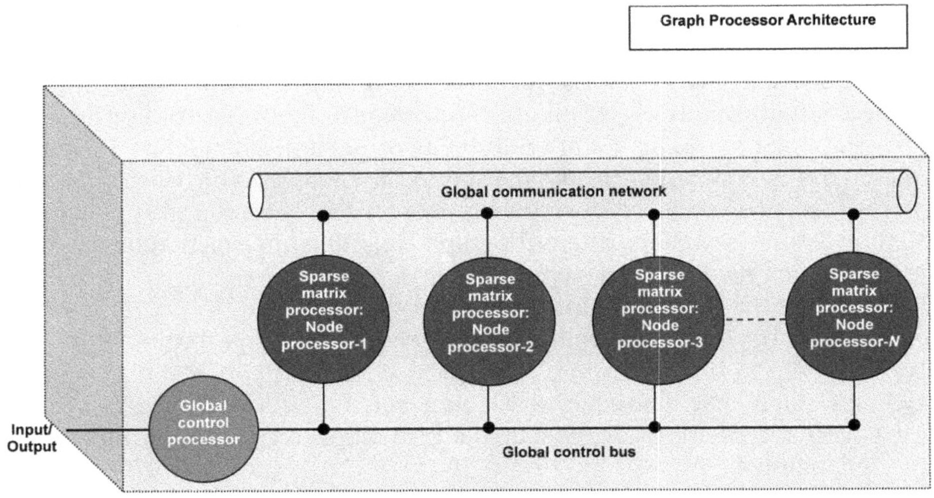

Figure 11.4. Parallel architecture of a graph processor. The constituent processors and their interconnections are shown. The graph processor is formed by combination of N node processors. These node processors are labelled as node-1 processor, node-2 processor, ..., node-N processor. A global communication network is the link joining all the node processors at their top ends. The bottom ends of the node processors share a global control bus. This bus terminates at one end of a global control processor. For interfacing with an input/output system, the given system is tied with the free opposite end of the global control processor.

communicate among themselves swiftly, thereby enabling coordination of their actions to work towards achieving a common objective. Thus, the graph processor operates on a tremendously parallel computing architecture with a distributed memory on the processor chip, and a very high bandwidth internode processor communication channel.

At the first sight, and to all intents and purposes, the generic high-level architecture of a graph processor ostensibly appears to display resemblance to a usual multiprocessor system. However, this speculation is not true. On deeper thought it is realized that the graph processor architecture is significantly different from the parallel architecture enactments. A key difference is the instruction set of the graph processor which markedly diverges from the instruction sets of common-place computers. This is expected because the instruction set of the graph processor is based on sparse matrix algebra operations, rather than full-matrix functions. A few instruction-kernels germane to the discussion may be mentioned. These apply to instructions for sparse matrix-matrix multiplication operation. Operations to add, subtract and divide are included too. The kernel is a computer program at the core of the operating system (OS) of the computer, so-named because it is like a spore in a hard shell which lies in the OS and controls all its chores. Sparse matrix formulation not only influences the software, it also impacts the hardware, as we shall explain below in section 11.4.2. In sections 11.4.4 and 11.5.3, we shall see that the data-intensive demands of graph processing make it necessary to build a specialized communication network for this processor having a large bandwidth and therefore higher data rates and higher network capacity, that works with low power consumption and is supplemented by a routing algorithm with small message sizes. These aspects noticeably set this processor apart from a normal multiprocessor.

11.4.2 Advantages accrued by graph algorithm processing in sparse matrix formalism through the instruction set created on a sparse matrix algebra premise

Sparse matrix algebra is the mathematical branch dealing with properties of sparse matrices using which the wastage of time in addition/subtraction or multiplication by zero is avoided. The benefits of representing graphs by sparse matrices and running graph algorithms with instruction sets framed in terms of sparse matrix algebra are succinctly explained below:

  (i) Shortening of the code length and enhancement of software development efficiency: Traditional software for graph algorithms uses lengthy instruction sets. In the sparsity notion-based approach, the number of lines of code is much smaller than the amount of code written for traditional software. Such shortening of code length improves the overall competence of software development. It must be noted that the code-conciseness advantage-induced upliftment of software development efficiency does not necessarily result in a higher throughput of computation in the time-honored processors. Consequently, the processor must be redesigned afresh in the light of sparse matrix operations.

  (ii) Convenience of simplified instruction set for sparse matrix operations than non-sparse operations on graph algorithms: The instruction set of sparse matrix-based graph algorithms requires surprisingly few base instructions. So, it can be enormously simplified.

  (iii) Easier design of necessary parallel processor and facilitation of sparse matrix processor architecture design: Sparse matrix operations help in the designing of the processor architecture. The design aid comes in the form of ease of conjuring up a picture in the mind's eye rendering the computational flow and data movement of sparse matrix operations running on parallel processors. This paves the way towards simpler and straightforward design of a parallel processor that computes sparse matrix operations in preference to general graph algorithms. The benefit reaped assists processor innovators to think up and devise novel architectures and unprecedented designs almost effortlessly.

11.4.3 Dissimilarity between the accelerator-based architecture of individual node processor and the von Neumann/Princeton model

Consider the situation that transpires when a common microprocessor following the von Neumann/Princeton architecture is used for accomplishing all the subtasks dealing with signal processing for graphs such as memory access to input instructions/data, communication-related processing with a bus interface unit for external transfers, performing arithmetic operations (addition/subtraction, multiplication/division), executing logical operations (AND, OR, NOT, Exclusive OR, etc), and chartering control functions (supervision of other devices and decision-making). It is common knowledge that serial processing of tasks by this processor consumes many clock cycles for completion, thereby lowering the overall computational throughput.

Now we concentrate our attention on the graph processor. The architecture of the graph node processor marks the pinnacle of a thorough re-examination and reworking of basic computer architecture, keeping in mind the essential and peculiar requirements of graph processing. For designing the graph node processor, the various computing tasks are decomposed into subtasks for parallel execution by distribution across numerous modules. Each subtask is allocated to an individual accelerator module. This accelerator module is referred to as a node processor (Song *et al* 2013, Song 2014). Figure 11.5 displays the architecture of a node of the processor.

Each separate processor node in a node processor module is equipped with a resident cacheless memory dwelling adjoining it. It is called the local memory. This cacheless memory offers the convenience of transferring the necessary information directly to the processor without requesting for help to an intermediary fast-reclamation storage basket. The cacheless memory represents a significant demar-cating feature of the node processor architecture from the regular von Neumann computer.

The main guiding principle underlying the design of modules of the node processor is to achieve as fast throughput as possible. Custom-made architectures are used to this end. Highly parallel pipelined computations are enabled by this

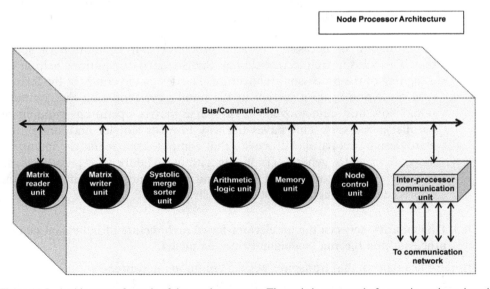

Figure 11.5. Architecture of a node of the graph processor. The node is composed of a matrix reader unit and a matrix writer unit. Other vital components of the node are a systolic merge sorter unit, an ALU, and a memory unit. All these units are connected by two-way links to a common bus. A node control unit and an internetwork communication unit coordinate the activities of the node. Two-way signal traffic flows between the internetwork communication unit and the communication network.

design sowing the seeds to engender the fostering of specialized accelerator modules. These dexterous modules are capable of yielding much higher output than the old-style architectures.

In the ideal situation, fastest data rate viable should be targeted during design. To achieve the fastest rate, one matrix element or one partial product needs to be processed within a single clock cycle. Effective throughput is thereby obtained. A multiplicity of parallel versions of these modules further expedites the computing by opening the door for processing of a larger number of matrix elements or partial products per clock cycle

11.4.4 Communication network with large bandwidth capability and low power requirement

Noteworthy features of the communication network of the graph processor are its large bandwidth which aids the transferring of a large amount of data in a given interval of time and its low power loading which drains less electrical energy. To make clear, bandwidth is the maximum rate of transference of data. It governs the theoretical upper boundary to the volume of data that is transferable. The network can be envisioned as a water pipe. Water flowing through it represents data. The wider the pipe, the greater the quantity of water flowing through the pipe in a given timeframe. When the bandwidth is narrow, the packets of data passing through the network experience difficulty as if they are moving in a traffic congestion on a crowded road so that the time for reaching the destination is lengthened. Obviously, insufficient bandwidth leads to slower data traffic. A related parameter to bandwidth is speed. The network speed is the actual rate of data transference at a certain instant of time. It is governed by several factors besides bandwidth including the number of processing devices working on the network, the type of hardware and the operating system. Any delay incurred during transference of data is measured by latency.

The communication network of the graph processor attracts close attention by its inimitable virtue of being designed for transmitting small messages. In accord with the large bandwidth, six dimensional (6D) toroidal communication network, the necessary optimization of the parallel graph processor hardware has also been carried out. The reader may be aware that the torus is a surface created by revolving a circle in 3D space about an axis that is coplanar with the circle. The generalized version of torus is an n-dimensional torus, which is the product space of n circles.

Ordinarily, the content of a message is one matrix element or one partial product. Data value along with row and column indices are the principal components constituting the message. This practice deviates from the usual trend characterized by maximization of message sizes. Maximization is desperately needed in order to reduce the overhead allied with data movement to the bare minimum attainable limit.

A statistical routing algorithm with small message sizes has also been formulated to take care of routing issues in the network. This algorithm ameliorates the communication throughput for graph processing. Furthermore, as already said, the network hardware itself has extensively wider bandwidth than that of regular parallel processors. The cooperative action of the algorithm and the hardware is able to satisfactorily meet the condition of the large bandwidth need that must be compulsorily imposed for judicious management of the exigencies of graph processing.

11.5 Exclusively developed modules of the graph processor

11.5.1 Modules for matrix reading and writing: the matrix reading/writing modules

11.5.1.1 Data formats used for matrices

Data formats determine the structuring of data for its usage, analysis and storage to secure its faultless processing and interoperability between systems. They act as a common language of exchanging data between different parts of a device or system. The structuring of data is defined by rows and columns. The following formats are used for storing the matrices economically:

    (i) Coordinate list (or tuple) format: In this format of matrix, also called the transaction format and abbreviated as COO, the repository of the data, the row index, and the column index is actualized in the form of triples. The matrix is represented as a set of triples (i, j, x) where x is the number found in row i and column j; the symbols i and j denoting the row and column indices are row and column numbers, respectively, indicating their positions in the matrix.

    (ii) Compressed sparse row (CSR) format: Also called the compressed row storage (CRS) or Yale format, in this format, the sparse matrix is squeezed by storage of the non-zero elements and the matching row indices. Three one-dimensional arrays constitute its structural parts, viz., the data, index and index pointer. The data part stores the non-zero values, the index part stores the column indices of elements in the data array, and index pointer part stores the index pointers to the beginning of each row in the data array. Thus, the element data and column index are stockpiled under the guise of pairs in an array format. This kind of storage offers two advantages, firstly helping in preserving memory space and secondly, improving the data lookup and foraging. As the row onset address for each column is stored in an extra array, these pointers facilitate in tracking down the accurate memory locations in which the rows are deposited.

    (iii) Compressed sparse column (CSC) format: The CSC format is analogous to the CSR format. Like the CSR format, the three one-dimensional arrays constituting its structural parts are the data, index and index pointer. The data part stores the non-zero values, the index part stores the row indices of elements in the data array, and index pointer part stores the index pointers to the beginning of each column in the data array. Obviously, the primary differentiating feature of this format with respect to the CSR format is that compression of columns is done in this format in place of the rows.

Comparison between the CSR and CSC formats: The CSR format keeps the non-zero elements by row, and preserves the row indices and column pointers in a contiguous block. It is favored for a matrix which has a large number of non-zero elements in each row, and is befitting for row-based access.

The CSC format keeps the non-zero elements by column, and preserves the column indices and row pointers in a contiguous block. It is more efficient for a matrix which has a large number of non-zero elements in each column, and is befitting for column-based access. The characteristics of the sparse matrix together with those of the application dictate which of the two formats must be used, CSR or CSC (Massed Compute 2024).

11.5.1.2 Design considerations beneficial to matrix reader and writer
These units are geared towards efficient reading and writing of the matrix data from the memory. The design double-checks that all the overhead operations are performed without engaging the human operator. Additionally, feeding of any supplementary instructions is abjured. These operations are:
  (i) arranging the matrix element data and indices in the accepted formats for writing the data,
  (ii) producing the pointer arrays for CSC and CSR formats for writing the data, and
  (iii) creating the requisite indices of matrix elements for reading the data.

As a consequence of taking these measures into account, the intricacies accompanying the sparse matrix read and write operations are significantly curtailed. Evidently, the operations taking place at the interface of memory with the remaining system are hastened appreciably resulting in quicker interactions.

11.5.2 Module for arithmetic and logic operations: the ALU module

Interestingly, the ALU (arithmetic-logic unit) module does not operate with a register file as one is tempted to think in traditional terms. This attribute of the ALU module sharply contrasts with prevalent processor architectures. In distinction to these architectures, the ALU module is designed to operate on a stream of sparse matrix elements or partial products. The streaming approach gets rid of the delays encountered due to register load operations. By avoiding these delays, the computational throughput is much elevated.

Clarifying further, the designated arithmetic or logical operations are carried out on a data stream, depending on the indices. As an illustration, the ALU module accumulates consecutive matrix elements only on the occurrence of precise matching of the element indices. Execution of computations by these matrix operations solely for the selected situations conforming to exact matching of the indices is a significant advantage. It greatly contributes towards the appropriateness of this feature for sparse-matrix add and multiply operations.

11.5.3 Module for communication between processor nodes: the communication module

This module oversees communication between processor nodes. A communication message is written by taking a particular matrix element or partial product. This message consists of two parts: the matrix element in coordinate format, and a header containing the destination processor address. Apart from these, it includes the error detection and correction bits and other pertinent information, notably the precedency of the message. After transmission of a communication message to the global communication network, it is dispatched to the destination nodes.

Other obligations of the communications module include:

   (i) deciphering the received messages,
  (ii) performing the correction of errors, and
 (iii) delivering the matrix element or partial product into the node in the desired coordinate format.

The communication network is designed as a network that groups data into packets for transmission from their source to destination. Such a network is known as a packet routing network. Its optimization is done for supporting small packet sizes. The degree of packet trimness is contemplated as keeping the packet sizes as small as a single sparse matrix element. Overcrowding in the network is prevented by orchestrating the network scheduling and protocol in such a fashion that sequential communication packets from a node will have stochastic termini. The optimization performed for minuscule message sizes together with scattered destination routing, abates message contentions. It boosts the overall communication throughput of the network.

The design strategy adopted here differs from a run-of-the-mill multiprocessor message-routing scheme. We know that the stereotyped scheme is based on much larger message sizes and globally reconciled routing. This is done essentially to reduce the message-routing expenditure, which does not directly contribute to service. Notwithstanding, large message-based communications are often problematic to route. They can have a relatively high message disputation rate. The high rates are caused by the long intervals of time during which the entwined communication links are tethered.

Curiosity is aroused to seek a comparison of randomized destination communication with unique destination communication. When such a comparison is made for the case when both routing methods are based on small message sizes, it is discovered that the unique destination routing has a roughly similar message contention rate as the widespread routing algorithms that rely on large message sizes. However, the randomized destination routing proves its utility by provision of a much higher data transfer rate and efficiency of utilization of the network when a similar network is used.

11.5.4 Module for data storage: the memory module

There are various feasible options from which one can choose the type of memory device for implementation of the node processor memory. The available options are: static random-access memory (SRAM), dynamic RAM, and synchronous DRAM.

SRAM, DRAM and flash memory technologies have different pros and cons. SRAM is flip-flop based while DRAM and flash memories use capacitors for bit storage. Both SRAM lose data when power is switched off but flash memory is permanent with a limited lifespan. SRAM is faster and more expensive than DRAM. Flash memory is slower than RAM. But it is less costly and consumes less power. For data storage over a lengthy period of time, a nonvolatile memory, such as a flash memory is useful, and becomes a backup asset for occasions where the storage requirement is high.

11.5.5 Module for controlling the node processors: the node controller module

Establishment and coordination of the sparse matrix operations are two activities which fall under the dominion of the node controller module, both its hardware and software. It acts a single entity that performs the key control functions for smooth functioning of the system. Prior to a sparse matrix operation, the controller module has to execute a preparatory step. In this preliminary step, it loads the control variables into the control registers and control memory of the accelerator modules. For this job, it uses a local control bus. These control variables comprise:

(i) the sparse matrix operations that are planned to be carried out,
(ii) the address locations of the matrix memory storage,
(iii) the mapping delineating the matrix distribution, and
(iv) other applicable information of relevance to the concerned operation.

The operations of timing and controlling the nodes too fall under the purview of the controller module. Apart from the above activities, processing tasks that are not well espoused by the accelerator modules are also delegated to the node controller. Amongst these tasks, mention may be made of the duty of creation of an identity matrix and steering the desired verification to ascertain whether a matrix is vacant/unfilled across all the processor nodes. The controller module is appended to the global control bus. The global control bus attends to the functions of loading of data and programs to and from the nodes. The global computation process is maneuvered by the same bus.

The services of a common general-purpose microprocessor are availed to realize the function of the node controller module. This processing is not specialized, rather a set of actions to be followed, largely of a regimen nature. So, this microprocessor does not have any fastidious specifications. It may also have a cache.

11.5.6 Module for systolic merge sorting: the sorter module

The sorter module is a module of paramount importance in graph processing (Song 2012). The reason is that a large proportion amounting to more than 95% of the units of information that the system processes in a given time, measured by its computational throughput, are concerned with the index sorting. This module is used for two purposes:

(i) for categorization of the matrix element indices oriented towards their storage, and
(ii) for discovering the matching element indices during matrix operations.

Cogitating over what exactly are the sparse matrix and graph operations, we notice that they principally involve ascertaining which specific element (s) or partial product (s) must be worked upon. In opposition to this, comparatively few actual element-level operations are executed. A systolic k-way systolic merge sorter architecture is designed. It attains strikingly higher computational throughput than the orthodox merge sorters.

Merge sort is an optimized sorting algorithm, which is in vogue in the industry due to its stability and efficiency. It is particularly advantageous when a stable sort is a priority. Maintenance and preservation of the initial order of equivalent elements in the sorted output is guaranteed by a stable algorithm. Hence, whenever a stable algorithm is necessary, merge sort algorithm is the favorite choice.

Proposed by J von Neumann in 1945 (Knuth 1998), the merge sort algorithm is inspired by the intuition that a problem is easily solved by decomposition into smaller sub-problems, solving the smaller sub-problems parallelly in chorus, and re-composition of the sub-solutions obtained to the smaller sub-problems to get a final answer in less time.

The k-way merge sorter compartmentalizes long sequences of numbers. The compartmentalization is realized by following a recursive 'divide and conquer' method. The step-by-step procedure followed in the merge-sort algorithm is (figure 11.6):

    (i) Divide step: The sequence is subdivided either into k shorter sequences of equal lengths, or sequences of as closely equal lengths as attainable. These shorter sequences are termed subsequences.

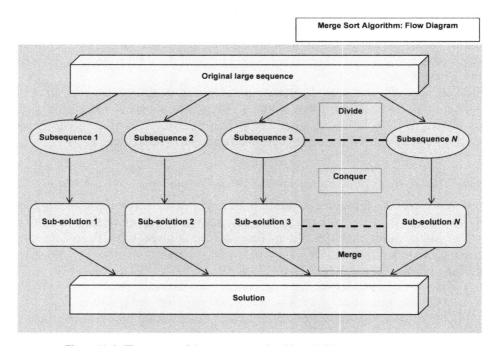

Figure 11.6. Three parts of the merge sort algorithm: divide, conquer and merge.

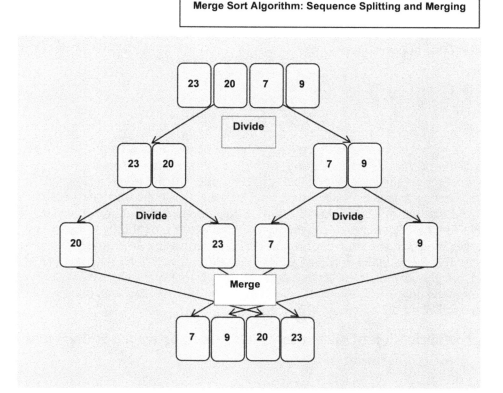

Figure 11.7. Cutting down a sequence into multiple subsequences until each subsequence has only one item, and then merging together the subsequences to obtain a sorted sequence.

  (ii) Conquer step: The k shorter subsequences are sorted individualistically.
  (iii) Merge step: Eventually, the sorted sequences are combined together in the sorted order (figure 11.7) to generate the final result of sorting. Since the subsequences are merged independently, the algorithm is expedient for parallel processing.

Further segmentation of subsequences is also possible. The sorting of k shorter subsequences sequences can be segmented to a greater extent by splitting into k even shorter subsequences, which are essentially smaller sub-subsequences made by breaking the bigger subsequences. These are subjected to recursive sorting by application of the same merge sort algorithm. Recursive repetition of this process is continued relentlessly, and ceases only when the length of the divided sequence becomes unity, which is called atomic length. The number N of memory cycles during which the sorting process takes place is given by the equation

$$N = n \log_k n \tag{11.3}$$

where n is the sequence length for the k subsequences. When $k > 2$, the k-way merge sort is $\log_2 k$ times quicker than the two-way merge sort process, e.g., for $k = 64$, the k-way merge sorter has six times greater sorter throughput than the two-way merge sorter. This happens because

$$\log_2 k = \log_2 64 = 6 \tag{11.4}$$

since

$$2^6 = 64 \tag{11.5}$$

The main hurdle to a k-way merge sorter implementation in a hackneyed processor needs to be underscored. During each step of the merge sorting process, this processor goes through many clock cycles to find out the minimum or maximum value among k entries, whereas the requirement is just the opposite because it is desired that the smallest value of k should be computed within one processor clock cycle for the peak sorter throughput. The cent percent or totally efficient systolic merge sorter achieves this performance requirement using k linear cells of systolic array. It is amenable to easy realization as a field-programmable gate array (FPGA) and integrated circuit (IC) device. Its constitution from repetitive systolic cells communicating with adjacent neighbors only endows the capability for such amenability on the sorter.

11.6 Fabrication of specimen graph processor device and evaluation of its operation

Lincoln Laboratory, Massachusetts Institute of Technology, Lexington, has technologically validated an FPGA version of the graph processor. Mainstream FPGA boards are used for making this model. Each board has one large FPGA memory bank. It also has two 4-GByte DDR3 memory banks. Two graph processor nodes are launched inside it (Song *et al* 2016). An eight-node graph processor knotted together with 1D toroidal network is put into service in a small four-board chassis. The performance of the FPGA prototype processor was critically assessed. The appraisal is done using sparse matrix-matrix multiply operations on power law matrices. The FPGA prototype at Lincoln laboratory displayed momentous performance enhancement over standard processors in graph computational throughput. This processor was measured to be 10 times faster than congruent computers at the 100 Watt scale. It is anticipated to exceed 100 times faster speed at the 1 Megawatt scale (Song *et al* 2010a, 2010b, 2013, 2016, Song 2012, 2014).

11.7 GraphCore intelligence processing unit

An intelligence processing unit (IPU) is a chip specifically designed for graph-related applications by GraphCore Limited, a British semiconductor company. The chip design uses graph as a representation of knowledge data model. This model is for AI-related algorithms, including neural network, Bayesian network, Markov field, etc. The terms 'Bayesian network' and 'Markov field' need explanation. A Bayesian network (BN) is a probabilistic graphical model. In this model, a directed acyclic graph (DAG) is used for representation of a set of variables and their conditional dependencies. The nodes of the DAG indicate variables while its edges illustrate the causal dependences among the variables. A Markov random field (MRF) is a graphical model appertaining to a joint probability distribution. This distribution shows a set of random variables. The interdependency relationship of these variables is represented by an undirected graph, which is a symmetric neighborhood matrix. Thus, in both BN and MRF, the data model of knowledge is framed as graphs. In these graphs, every vertex measures the probability of a particular feature. The edges represent correlation or causation between features (Selwood 2017). As each vertex links to only a few others, the graph is sparse. The classical problem of bundle adjustment (BA) in computer vision concerned with refining a visual reconstruction for estimating the structure and viewing parameters of a scene is shown to be solved extremely fast on an IPU (Ortiz *et al* 2020).

The IPU chip is a memory-centric design. It supports both training and inference (Xie 2018). Memory-centric approaches will be discussed in chapter 12.

11.8 Intel's PIUMA architecture processor

The Programmable Integrated Unified Memory Architecture (PIUMA) is developed by Intel for the Defense Advanced Research Project Agency's (DARPA) Hierarchical Identify Verify Exploit (HIVE) program for advancement of graph analytics (Aananthakrishnan *et al* 2020, 2023). In similarity to the processor of Song *et al* (2016), it features efficient communication and memory accesses, and absence of cache. But in disparity to the same, it has a flexible instruction set architecture (ISA), not restricted to algorithms using sparse matrix algebra.

The PUMA layout shown in figure 11.8 consists of:
   (i) Single-threaded cores (STCs): These are used for single thread performance-sensitive tasks.
   (ii) Multi-threaded cores (MTCs): These are multi-threaded pipelines based on:
      (a) Round robin algorithm in which time slices or time quanta are allocated to different processes in equal portions in a circular order without prioritizing any process.
      (b) In-order pipelining in which instructions are executed in the order of their fetching.
   (iii) A small data cache (DCe) and instruction cache (ICe) and a register file (RF): These are attached to each core.
   (iv) Scratchpad (SPAD): This is a large local high-speed memory to reduce latency by holding rapidly small data items for rapid retrievability.
   (v) Block offload: The STCs and MTCs are combined into blocks, and embellished with an offload engine for core offloads of the background memory operations as the cores handle the principal computing steps.
   (vi) DRAM.

The first PIUMA version is an upfront implementation of sparse matrix dense vector multiplication (SpMV), a computational operation in which a sparse matrix is

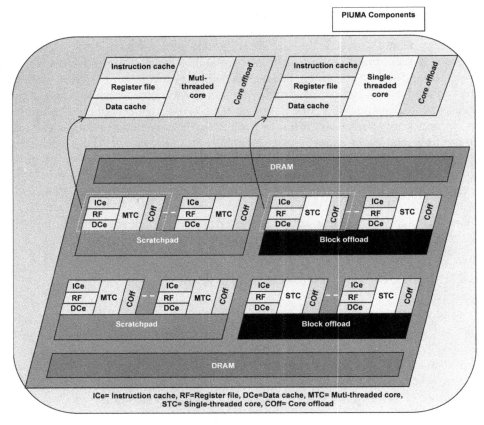

Figure 11.8. Structural unit components in the Intel's PUMA layout representation showing the single-threaded and multi-threaded cores, data and instruction cache, register file, the scratchpad, block offload and DRAM.

multiplied with a dense vector. The ensuing version works with selective caching wherein caches are bypassed during sparse accesses to the vector but the accesses to the matrix values are cached. Performance is further improved by direct memory access (DMA) for fetching the elements of the dense vector required for the recent row from memory to the local scratchpad. PIUMA overtakes the prevailing high-end processors for usual graph workloads. It is expected that PIUMA node will surpass a typical compute node by one-to-two orders of magnitude.

11.9 Summary and the way forward

11.9.1 Highlights of chapter 11 at a glance

(i) Graphs are valuable tools for real-world data modeling. The abstraction of graphs brings about a great simplification of the presentation and analysis of pairwise or dependency relationships among items of a certain form.

(ii) Graph algorithms are a set of instructions that traverse the nodes of a graph to determine specific nodes or paths between two nodes. The design, analysis, and implementation of graph algorithms helps in taking advantage of relationships within a dataset to uncover hidden connections and reveal crucial influencers for developing intelligent solutions.

(iii) Problems faced by common processors in handling graph algorithms were elucidated. Multi-core and networked parallel processors are able to handle these algorithms with limited success. So, the evolution of the graph processor is traced as a convenient solution to meet these situations.

(iv) The graph processors were specially developed by Song and co-workers for efficient processing of graph algorithms (Song *et al* 2016).

(v) The parallel architecture of the graph processor was emphasized and the advantages of implementation of graph algorithms as sparse matrix operations on a sparse matrix algebra instruction set were mentioned.

(vi) An important module in the graph processor is the matrix reader and writer module. Matrix data formats and reader/writer design considerations were outlined.

(vii) Other modules of the graph processor include the ALU module, the communication module, the node controller module and the systolic merge sorter module.

(viii) A high-bandwidth, low-power communication network is used in this processor.

(ix) The graph processor of Lincoln laboratory is prototyped on an FPGA.

(x) As a commercial example, the features of the GraphCore intelligence processing unit were described (Selwood 2017). A prominent architecture, the Intel PIUMA was discussed.

(xi) The prospective applications of graph processors are propitious. Processors for graph algorithms help organizations to optimize processes and make data-driven decisions in solving multifaced practical problems in social relationships and network dynamics. In social media platforms such as Meta, *X* and LinkedIn, the vertices of a graph are the users while their edges are relationships, e.g., friendships or followings. In a telecommunication network, the vertices are the user phones whose graphical study helps companies in contriving strategies for marketing (Coimbra *et al* 2021).

(xii) The graph processors are leveraged in recommendation systems to provide personalized and relevant exhortation to users. Reduced costs, improved delivery times, and better resource allocation are ensured in logistics and transportation by route optimization using graphical methods.

(xiii) Graph processing assists in identifying patterns in network traffic on the web to detect anomalies, and mitigate security threats. These applications make them crucial in cybersecurity for fraud detection.

(xiv) Graph algorithm processing facilitates semantic analysis and natural language processing tasks by extraction and representation of relationships between words, concepts, or entities. Text summarization, and sentiment analysis are thereby made comparatively easier.

(xv) Deciphering patterns in protein–protein interaction and genetic networks through graphical techniques aids in drug discovery and personalized medicine. So, the role of graph processors in interpreting real-life situations and solving problems is worthy of admiration and esteem.

(xvi) Keywords for this chapter are: sparse matrix operations, systolic merge sorter FPGA, IPU, PIUMA.

11.9.2 Getting ready to begin chapter 12

In the next chapter, we move to a spectacularly different approach of building a processor for performing calculations directly in its data memory. In this in-memory computing method, all the processing is done exclusively on the data stored in the RAM to eliminate any slow data accesses and transfers.

References

Aananthakrishnan S *et al* 2020 PIUMA: programmable integrated unified memory architecture arXiv:2010.06277v1

Aananthakrishnan S, Abedin S, Cavé V, Checconi F, Bois K D and Eyerman S 2023 The intel programmable and integrated unified memory architecture graph analytics processor *IEEE Micro* **43** 78–87

Balaban A T 1985 Applications of graph theory in chemistry *J. Chem. Inform. Comput. Sci.* **25** 334–43

Boldi P and Vigna S 2004 The webgraph framework I: compression techniques *WWW'04: Proc. of the 13th International Conference on World Wide Web (New York, May 17–20)* ed S I Feldman, M Uretsky, M Najork and C E Wills pp 595–602

Coimbra M E, Francisco A P and Veiga L 2021 An analysis of the graph processing landscape *J. Big Data* **8** 41

Cui P, Liu H, Tang B and Yuan Y 2014 CGgraph: an ultra-fast graph processing system on modern commodity CPU-GPU co-processor *Proc. of the VLDB Endowment, 50th Int. Conf. on Very Large Data Bases(August 26–August 30) (Guangzhou) vol 17 pp 1405–17*

Deo N 2016 *Graph Theory with Applications to Engineering and Computer Science* (Mineola, NY: Dover) p 496

Knuth D E 1998 Sorting by merging *The Art of Computer Programming, Vol. 3: Sorting and Searching* 2nd edn (Reading, MA: Addison-Wesley) sec 5.2.4 pp 158–68

Liu H and Huang H H 2019 SIMD-*X*: programming and processing of graph algorithms on GPUs *USENIX ATC'19: Proc. of the 2019 USENIX Conf. on Usenix Annual Technical Conf.* pp 411–27

Massed Compute 2024 *Differences between CSR and CSC Sparse Matrix Formats* (Massed Compute LLC) p 265

Needham M and Hodler A E 2019 *Graph Algorithms: Practical Examples in Apache Spark and Neo4* (Sebastopol, CA: O'Reilly Media Inc) pp 1–14

Ortiz J, Pupilli M, Leutenegger S and Davison A J 2020 Bundle adjustment on a graph processor *IEEE/CVF Conf. on Computer Vision and Pattern Recognition (CVPR)(13–19 June) (Seattle, WA)* pp 2413–22

Selwood D 2017 IPUs—a new breed of processor: machine learning platforms for AI, Feature article *Electron. Eng. J.* https://eejournal.com/article/20170119-ipu/

Shao C, Guo J, Wang P, Wang J, Li C and Guo M 2022 Oversubscribing GPU unified virtual memory: implications and suggestions *ICPE'22: Proc. of the 2022 ACM/SPEC on Int. Conf. on Performance Engineering (April 9–13) (Beijing)* (New York: Association for Computing Machinery) pp 67–75

Shi X, Zheng Z, Zhou Y, Jin H, He L, Liu B and Hua Q-S 2018 Graph processing on GPUs: a survey *ACM Comput. Surveys (CSUR)* **50** 1–35

Song W S *et al* 2010a 3-D graph processor *Workshop on High Performance Embedded Computing (15–16 September) (Lexington, MA)* (Lincoln Laboratory, Massachusetts Institute of Technology) p 2 https://archive.ll.mit.edu/HPEC/agendas/proc10/Day2/S4_1335_Song_abstract.pdf

Song W S *et al* 2010b 3-D graph processor, Workshop on High Performance Embedded Computing, HPEC 2010 16 September 999999-1 to 999999-21, https://archive.ll.mit.edu/HPEC/agendas/proc10/Day2/S4_1335_Song_presentation.pdf

Song W S 2012 Systolic merge sorter *US Patent* US 8,190,943 B2 pp 1–24

Song W S, Kepner J, Gleyzer V, Nguyen H T and Kramer J I 2013 Novel graph processor architecture *Linc. Lab. J.* **20** 92–104

Song W S 2014 Processor for large graph algorithm computations and matrix operations *US Patent* US 8,751,556 B2 Sheet 1–14 pp 1–32

Song W S, Gleyzer V, Lomakin A and Kepner J 2016 Novel graph processor architecture, prototype system, and results *2016 IEEE High Performance Extreme Computing Conf. (HPEC) (13–15 September) (Waltham, MA)* pp 1–7

Xie Y 2018 A brief guide of xPU for AI accelerators, ACM SIGARCH, computer architecture today *Comput. Arch. Today* **Apr. 26** https://sigarch.org/a-brief-guide-of-xpu-for-ai-accelerators/

Zerbino D R and Birney E 2008 Velvet: algorithms for de novo short read assembly using de Bruijn graphs *Genome Res.* **18** 821–9

IOP Publishing

AI-Processor Electronics

Basic technology of artificial intelligence

Vinod Kumar Khanna

Chapter 12

Associative processing unit

Computing in memory has ushered a new era in processors. It works on the content-addressable memory (CAM) concept by adopting a data-centric computing strategy to cope with the von Neumann bottleneck. CAM is differentiated from the conventional random-access memory (RAM). The parts, circuit configuration and operation of a CAM cell are delved into. Internally, the 10T CAM cell consists of a 6T static random-access memory (SRAM) cell. Operating modes of this SRAM cell, and the data matching procedure are elaborated. Construction of a generic associative in-memory (AiMP) processor is illustrated. Operation cycles of AiMP include the comparison and writing cycles. The AND operation is executed on an AiMP by preparing a lookup table (LUT) for this operation. Similarly, other logic operations as well as arithmetic operations are carried out on the AiMP. The AiMP outperforms several available computing platforms in data-intensive workloads, and can be implemented in standard complementary metal-oxide semiconductor (CMOS) as well as emerging memory technologies. An in-memory processor for big data analysis is described.

12.1 Introduction

In this chapter, we study the whys and wherefores of the operational mechanisms of the associative processing unit. Its main characteristic feature is that its working represents a deviation from the time-honored concept of computing based on data transfers between processor and memory. It begins with a root-and-branch rejection of the von Neumann approach, and replaces this customary practice with a radically different functioning style based on data accessing by content and processing directly in place in the memory array without crossing the input/output. This method increases the performance-over-power ratio by several orders of magnitude compared to methods using central processing unit (CPU) and graphical processing unit (GPU), along with dynamic RAM (DRAM) (GSI Technology 2024).

12.2 Adopting a data-centric computing strategy to cope with the von Neumann bottleneck

AiMP is a non-von Neumman architecture. It can help in overcoming the memory and power barricades that have put the brakes on the progress of the CPU. Data transfers between the memory unit and the processor constitute a major chunk of all computing today. Problematic issues at the focus of attention are:

    (i) the time taken in accessing data,
    (ii) the limitations on throughput of data, and
    (iii) the enormous energy needed for data transfers.

All these are wasteful of time, power and cost. So, an approach devised by diverging from the usual data transfer practice is highly desirable. Necessity is the mother of invention. In lieu of this run-of-the-mill data transferring process, concentrating on computation directly at the location of the data should prove more beneficial (Yau and Fung 1977, Grosspietsch 1992, Krikelis and Weems 1994). Both structured data and unstructured data are retrievable by the naturally parallel and scalable associative processing technique. The AiMP works in an induction-like style on a deep processing in-memory (PIM) principle. It uses arrays of memory cells fabricated with the proven CMOS as well as upcoming memory resistive RAM (ReRAM) technologies. Benefitting from these improvements, it greatly minimizes data movement (Yavits et al 2015, Yavits 2023).

12.3 Content addressable memory

12.3.1 Difference between CAM and RAM

The conventional RAM working with reference to data addresses has an inherently sequential nature following a serial order of events. This manner of working is the root cause rendering it unsuitable for big data and AI. High speed and low-power, large data capacity memory are essential preconditions for achieving success in these fields (Kim et al 2022). Therefore, the central component acting as the keystone of the AiMP is the CAM. It is also called associative memory or associative storage. It is a kind of computer memory adept at comparing the input data known as the search data with a table of stored data. As the outcome of this comparison, it returns the address of any data found in the table of data that matches with the search data (Yavits 1994).

12.3.2 Parts of a CAM

The five main parts of a CAM are (figure 12.1):

- Key register: This stores the value that is written or against which the comparison is made.
- Mask register: This indicates the bit or bits which are activated during comparison or writing.

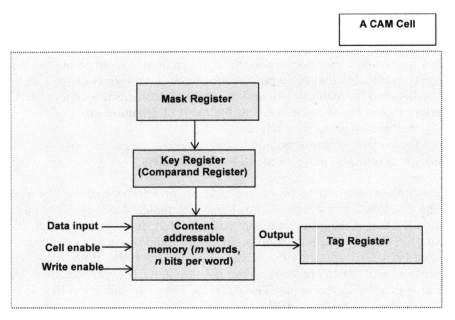

A CAM Cell

Mask Register

**Key Register
(Comparand Register)**

Data input →

Cell enable →

Write enable →

**Content
addressable
memory (*m* words,
n bits per word)**

Output →

Tag Register

Figure 12.1. The fundamental CAM cell consisting of registers (mask, key and tag) and CAM together with data input, cell enable, output and write enable terminals.

- Tag register: In this register, the rows matched by the comparison or writing are marked. Tagging a row with logic-1 means that this particular row has been found to be matching with the given key and mask value.
- Content-addressable memory unit as a two-dimensional data array: The CAM stores the data to be searched against or processed in a 2D matrix organization spoken of as a 2D array. The data is arranged in a grid format as a table made of several rows and columns of memory cells. Each cell of CAM is accessible by activation of the corresponding search line (SL) and its complementary, the search line bar $\overline{SL}$.
- Data input, cell enable, output and write enable terminals: The CAM performs data storage in a similar manner to the SRAM. In addition, it has a comparison functionality. Its terminal connections are:
  1. Data input: The data to be searched are entered through the data input port.
  2. Cell enable: If cell enable signal is high, then a comparison process between input data and the stored data in the cell will indicate either a matching or no matching status at the output. An XNOR operation is performed during the comparison.
  3. Output: It gives the result of the matching operations carried out.
  4. Write enable: If the write enable is high, the input data are written to the cell.

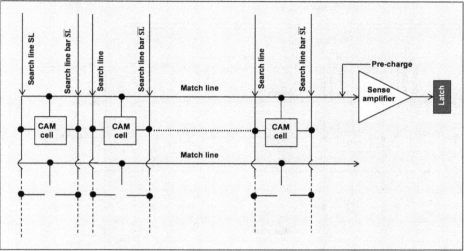

Figure 12.2. The circuit configuration of a word stored in CAM. The diagram shows the CAM cells, the SL, search line, $\overline{SL}$, search line bar, the ML, match line, sense amplifier, pre-charge point and the latch.

12.3.3 Circuit configuration of CAM

Figure 12.2 shows the circuit configuration of a CAM word. Each bit of the word is stored in one CAM cell. An SL drawn in the vertical direction is connected to each CAM cell in a column. A complementary $\overline{SL}$ is drawn in the vertical direction parallel to the SL. A common match line (ML) is drawn horizontally. It is connected to all the CAM cells in a row. It terminates in the sense amplifier. Thus, every CAM cell is connected to an SL on the left, an $\overline{SL}$ on the right and an ML at the top. Hence, the SL is shared by all the CAM cells in a row. Inversely, the ML is shared by all the CAM cells in a column.

12.3.4 Operation of a CAM

The data to be compared are loaded and stored in the key register (figure 12.3). All the match lines are pre-charged with a high voltage. The key register is applied on the SL and $\overline{SL}$ search line bar. The mask register is used to adjust the SL and $\overline{SL}$ of those columns that are not to be compared, to a low value of voltage. If all the chosen bits of a row match with the given data, the associated pre-charged word line (WL) maintains a high voltage. This high voltage is captured by the sense amplifier. It is stored in the tag latch. In contrast, if there is a mismatching of data, the current in that memory cell leaks away. Therefore, the voltage of the WL falls. All the selected bits of each column are compared with the input data in parallel. Their comparisons are completed in one cycle.

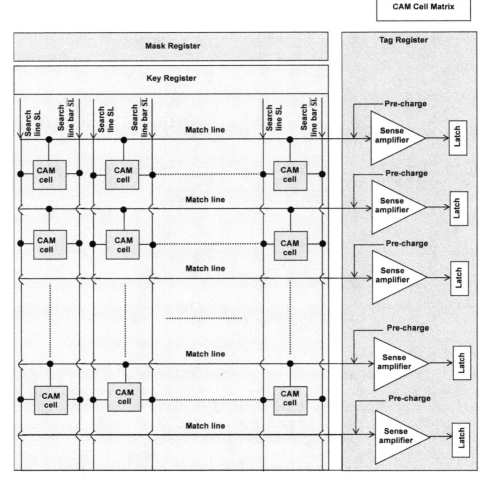

Figure 12.3. Organization of the constituent memory cells and registers in a CAM system. The mask and key registers are placed on the top side. The tag register containing its sense amplifiers, pre-charge points and latches is placed on the right side. The CAM cells are mounted along horizontal rows with their upper ends tied to the match lines. Their left and right sides are joined to vertical lines. To these vertical lines, the search line and the $\overline{SL}$ line are connected at the top.

12.3.5 Internal structure of a CAM cell

The 10T CAM cell shown in figure 12.4 consists of a 6T SRAM cell. It comprises the four memory cell transistors (M_1, M_2, M_3, M_4) constituting a pair of cross-coupled inverters (inverter-1 and inverter-2) that are used for bit storage. Furthermore, it has two access transistors A_1, A_2 for performing read/write operations. It also has four comparison transistors C_1, C_2, C_3, C_4 for matching the searched data.

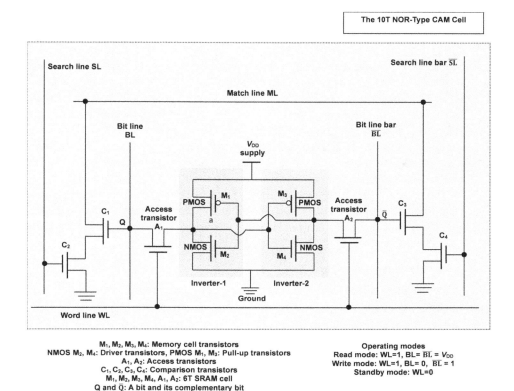

Figure 12.4. A 10-transistor CAM cell with 6T SRAM cell. The 10-transistor CAM cell consists of the transistors: M_1, M_2, M_3, M_4, A_1, A_2, C_1, C_2, C_3, C_4. The 6T SRAM cell comprises the transistors: M_1, M_2, M_3, M_4, A_1, A_2. The M-transistors are the memory transistors. The A-transistors are the access transistors. The C-transistors are the comparison transistors. The PMOS transistor M_1 and NMOS transistor M_2 constitute one inverter circuit. The PMOS transistor M_3 and NMOS transistor M_4 constitute another inverter circuit. The bit line (BL), bit line bar $\overline{BL}$, ML and WL are shown. The V_{DD} supply and ground terminals are also indicated.

12.3.6 Operating modes of SRAM cell

The SRAM cell has three different modes of operation (Apostolidis *et al* 2016):

Mode 1: Standby mode: Here the WL is not asserted, hence WL = 0. Then access transistors A_1, A_2 are in the OFF state. The memory cell is inaccessible. The two inverters: inverter-1 and inverter-2 continue to feed each other through feedback signals in a positive feedback loop. This feeding continues as long as the V_{DD} supply is ON. Therefore, the data inside the memory cell is preserved, and not lost.

Mode 2: Read mode: This operation begins with the pre-charging of the BLs to keep them high, BL = $\overline{BL}$ = V_{DD}. The BLs are maintained at a floating potential while the WL is asserted: WL = 1. Both the access transistors A_1, A_2 are turned ON. The data in the nodes are driven onto the BLs to create a potential difference between the BLs. The potential difference produced is detected by a sense amplifier, thereby reading the value of data stored in the cell.

Mode 3: Write mode: For performing a write operation, the BLs are pushed to complementary voltage levels corresponding to the bits Q, $\overline{Q}$: BL = 0, $\overline{\text{BL}}$ = 1. Then the access transistors A_1, A_2 are turned ON by activating the WL (WL = 1). The memory cell (M_1, M_2, M_3, M_4) is appended to the BL (and $\overline{\text{BL}}$). Therefore, the data to be written are driven into the BLs. One of the storage nodes is discharged through the access transistor. The voltage of the opposite storage node is raised by the action of inverters. The memory cell is latched with the new data, overpowering the inverters. The SRAM circuit holding the state is sufficiently feeble to be overpowered during a write cycle; this property is called writability. Yet it is adequately strong so that it is not disturbed during a read cycle; this capability is known as read stability.

12.3.7 Data matching procedure

The complementary stored bits, Q (and $\overline{Q}$), are compared with the complementary search data on the complementary search line, SL (and $\overline{\text{SL}}$). Four transistors C_1, C_2, C_3, C_4 are used for making these comparisons. The transistors control the pulldown path of a dynamic XNOR logic gate from the ML with the inputs SL and Q. A mismatch of SL and Q activates at least one of the pulldown paths, thereby connecting ML to ground. A match of SL and Q disables both pulldown paths, thus disconnecting ML from ground. The ML remains in the pre-charged state. Remaining in pre-charged state is an indication showing that a matching has taken place.

A CAM word is formed by connecting multiple cells in parallel. This is done by shorting the ML of each cell to the ML of adjacent cells. In this alliance, the pulldown paths link up in parallel. A match condition occurs on a given ML only if every individual cell in the word has a match (Pagiamtzis and Sheikholeslami 2006).

12.4 Construction of a generic AiMP

The generic architecture of AIMP is exhaustively described by Yantir (2018), and by Fouda *et al* (2022) in a tutorial brief. The AiMP consists of an instruction cache; the mask, key and tag registers; a controller; a CAM, and an interconnection matrix (figure 12.5):

  (i) Instruction cache: This stores the instructions to be performed by the content-addressable memory.
  (ii) Mask, key and tag registers.
  (iii) Controller: This manages all the processing operations, e.g.,
      (a) issuing instructions, producing the key and mask values for the instructions, and hence setting the key and mask registers;
      (b) handling control sequences and executing read requests; and
      (c) acting as a data buffer.

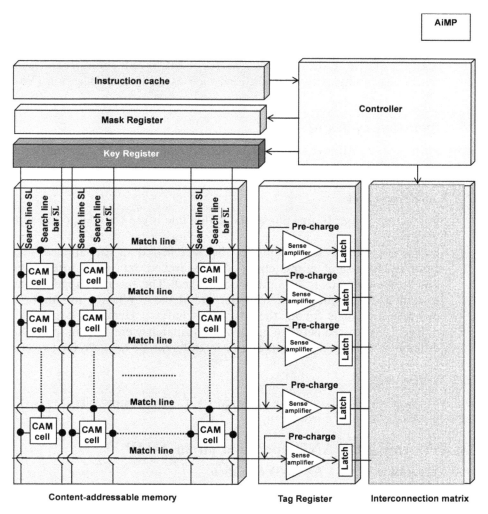

Figure 12.5. AiMP architecture. The AiMP is formed by attaching an instruction cache and a controller to the CAM with its mask, key and tag registers. The coupling is done through an interconnection matrix. The interconnection matrix is a changeover or exchange matrix enabling parallel communication among rows of the AiMP.

(iv) CAM: This consists of the memory storage unit, and the accompanying mask, key, and tag registers. The compared data are prefetched into the key or comparand register. A binary pattern is stored in the mask register. This pattern determines whether or not the corresponding bits in the word are to be concealed from further comparing and writing steps (Li 2005).

(v) Interconnection matrix: This is a switching matrix. It bears the responsibility of permitting parallel communication among the rows of the associative processor, either bit-wise or word-wise. Two types of interconnection matrices are the CMOS-based reduction tree and the memristive crossbar. Amongst these options, a higher density together with a better functionality is achieved using the memristive switching matrix. The underlying reason is that the disallowance of intercommunication between the associative processor rows in the CMOS reduction tree structure introduces complexity in operations such as array multiplication (Yantir 2018).

12.5 Operation cycles of AiMP

The working of an AiMP involves different operational cycles which are outlined in the subsections below.

12.5.1 Comparison cycle

   (i) Pre-charge phase: In this phase, the match line is pre-charged to the supply voltage.
   (ii) Evaluation phase: In this phase, the searched data are applied to the columns. In case of mismatching of the searched data with stored data, a low resistance path to ground is established. The reverse happens during matching of searched data with the stored data. When a minimum one mismatched cell is found, the ML is discharged. Hence, a large voltage difference is created between match and mismatch events. From this voltage difference, the sensing circuits can easily differentiate between the two cases.

12.5.2 Writing cycle

The CAM cell is updated based on a prepared LUT. The operand bits are rewritten if an in-place operation is carried out.

12.6 Performing the AND operation on an AiMP

12.6.1 Preparation of LUT for AND operation

An AiMP performs computation with the help of LUTs of logical operations (Yantir *et al* 2019). An LUT for the AND operation is made as shown in table 12.1 (Fouda *et al* 2022).

   The comment column in the table lays down the running order of this key combination. 'No change' means the CAM content is not affected by the given input combination. Note that the AND operation involves a single pass in which a single

Table 12.1. LUT for AND operation.

B	A	R (Result)	Remarks
0 or OFF	0 or OFF	0 or OFF	No change
0 or OFF	1 or ON	0 or OFF	No change
1 or ON	0 or OFF	0 or OFF	No change
1 or ON	1 or ON	1 or ON	First pass

comparison cycle is required to search for '11' and a single writing cycle is needed in which the output cell is updated. The AiMP moves over the LUT in a defined order. This orderly movement ensures avoiding any cycles in which the outcomes of preceding passes may be overwritten. Therefore, adherence to prescribed pass order is essential.

(i) The masks are produced for implementation of comparison and writing cycles. The controller sets the mask and key values in agreement with the LUT.

(ii) In a 3 × 3 CAM, the first two columns are kept for the inputs A and B. The third column is reserved for the result R (figure 12.6). A capacitor is

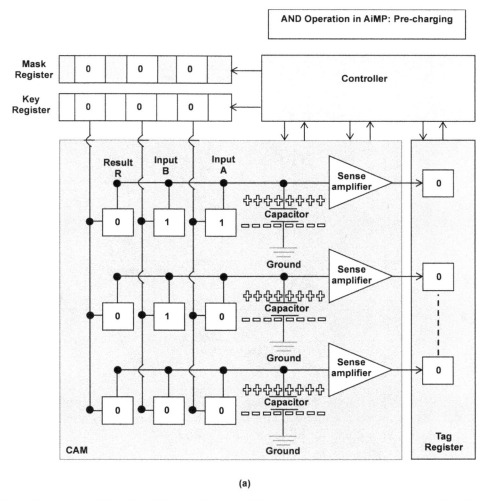

(a)

Figure 12.6. Accomplishing the AND operation on AiMP: (a) pre-charging, (b) evaluation and (c) writing. Pre-charging is done during the compare operation. In pre-charging, all the rows of CAM are pre-charged. Evaluation phase involves searching the A and B columns for '11'. The mask bits related to A and B are activated and set to '11' while the R mask is set to '0'. In the write cycle, a write signal is produced for setting the R cell to '1' for the matched rows.

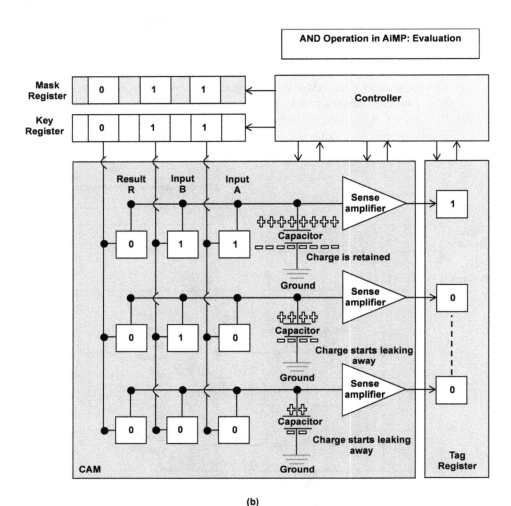

(b)

Figure 12.6. (Continued.)

incorporated in the circuit at a point adjoining the sense amplifier to store charge. In the beginning, all the cells of the CAM are set to zero. The status of the tag column for all rows is: 0, 0 and 0.

(iii) For carrying out the LUT operation in the CAM, a search is made for finding '11' in the A and B columns of the CAM. A logic-1 is written in the R column of the row which matches with '11'.

12.6.2 Performing the AND operation

The operation is performed through comparison and writing cycles:

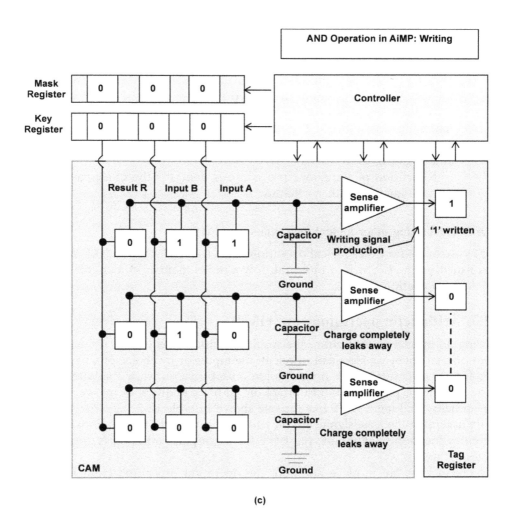

Figure 12.6. (Continued.)

(i) Comparison cycle: Here, the key and mask fields are set. They are compared with CAM content according to the left side of the LUT.

(a) Pre-charging is done on all rows of the CAM. The status is: R column: 0, 0, 0 and tag column: 0, 0, 0.

(b) During evaluation, the A and B columns are searched for '11'. The mask bits for A and B are set to '11' and R mask is set to 0. The LUT of the AND function is imposed on the CAM with the key bits of columns A and B set to 1. Matching of the key and mask values is observed only with the first row. Hence, the charge is retained on the capacitor. The same does not happen with the second and third rows. So, the charges stored in the capacitors in these rows start to leak away. The sense amplifier compares the capacitor's voltage for the

ML of each row to a threshold value. In case of matching, the tag bit is set to 1. For a mismatching case, it is set to 0. Now the status is: R column: 0, 0, 0 and tag column: 1,0, 0.

(ii) Writing cycle: The right side of the LUT is observed to set the mask and key values. Also, the values in the tagged rows are changed in this cycle.

A write signal is produced to set the R cell to 1 for the row in which matching was noticed. The charges stored on the capacitors in the other rows completely leak away. Finally, the status is: R column: 1, 0, 0 and tag column: 1, 0, 0.

Repetition of the above procedure is done for the other passes on the same columns or other columns.

12.6.3 Performing other logical operations on AiMP

LUTs are made for other logical operations such as NOT, OR, and XOR. They are performed on the CAM in a bit-serial, row-parallel manner in identical fashion to the AND operation.

12.7 Arithmetic operations on AiMP

During addition or subtraction, the result is written into one of the two possible locations, either by replacement of one of the inputs or by choice of a new location. The former option is called in-place addition/subtraction while the latter option is known as out-of-place addition/subtraction. In-place addition of two vectors A and B is done by making a LUT like the one shown in table 12.2.

By accepting the least significant bit and most significant bit locations along with the carry location C as inputs, the LUT for an in-place addition is applied to the vectors A and B.

Note that for an in-place addition, the operation order and truth table order differ. Entry number 4 is run before entry number 2. Suppose the '001' (entry

Table 12.2. LUT for addition of vectors A and B.

	Comparison operation			Writing operation		
Entry number	Carry	Vector B	Vector A	Carry	Vector B	Remarks
1	0 or OFF	0 or OFF	0 or OFF	0 or OFF	0 or OFF	No change
2	0 or OFF	0 or OFF	1 or ON	0 or OFF	1 or ON	Second pass
3	0 or OFF	1 or ON	0 or OFF	0 or OFF	1 or ON	No change
4	0 or OFF	1 or ON	1 or ON	1 or ON	0 or OFF	First pass
5	1 or ON	0 or OFF	0 or OFF	0 or OFF	1 or ON	Third pass
6	1 or ON	0 or OFF	1 or ON	1 or ON	0 or OFF	No change
7	1 or ON	1 or ON	0 or OFF	1 or ON	0 or OFF	First pass
8	1 or ON	1 or ON	1 or ON	1 or ON	1 or ON	No change

number 2) is searched first for Carry, B, A, respectively. Then the B values of tagged rows are changed to logic-1. Subsequently, when '011' (entry number 4) is searched as the second item in the progression, the same rows will be tagged again, and altered twice. Errors arising from such causes are avoidable by correct sequencing.

12.8 In-memory processor for big data analysis

Big data is a term for a blended structured/semi-structured/unstructured database. The data are collected by organizations for analyses and mining for extracting useful information and insights. Such data are identifiable by their three distinguishing features: a large volume, a wide variety and a high velocity of data generation. These are known as the three V's of big data.

The CPU is easily fatigued in performing simple search tasks on these data. As its local memory overflows, the data are stored in the larger on-chip memory of the CPU. The data transmission rate from this larger memory to processor cores being time-consuming, a bottleneck quickly ensues and obstructs functioning of the CPU.

A chip classified as an associative processing unit is designed to handle such data by combining memory and computation units, and performing the computation near the data (Gwennap 2020). The chip contains millions of processors. The memory is divided into a large number of subunits to feed the different processors. Each processor can load data directly from memory. Thus, data flow is increased, and much greater data flow is obtained. Moreover, this on-chip connection between memory and processor is extremely short. Hence, the power used in transferring the data is low.

The large memory bandwidth in conjunction with smaller power consumption enable the chip to outperform standard processors by $\geqslant 100\times$ on certain big-data workloads. At the same time, the power consumption is decreased by 70%. Even further, the chip deployment is simplified by providing the chip as a part of a peripheral component interconnect express (PCIe) board known as Leda-G, for direct plugging into standard servers. Also supplied is a software stack for helping customers in running different search algorithms using the chip. The chip finds applications in image-recognition, molecular-identification, e-commerce, natural-language processing and visual search.

12.9 Summary and the way forward

12.9.1 Highlights of chapter 12 at a glance

    (i) Computing in memory has ushered a new era in processors. It works on the CAM unit by adopting a data-centric computing strategy.

    (ii) The CAM was differentiated from the conventional RAM. The parts, circuit configuration and operation of a CAM cell were delved into. Internally, the 10T CAM cell consists of a 6T SRAM cell. Operating modes of this SRAM cell, and the data matching procedure were elaborated.

    (iii) Construction of a generic AiMP processor was illustrated.

    (iv) Operation cycles of AiMP include the comparison and writing cycles.

(v) The AND operation is executed on an AiMP by preparing an LUT for this operation. Similarly, other logic operations as well as arithmetic operations are carried out on the AiMP.

(vi) An in-memory processor for big data analysis was described.

(vii) The popularity of the emerging non-von Neumann computational paradigm of in-memory processing stems from the speed gained over outside-the-memory processing. In the slow data reading/writing operations necessary in the latter technique, a lot of time is wasted in shuttling the data back-and forth between the processor and the memory.

(viii) The sluggish operation of stored program computers often creates a data bottleneck known as input/output bound (Mannocci *et al* 2023). Therefore, in-memory processing has carved its niche in real-time processing to cope with von Neumann bottleneck in several areas such as for financial businesses and e-commerce as well as for personalized recommendations and inventory management.

(ix) Fraudulent activities can be prevented if banks and stock exchanges rely on in-memory computing for lightning-fast transaction processing. In-memory computing is also helpful for healthcare by providing quicker doctor intervention in critical patient monitoring.

(x) The AiMP outperforms several available computing platforms in data-intensive workloads. It can be implemented in standard CMOS as well as emerging memory technologies (figure 12.7). CPUs and GPUs are

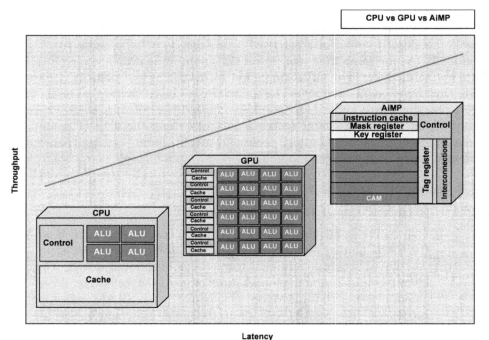

Figure 12.7. Comparison among CPU, GPU and AiMP.

designed with the aim of minimization of latency. In striking dissimilarity to CPUs and GPUs, AiMPs are optimized for throughput maximization. Their single memory structure and simpler control unit are the main reasons for this approach. Hence, the AiMPs excel over the common processors based on the single-instruction, multiple-data (SIMD) instruction set.

Applications such as matrix multiplications and fast Fourier transform calculations with an intrinsic SIMD pattern of computation are the main beneficiaries of AiMP processing (Fouda *et al* 2022).

(xi) Keywords for this chapter are: AiMP, CAM, SRAM, non-von Neumann architecture.

12.9.2 Getting ready to begin chapter 13

From the next chapter, we are bidding good-bye to classical computing and moving to an altogether different technology model. This paradigm is known as quantum computing.

References

Apostolidis G, Balobas D and Konofaos N 2016 Design and simulation of 6T SRAM cell architectures in 32 nm technology *J. Eng. Sci. Technol. Rev.* **9** 145–9

Fouda M E, Yantır H E, Eltawil A M and Kurdahi F J 2022 In-memory associative processors: tutorial, potential, and challenges *IEEE Trans. Circuits Syst. Express Briefs* **69** 2641–7

Grosspietsch K E 1992 Associative processors and memories: a survey *IEEE Micro* **12** 12–9

GSI Technology 2024 GSI's solution for search and next generation high performance computing https://gsitechnology.com/compute/

Gwennap L 2020 *In-Memory Acceleration for Big Data* (The Linley Group, Inc.) pp 1–6 https://gsitechnology.com/wp-content/uploads/2023/01/GSIT-Gemini-WP-Final-Linley.pdf

Kim H, Cho M, Lee S, Kwon H S, Choi W Y and Kim Y 2022 Content-addressable memory system using a nanoelectromechanical memory switch *Electronics* **11** 1–15

Krikelis A and Weems C C 1994 Associative processing and processors *Computer* **27** 12–7

Li J F 2005 Testing priority address encoder faults of content addressable memories *IEEE Int. Conf. on Test, Paper 33.2 (8 November) (Austin, TX)* pp 1–8

Mannocci P, Farronato M, Lepri N, Cattaneo L, Glukhov A, Sun Z and Ielmini D 2023 In-memory computing with emerging memory devices: status and outlook *APL Mach. Learn.* **1** 1010902-1–25

Pagiamtzis K and Sheikholeslami A 2006 Content-addressable memory (CAM) circuits and architectures: a tutorial and survey *IEEE J. Solid-State Circuits* **41** 712–27

Yantir H E 2018 *Efficient Acceleration of Computation Using Associative In-memory Processing* (Irvine: UC Irvine Electronic Theses and Dissertations, Doctor of Philosophy in Electrical and Computer Engineering, University of California) pp 1–150

Yantir H E, Guo W, Eltawil A M, Kurdahi F J and Salama K N 2019 An ultra-area-efficient 1024-point in-memory FFT processor *Micromachines* **10** 1–16

Yau S S and Fung H S 1977 Associative processor architecture—a survey *ACM Comput. Surv.* **9** 3–27

Yavits L 1994 Architecture and design of an associative processor chip for image processing and computer vision *Project Thesis* Master of Science in Electrical Engineering, The Technion Israel Institute of Technology, AV5754, Haifa p 25

Yavits L, Kvatinsky S, Morad A and Ginosar R 2015 Resistive associative processor *IEEE Comput. Archit. Lett.* **14** 148–51

Yavits L 2023 Will computing in memory become a new dawn of associative processors? *Memories—Mater., Devices, Circuits Syst.* **4** 1–10

IOP Publishing

AI-Processor Electronics
Basic technology of artificial intelligence
Vinod Kumar Khanna

Chapter 13

Quantum computing principles and devices

Present-day computers are essentially non-quantum mechanical in operation. The logic gates of a classical computer are irreversible in nature. For easy understanding, elementary quantum mechanics is reviewed including the Schrödinger's equation, the particle-in-a-box problem, state superposition and collapsing. Bloch sphere representation of a quantum bit is explained. The qubits display interesting properties of quantum superposition, entanglement and interference, characteristic of atomic and subatomic particles. Measurements carried out on qubits cause their wave functions to collapse forcing them into a definite 0 or 1 state. Qubit decoherence on interaction with the environment is a major problem in quantum computing Three types of quantum computing devices are discussed including silicon, superconducting and trapped qubit devices. Among silicon devices, gate-controlled silicon quantum dots and phosphorous donor implanted Si substrates are considered. Superconducting qubit archetypes and their realization techniques are explained. The transmon qubit design is preferred for charge noise mitigation. Under the trapped ion qubit devices, construction and operation of Paul ion traps is discussed, and the types of trapped ion qubits are elucidated.

13.1 Introduction

Classical computing has been accepted, widely adopted over the years and now has become firmly established, omnipresent and commonplace as a benchmark of computing. Thanks to the painstaking research by many engineers and scientists. But, of late, quantum computing is gaining visibility although it is still in infancy. It is the rising star in the field of computing. It may take several years for its maturing to attain its full brightness and widespread implementation because more regulated physical conditions are imperative for smooth functioning of quantum computers.

Quantum computing works on different types of bits and follows a different set of rules from classical computing which has a deterministic nature producing either a 0 or 1 output. Quantum computing has a probabilistic nature which means that a

doi:10.1088/978-0-7503-6259-7ch13

probability value is associated with each possible output, allowing for the exploration of multiple solutions. In this respect, it differs from its classical contestant to the apex position.

In opposition to classical computing which operates on Boolean logic and algebra, quantum computing uses linear algebra. Linear algebra is the branch of algebra devoted to the study of linear equations and functions, and their matrix/determinant representations. Further, quantum computing is based on quantum mechanical theory. In quantum mechanics, representation of the state of a physical system is done with the aid of a vector in the space known as the Hilbert space. The Hilbert space is defined as a vector space equipped with an inner product. The inner product is the analog of the dot or scalar product in an arbitrary vector space. The concept of Hilbert space applies 2D and 3D mathematics to describe outcomes in more than three dimensions. It thus allows the generalization of the methods of linear algebra and calculus from Euclidean vector space to an infinite–dimensional space.

In this chapter, we embark on a journey of quantum computing principles and the enabling device technologies to familiarize ourselves with the differences between the classical and quantum worlds in the context of computing.

13.2 Peregrination from classical to quantum computing

13.2.1 Non-quantum-mechanical nature of the present-day computers, and the theoretically-driven curiosity for quantum computing

Contemporary computers are essentially a conglomeration of semiconductor technology-enabled devices called transistors wired together to execute logic and arithmetic operations. The properties of semiconductors and the working of transistors are explained by quantum-mechanical theory. So, the present-day computers should be quantum-mechanical computers. It is indeed true but only partially (Vedral and Plenio 1998).

No doubt, semiconductor device operation is based on the core principles of quantum physics but the computers made from them are not fully quantum-mechanical in operation because the information is recorded in these computers in macroscopic two-level systems. A bit of information storage depends on the two states defined as current flowing in a wire for logical 1 state and no current flowing in the wire as the logical 0 state.

13.2.2 A pragmatic approach towards quantum computing

Gordon Moore's observation that the number of transistors in an integrated circuit chip and hence the complexity of computers doubles every year with minimal rise in cost leads us to believe that the relentless downscaling of transistor features sizes will push the miniaturization to the scale that a bit of information will be ultimately encoded into a single atom. So, the role played by quantum-mechanical effects in computing will greatly predominate as computers march forward unabated. Thus, the influence of quantum effects on computation will become increasingly significant day after day. So technological progress clearly signals that quantum computing is already on the horizon, and its imminence can no longer be ignored.

13.2.3 Irreversible nature of gates in a classical computer, and the consequent dissipation of energy

A classical computer is viewed as a collection of bits and gates in which the action of gates on the initial states of bits results in final states of bits. Considering a simple two-input OR gate of this computer, its output C is 1 when either A is 1 or B is 1. Let us operate it backwards. Given that the output C is 1, is it possible to tell whether A is 1 or B is 1? No. The answer is ambiguous because:

    (i) When A is 1 but B is 0, then C is 1, i.e., for input 10, the output is 1.
    (ii) When B is 1 but A is 0, then C is 1, i.e., for input 01, the output is 1.
    (iii) When A = B = 1, then also C is 1, i.e., for input 11, the output is 1.

Thus, the input values are not uniquely determined from the output value. They are either 10, 01 or 11, when C is 1, or better to say, the inputs are simply undetermined. Obviously, this gate cannot be run backwards. Hence, it is an irreversible gate. Construction of any other gate is possible using OR, AND and NOT gates. So, they are the fundamental gates.

A noteworthy feature of all existing real-world computers built using irreversible gates is the unavoidable dissipation of energy in the form of heat. This energy dissipation occurs when the information is deleted. It must also be highlighted that the aforementioned OR gate can be made reversible by allowing the first input bit as an additional output bit. This modified OR gate has two outputs and two inputs. Thus, by saving the necessary input at the output, an irreversible gate can be made reversible. This statement paves the way toward making the entire irreversible computing a reversible dissipation-free process. Although technically feasible, it is an impractical proposition because it adds significant overheads, complexity and costs due to the extra circuitry needed to store the garbage bits. Quantum gates are inherently reversible because they are represented by unitary matrices which are guaranteed to be invertible allowing the gates to return to their original states thereby ensuring reversibility (section 14.2).

13.3 Concise recapitulation of quantum mechanics

13.3.1 Schrödinger's equation

Quantum entities exhibit wave–particle duality. They show wave nature or particle behavior depending on the experimental circumstances. The one-dimensional, time-independent Schrödinger equation, the foundational phenomenological model of quantum mechanics for the energy E of an electron of mass m moving in the potential $V(x)$ due to an applied electric field is used to formulate the famous Young's experiment (figure 11.1) on interference diffraction pattern generated by electrons passing through parallel slits (Webb 2024).

In initial experiments, sunlight or a laser beam was used as a coherent source of light, and later atomic-scale entities, e.g., electrons, were used to illuminate/bombard

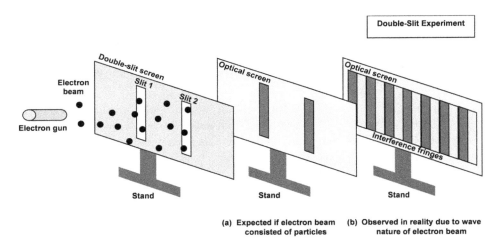

Figure 13.1. Enacting the double-slit experiment performed by Thomas Young with an electron beam: (a) the pattern that would be seen on the optical screen if electrons behaved like particles, (b) the pattern actually observed because of the wave nature of electrons. The diagram shows electrons emitted by an electron gun striking a screen with parallel slits 1 and 2. Part (a) shows the likely pattern if electrons were particles but this pattern is not actually obtained. Part (b) is the pattern registered experimentally on the screen revealing that electrons behave like waves, not particles.

a double-slit screen pierced by two parallel slits: slit 1 and slit 2. The light/electron beam emerging from the slits strikes an optical screen placed behind the double-slit screen. If the light/electron beam was particulate in content, a pattern containing two parallel lines for the two slits would be observed, as shown in figure 13.1(a). But actually, the light/electron beams passing through the two slits interfere between themselves. This wave phenomenon displayed by electron beams results in the creation of a pattern of alternate dark and bright bands known as interference fringes, as shown in figure 13.1(b). The experiment demonstrates the wave nature exhibited by light/electron beam. The phase difference between the two light/electron waves, caused by the path difference between the beams from the two slits is the cause of appearance of the interference pattern.

Schrödinger's equation was proposed in a way resembling the classical equation for the mechanical energy of a body as the sum of its kinetic and potential energies. It is expressed as

$$\frac{\hbar^2}{2m}\frac{d^2\psi(x)}{dx^2} + V(x) = E\psi(x) \qquad (13.1)$$

where $\psi(x)$ is a mathematical function called the wave function and $\hbar$ is the reduced Planck constant.

13.3.2 Particle-in-a-box or infinite potential well problem

A simple problem providing an intuitive understanding of quantum mechanics is the infinite potential well problem (figure 13.2). Schrödinger's equation is applied to a

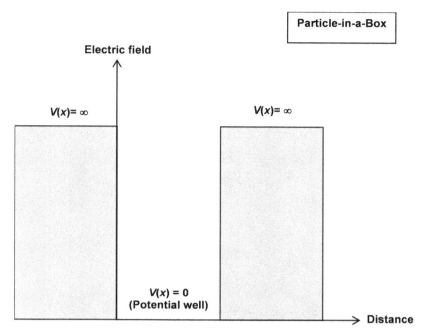

Figure 13.2. The hypothetical infinite potential well model of a free particle confined in a region of space surrounded by two impenetrable barriers. The model illustrates the differences between classical and quantum mechanical systems.

one-dimensional infinite potential well. This kind of potential well is defined as a hypothetical space at zero electric potential that is enclosed between two walls kept at an infinite potential of width L. For this well, Schrödinger's equation reduces to

$$\frac{\hbar^2}{2m}\frac{d^2\psi(x)}{dx^2} = E\psi(x) \tag{13.2}$$

The equation written down for the infinite potential well is subject to two boundary conditions:

$$\psi(x) = 0 \ \text{ at } \ x = 0 \ \text{ and } \ \psi(x) = 0 \ \text{ at } \ x = L \tag{13.3}$$

or,

$$\psi(0) = 0 \ \text{ and } \ \psi(L) = 0 \tag{13.4}$$

Evidently, this is a second order differential equation having general solutions of the form

$$\psi(x) = A\sin(\lambda x) + B\cos(\lambda x) \tag{13.5}$$

where A and B are constants, and

$$\lambda = \sqrt{\frac{2mE}{\hbar^2}} \tag{13.6}$$

On applying the boundary conditions to the solutions, we get

$$\psi(0) = A \sin(\lambda \times 0) + B \cos(\lambda \times 0) = 0 \qquad (13.7)$$

giving

$$B = 0 \qquad (13.8)$$

and

$$\psi(L) = A \sin(\lambda \times L) + B \cos(\lambda \times L) = 0 \qquad (13.9)$$

or,

$$A \sin(\lambda L) = 0 \qquad (13.10)$$

which implies

$$\lambda = \frac{n\pi}{L} \qquad (13.11)$$

where

$$n = 0, 1, 2, 3, \ldots. \qquad (13.12)$$

Putting the value of λ in equation (13.6)

$$\frac{n\pi}{L} = \sqrt{\frac{2mE}{\hbar^2}} \qquad (13.13)$$

or,

$$E = \frac{\hbar^2}{2m}\left(\frac{n\pi}{L}\right)^2 \qquad (13.14)$$

which is rewritten as

$$E_n = \frac{\hbar^2}{2m}\left(\frac{n\pi}{L}\right)^2 \qquad (13.15)$$

Thus, the valid solutions of Schrödinger equation are obtained only for certain values of E. These solutions constitute a discrete set containing one solution for each value of E. Discreteness of energy states is the foundational principle of quantum mechanics. The permissible values of energy E for the electron are separated. So, the electrons cannot move continuously from one energy level to another as expected from our customary experience with the macroscopic world. Instead, they can only leap from one energy level to another. Each energy level of an electron is called its energy state. Another noteworthy feature of Schrödinger's equation is that its solutions are orthogonal to each other. Orthogonality means that they do not overlap so that the integral of their product is zero.

It is re-emphasized that the wave functions represent the solutions of the Schrödinger's equation. The square of the wave function represents the probability of finding an electron at various locations in a given region around the atomic nucleus. It is called the energy probability density or probability density function.

Besides their distinct energy levels, electrons also have angular momentum. This means that that the electrons spin around an axis passing through their centers. The direction of spinning can only be either clockwise or counter-clockwise.

13.3.3 State superposition and collapsing

Calculations of the possible energies in which an electron can exist and the corresponding wave functions were carried out. These calculations yield the probability of finding the electron at certain positions at each energy level. The question arises: at which energy level will the electron be actually found? Linear combinations of the solutions for the differential equations are also solutions of the equation. This statement forms the basis of state superposition. The electron can be simultaneously present at any of the several possible energy states. There is a certain probability of its presence for each state. The actual state cannot be exactly predicted. It is only known after a measurement is made.

When the energy level of the electron is measured, the states collapse into the present state. In a quantum computation, a circuit is used to solve a problem by operating on its inputs. But the outputs are found in all possible states albeit in such a way that the correct answer is the state with highest probability. Upon reading the system, all the states collapse into the correct answer for the inputs that were fed.

13.4 Qubits and their properties

13.4.1 Qubits and quantum superposition

A quantum bit or qubit is a fundamental unit of quantum information. The concept of qubit and its computational perspective emerges if we choose to represent the two states for storing information as the ground state and the excited state of an atom, namely the two states of an atom:
   (i) the state in which an electron in the atom is at its lowest energy state, and
   (ii) the state in which the electron in the atom is at a higher energy level, where it reaches by absorbing energy from an external source such as heat, light, etc.

This point marks the beginning of the concept of a quantum bit or qubit, the quantum-mechanical version of classical bit of binary data. A qubit is a linear superposition of the states $|0\rangle$ and $|1\rangle$ of a two-level quantum-mechanical system pronounced as ket 0 and ket 1 in accordance with the bra-ket notation of quantum states. Thus, a qubit $|\psi\rangle$ is represented as a linear combination of $|0\rangle$ and $|1\rangle$ by the equation

$$|\psi\rangle = \alpha\,|0\rangle + \beta\,|1\rangle \tag{13.16}$$

where α and β are complex probability amplitudes whose values are constrained by the equation

$$|\alpha|^2 + |\beta|^2 = 1 \tag{13.17}$$

which implies that the probability that the qubit is measured in the $|0\rangle$ state is $|\alpha|^2$, the probability that it is measured in the $|1\rangle$ state is $|\beta|^2$, and the total probability of observation of the system in either $|0\rangle$ state or $|1\rangle$ state is 1. The probabilistic nature of a qubit makes it possible for the qubit to acquire an infinite range of values.

Superposition is the ability of quantum entities to populate states combining seemingly conflicting realities. Whereas a classical coin has two faces, either head up or tail up, a quantum coin can exist as head up, as tail up, as both head and tail up or in a state that lies between head up and tail up. Thus, the qubits can occupy superposition states that are truly amazing and incredible in regular thinking (Celsi and Celsi 2024).

This scenario is vastly different from that for a classical bit which can acquire only one of the two values, 0 or 1. This is one strikingly different behavior of a qubit from a classical bit. More differences will appear as we progress further.

13.4.2 Bloch sphere representation of a qubit

This is a visual geometric representation of a qubit (figure 13.3) in a unit sphere with X, Y, Z axes by the equation (Sahu and Gupta 2023)

$$|\psi\rangle = \cos 0.5\theta \, |0\rangle + \exp(i\phi) \sin 0.5\theta \, |1\rangle \tag{13.18}$$

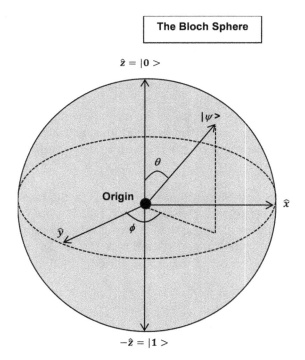

Figure 13.3. Block sphere representation of the qubit ψ. It is a unit sphere whose three axes represent the state of a qubit. Its surface defines all the possible qubit values.

where θ is the angle subtended by the wave function of the qubit with the z-axis and ϕ is the angle made by it with the x-axis. The z-axis denotes the probability of qubit measurement as a $|0\rangle$ or $|1\rangle$ state. The north pole of the sphere represents the $|0\rangle$ state and its south pole the $|1\rangle$ state. The x-axis signifies the real part of the state vector starting from the center of the sphere and terminating at a point with (x, y, z) coordinates.

13.4.3 Quantum entanglement

A group of qubits are said to be entangled when the quantum state of each qubit in the group cannot be described independently of the state of other members in the group, even when these qubits are far apart. If the spin of one of the qubits is measured on a certain axis to be directed clockwise, the spin of the other qubit around the same axis is counter-clockwise. It is a phenomenon counter to intuitive thinking, opening up the possibility to utilize wave–particle duality properties of qubits in certain circumstances for interaction with them by interference in quantum algorithms (Khan and Robles-Kelly 2020).

13.4.4 Quantum interference

Interference is a basic property of qubits arising primarily from wave–particle duality of electrons. The interference is constructive for in-phase electron waves and destructive for out-of-phase electron waves. In computation, interference influences the probability amplitudes. A popular example is Grover's algorithm used to search for values meeting specified criteria, where a diffusion operator works in collaboration with a problem-defining oracle. Constructive interference is applied for amplification of correct outcomes and destructive interference for minimization of all other possible solutions

13.4.5 Effects of measurements on qubits

A measurement is essential to read the result of a quantum computation. It extracts classical information from quantum information. The extracted classical information is processed by traditional algorithms on common computers. The measurement impacts the quantum state of a qubit owing to the inherent delicacy of the qubit. It causes a disruption of superposition of states and leads to a collapse of the wave function of the qubit from a superposition of multiple states into one definitive computational basis state or eigenstate, which is either $|0\rangle$ or $|1\rangle$. Which of these eigenstates appears on a measured qubit depends on the relative probabilities of the two states.

The measurement postulate states that a qubit remains in the superposition state until it is observed. It forms the basis of the no-cloning theorem, viz., the storage or recreation of an independent or identical quantum state to an arbitrary quantum state is impossible.

Laconically stated:

    (i) The superposition is a quantum-mechanical principle which allows a qubit to exist in multiple states simultaneously.

(ii) Entanglement of qubits is employed for superdense coding and quantum teleportation. Superdense coding is a communication procedure to transmit two classical bits from a sender to a user by using one qubit. Quantum teleportation is the transfer of the state of a qubit from one location to another without sending the qubit itself.

(iii) Interference of qubits affects the amplitudes of their measurement probabilities.

(iv) Measurement of qubits yields a classical bit from a qubit.

(v) The no-cloning theorem opposes the conventional methods of correcting errors, e.g., backup copies of a quantum state cannot be created during a computation, and subsequently used for correction.

13.4.6 Qubit decoherence and challenges in quantum computing

A serious challenge faced in quantum computing is the disturbance from the external environment. This difficulty is overcome by isolating the quantum system and cooling it to extremely low temperatures to obviate thermal noise perturbations.

Unsolicited deviations in the states of qubits are known as decoherence. They alter the state of the quantum system in an unpredictable and random manner. The time during which the qubit in a quantum system can store information reliably enforces a restriction on the time allotted to a computation. It is the decoherence time of the system.

Quantum error correction seeks to protect crucial quantum information by dispersing it across multiple qubits. Then the accidental interactions are identified and suitable corrections are done. The postliminary project is the realization of fault-tolerant quantum computers. These computers can work without terminating the program in case of environment-induced errors, and restarting the computation.

> *How quantum mechanics works fascinatingly?*
> *Wave–particle duality combined with the concepts of uncertainty and probability,*
> *Makes qubits behave counter-intuitively,*
> *Qubit superposition allows their existence across all possible quantum states simultaneously,*
> *Qubit entanglement permits manipulation of multiple qubits in a single operation rather than individually,*
> *Qubit interference enables their wave functions to interact constructively or destructively,*
> *Qubit measurement collapses their wave function to produce a definite state, alone and singly,*
> *Environmental noise causes decoherence impacting qubits adversely,*
> *Qubits are not binary bits, they act differently.*
> *Qubits and binary bits show obvious dissimilarities,*
> *Each has unmatched characteristics and peculiarities,*
> *It is like comparing apples and oranges in properties!*

13.5 Quantum computing devices

The semiconducting, superconducting, and trapped ion technologies form the deep-down groundwork on which the enormous edifice of quantum computing is being erected today. A certain level of maturity has been attained by quantum computing architectures based on two technology platforms, viz.,

(i) The superconducting qubits connected by microwave resonators.
(ii) The system of $^{171}Yb^+$ ions confined in a linear Paul trap and connected by laser-mediated interactions.

Superconducting qubits is a foremost quantum computing technology employed by several pre-eminent companies. Trapped ions offer a scalable opportunity for quantum computing by leveraging the intrinsic properties of ions as qubits. These two architectures have been experimentally compared by running a selected series of algorithms. Quicker gate clock speeds along with the benefits of a solid-state platform are provided by the superconducting qubits while the ion trap system displays superior qubits and reconfigurable connections (Linke *et al* 2017).

In photonics-based quantum computing, the photons produced by lasers propagate through optical components like beam splitters, phase shifters and detectors. Quantum computing is done by modulating individual photon states. Further options may be sought in the near future.

13.5.1 Silicon qubit devices

Here, the spin orientations of electrons or nuclei in silicon atoms are used as the two distinctive properties to encode the $|0\rangle$ and $|1\rangle$ qubit states. In order to achieve this aim, silicon quantum dots are made. The spin states of the electrons in silicon atoms of the quantum dots are utilized for encoding the quantum information. Alternatively, the spin states of the electrons or nuclei of single dopant atom impurities are used. The use of group V elemental donor impurities in a silicon wafer is an illustrious example of the donor method. The quantum dot and donor impurity-based silicon devices are fabricated by the standard complementary metal-oxide semiconductor technology.

13.5.1.1 Gate-controlled silicon quantum dots
Electrostatic control by properly designed and correctly biased Al or polysilicon gate electrodes enables a strong confinement of electrons in such a way that the lowest electronic orbital energy is far separated from the excited electronic energy states. Typically, the confinement length for electron in a Si quantum dot is $\sim$2 nm in the lateral direction and $\sim$10 nm along the vertical direction. For the donor electron, the confinement length is around 1.5 nm in all the three dimensions. Thus, the qubits available in the atoms of silicon quantum dots or a phosphorous donor implanted silicon substrates help in realizing the most compactly fabricated devices for quantum computing.

Two-qubit controlled-NOT (CNOT) gate operations are of paramount importance for quantum computing. A CNOT gate is fabricated by using electron spin qubits in Si quantum dots. It is applied to produce an entangled quantum state known as the Bell state. Compatibility with semiconductor electronics is an obvious advantage of this device (Zajac *et al* 2017).

The two-qubit device shown in figure 13.4 isolates two electrons in a silicon/ silicon germanium (Si/SiGe) quantum well heterostructure (Zajac *et al* 2017). The left and right gates control the energies of electrons trapped in the left and right quantum dots through voltages represented by the symbols V_{Left} and V_{Right}, respectively. The middle gate at a voltage V_{Middle} is used for controlling the exchange coupling. The charge occupancy of the double quantum dot (DQD) is

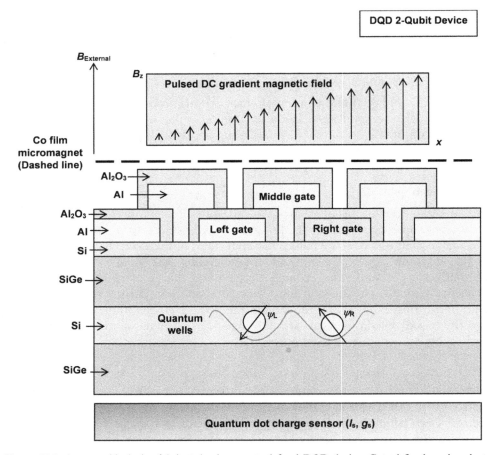

Figure 13.4. A two-qubit device fabricated using a gate-defined DQD device. Gate-defined semiconductor quantum dots are set up by lithographically fabricating nanoscale electrodes on the surface of or Si/SiGe semiconductor interfaces. These interfaces act as hosts to a two-dimensional electron gas at cryogenic temperatures. Application of voltages on the electrodes leads to the creation of electrostatic potential wells. These wells work as quantum dots for electrons. They have a well-defined spectrum of discrete electronic energy levels. Different characteristic parameters of quantum dots, such as the shape and depth of the quantum dots, the tunnel coupling between successive quantum dots, and the number of trapped electrons in them are accurately controlled by varying gate voltages.

detected by measuring either the current I_S or conductance g_S through a nearby quantum dot charge sensor.

A cobalt film is deposited on top of the device. An external magnetic field $B_{External}$ is applied for magnetization of the Co film to form a Co micromagnet. Spin control and manipulation for transitions between the quantum states $|\psi_L\rangle$ and $|\psi_R\rangle$ are done by a DC magnetic field gradient. The driving magnetic field B_z is suitably pulsed through application of microwave signals. An oscillating magnetic field is used. The frequency of the magnetic field is chosen to be equal to match the energy difference between the two qubit levels. The spin transitions take place at a rate dependent on the amplitude of the driving field.

13.5.1.2 Phosphorous donor implanted Si substrates

A buried donor charge qubit consisting of two dopant atoms in a semiconductor crystal is described (Hollenberg *et al* 2004). One of the dopant atoms is singly ionized. The charge qubit is shown in figure 13.5. The qubit is controlled with the help of surface electrodes. Fast readout is achieved with a RF single-electron transistor (SET). This transistor consists of a metallic island with a gate electrode separated by tunnel junctions from the source and drain leads. The gate electrode electrostatically controls the flow of electrons between the source and drain. Calculations reveal single gate times $\sim$50 ps or less. These are much shorter than the expected decoherence time of qubits.

A top-down nanofabrication, control and measurement process is developed for making a qubit consisting of two phosphorus dopant atoms $\sim$50 nm apart (Dzurak *et al* 2003). The phosphorus atoms are implanted in silicon using a low-energy $\sim$14 keV (kilo electron volt) $^{31}P^+$ ion beam through 25 nm diameter openings that are micromachined in a poly(methyl methacrylate) (PMMA) resist mask across 5 nm SiO_2 thin film to an average depth of 20 nm into Si. On-chip detector electrodes, integrated within the device structure enable single atom control by monitoring the current pulse resulting from electron–hole plasma created by the impacting ion. For device operation, voltage pulses are applied through surface electrodes to control the qubit. Dual SETs working close to the quantum limit deliver fast readout of signals.

13.5.2 Superconducting qubit devices

These are the quantum bits based on superconductivity. The phenomenon of superconductivity takes place in certain materials such as niobium, magnesium diboride and yttrium-barium-copper oxide at extremely low temperatures below the critical temperature T_C of the material. At such low temperatures, these materials show the ability to conduct direct current with zero electrical resistance. The vanishing of resistance allows qubits to store information without energy loss by resistive heating. Advantages of superconducting qubits are:

   (i) Scalability: The ability to grown in proportion to demand offers the possibility to build computers at a large scale.

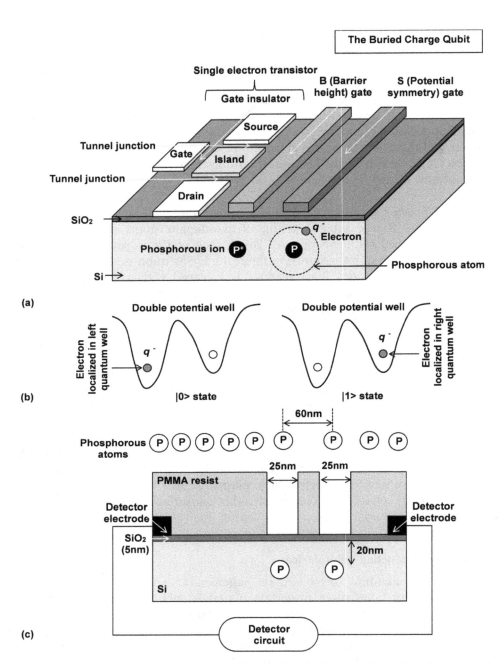

Figure 13.5. The donor charge qubit (a), its logical states $|0\rangle$ and $|1\rangle$, and (c) the implantation method for localization of phosphorous atoms buried at the desired locations. In part (a), there are two gates above the buried P-P$^+$ system: the B gate and the S-gate. The B-gate allows external control over the barrier height. The S-gate enables control over potential symmetry for manipulating localized qubit states, $|0\rangle = |L\rangle$ and $|1\rangle = |R\rangle$. Qubit initialization and readout are done using dual radio frequency (RF) SETs on the surface of the device. The SiO$_2$ film is a thin barrier layer used for isolation of the gates from the donors underneath.

In part (b), the quantum logical states $|0\rangle$ and $|1\rangle$ of a charge qubit are defined by the position of a single electron within a double well potential created by two P dopants. The quantum logic states are related to the lowest two states of the single valence electron localized in the double well formed by two donor phosphorus atoms. Part (c) depicts the technique for localization of individual phosphorus atoms at the desired qubit array sites. The array sites are defined in a nano-patterned ion-stopping resist, e.g., polymethyl-methacrylate (PMMA). The P dopants are implanted using a low-energy keV $^{31}P^{+}$ ion beam through the thin SiO_2 barrier layer to a mean depth of 20 nm into the substrate. On-chip electrodes connected to a detector circuit are used to monitor each ion strike, which creates an electron–hole plasma, producing a current pulse.

(ii) Long coherence times: Coherence time is the time duration for which the qubit can maintain its quantum state. A longer coherence time provides longevity of the qubits by prolonging their existence.

(iii) Compatibility with microelectronic fabrication technology: This feature enables integration of related electronics using the established semiconductor industry infrastructure.

13.5.2.1 Superconducting qubit archetypes

Superconducting qubit circuits contain a Josephson junction as the primary component. The other components of these circuits are the capacitors. In a Josephson junction, a non-superconducting tunneling barrier, e.g., a thin insulating aluminum oxide film is sandwiched between two superconducting aluminum layers; the critical temperature for aluminum (T_{CAl}) is 1.2 K.

The Josephson junctions act as non-linear inductors. Therefore, an inductance–capacitance oscillator is formed. Two parameters that determine the obtained electron energy levels in this device are the charging energy (E_C) of the superconducting island and the Josephson energy (E_J). The charging energy E_C is given by the well-known equation

$$E_C = \frac{q^2}{2C} \tag{13.19}$$

where q is the electronic charge and C is the total capacitance, directly or indirectly, in parallel with the junctions. The Josephson energy E_J is related to the Josephson inductance L and reduced Planck's constant $\hbar$ besides the electronic charge q as

$$E_J = \frac{\hbar^2}{4q^2L} \tag{13.20}$$

The ratio

$$\frac{E_J}{E_C} = \frac{\hbar^2}{4q^2L} \times \frac{2C}{q^2} = \frac{\hbar^2 C}{2q^4 L} \tag{13.21}$$

decides the type of qubit encoded by the Josephson junction. There are three primary qubit models, namely, the charge, flux and phase qubits (figure 13.6). Among these models,

(i) The charge qubits allow the fewest of Cooper pairs to tunnel through. A greater number of Cooper pairs are allowed tunneling by the flux qubits and still larger number by the phase qubits.

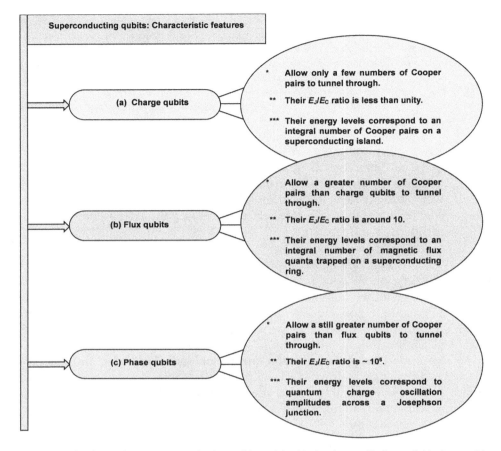

Figure 13.6. The three primary superconducting qubit models: (a) the charge, (b) flux and (c) phase qubits.

(ii) A charge qubit has a Josephson energy/charging energy (E_J/E_C) ratio < 1, a flux qubit has an E_J/E_C ratio ~10 while a phase qubit has an E_J/E_C ratio ~10^6.

(iii) Energy levels of a charge qubit relate to an integer number of Cooper pairs on a superconducting island, i.e., a small volume of semiconductor containing a controllable number of charge carriers. Energy levels of a flux qubit relate to an integer number of magnetic flux quanta trapped in a superconducting ring. Energy levels of a phase qubit relate to quantum charge oscillation amplitudes across a Josephson junction.

13.5.2.2 Archetype qubit realization

(i) Charge qubit: A small superconducting region called a superconducting island or Cooper pair box is formed between one lead of a capacitor C_G and a Josephson junction (figure 13.7). It is driven by a gate voltage,

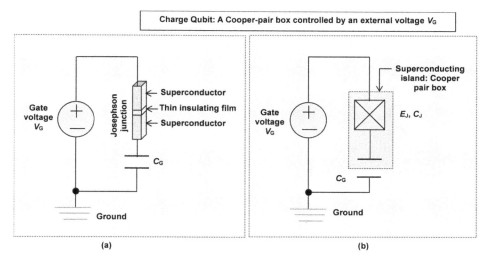

Figure 13.7. Superconducting charge qubit: (a) schematic and (b) equivalent circuit diagrams. The number of Cooper pairs is controlled by adjustment of gate voltage V_G.

$$V_G = \frac{q}{C_G} \tag{13.22}$$

through a capacitor C_G. Despite the equality of the energies of the classical states in which there are zero ($|0\rangle$) and one ($|1\rangle$) extra Cooper pairs in the box, quantum tunnelling through the Josephson junction forms two new quantum states. Of these, one state is formed by a symmetric superposition of the classical zero and one states ($|0\rangle + |1\rangle$). The other state is formed by an antisymmetric superposition ($|0\rangle - |1\rangle$). These two new quantum states differ in energy by $2E_J$. Superposition of these two states constitutes the charge qubit. The charge qubit is dynamically controlled by applying time-varying signals to the voltage gate.

(ii) Flux qubit: A Josephson junction is inductively coupled to a constant current source (figure 13.8). In the diagram, an external flux Φ_{External} originates from the magnetic field produced by a superconducting circuit on the left-hand side. It passes through the circuit containing the Josephson junction on the right-hand side. The magnetic flux embracing the super-conducting circuit on the right induces a current flow in that circuit. When the magnetic flux threading the loop, $\Phi_{\text{External}} = \frac{1}{2}$ the quantum of magnetic flux in a superconductor, the energy of the state with zero phase difference around the loop ($|0\rangle$) is = that of the state with a 2π phase difference ($|1\rangle$). One of these states pertains to a current circulating around the loop in the clockwise direction. The other state refers to a current moving in the loop in the counter-clockwise direction. Thus, two new quantum states are formed. The superposition of these states gives the flux qubit.

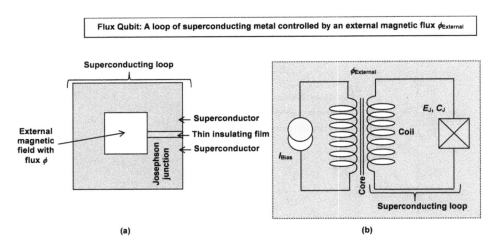

Figure 13.8. Superconducting flux qubit: (a) schematic and (b) equivalent circuit diagrams. The energy level structure of the qubit is controlled by adjustment of bias flux ϕ_{External}.

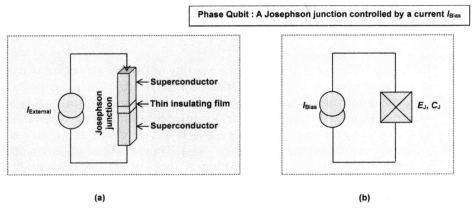

Figure 13.9. Superconducting phase qubit: (a) schematic and (b) equivalent circuit diagrams. The potential energy surface or energy landscape is tilted by adjusting the bias current I_{Bias}.

(iii) Phase qubit: This employs a single Josephson junction biased by a current I_{Bias}. (figure 13.9). The two energy levels $|0\rangle$ and $|1\rangle$ are defined by quantum oscillations of the phase difference between the electrodes of the junction. The superposition of their quantum states generates the phase qubit.

13.5.2.3 The transmon qubit design for charge noise mitigation
The coherence times of qubit archetypes are too short for error correction. A strategy for dephasing time enhancement is based on the reduction of linear noise sensitivity. This is possible by operating the qubits at optimal working points.

Transmon Qubit: Two Josephson Junctions Shunted by a Large Capacitance C_B

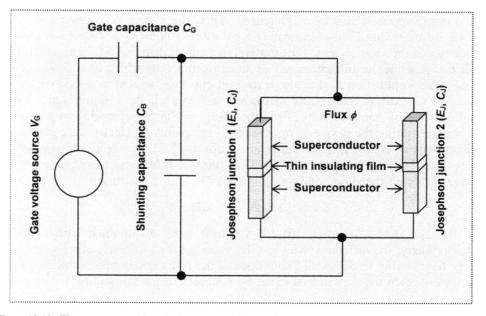

Figure 13.10. The transmon qubit: (a) schematic and (b) circuit diagrams. Transmon qubit measurement and control are performed by a microwave resonator. For coupling the qubits to the resonator, a capacitor is placed between the qubit and the resonator. The capacitor placement is done at the point of maximum electromagnetic field of the resonator.

By this approach it is possible to stretch the dephasing times by as much as three orders of magnitude. Hence, by proper quantum circuit design, the qubit performance can be significantly improved.

The transmon is the shortened form of 'transmission-line shunted plasma oscillation qubit' (Koch *et al* 2007). It closely resembles the charge qubit based on the Cooper pair box. The transmon consists of two superconducting islands (figure 13.10). Coupling between these islands is established through two identical Josephson junctions with capacitance C_J and Josephson energy E_J in a DC superconducting quantum interference device (SQUID) configuration. These junctions are shunted by a large capacitance C_B, which is matched by a gate capacitance C_G of comparable magnitude. The gate voltage source is represented by V_G.

13.5.2.4 Anharmonicity and charge dispersion of a Cooper pair box
Two vital performance metrics of a Cooper pair box are anharmonicity and charge dispersion of energy levels. If the anharmonicity is small, qubit operations will stimulate unwanted transitions. So, energy level anharmonicity should be high. On the other hand, a broad charge dispersion too is detrimental. This happens because it leads to a large change in qubit frequency whenever fluctuations take place in gate

voltage and environmental charge. So, the qubit becomes more sensitive to charge noise. Therefore, it is desirable for the qubit to exhibit less charge dispersion.

Both the anharmonicity and charge dispersion are governed by the E_J/E_C ratio of the qubit. Suppose the ratio is increased. Then the charge dispersion and noise sensitivity are reduced but at the same time the anharmonicity also decreases, which is counter to the requirement. Fortunately, the charge dispersion falls exponentially with E_J/E_C, while the anharmonicity declines algebraically with a slow power law in E_J/E_C. The implication is that the transmon can be operated at a comparatively higher E_J/E_C ratio to the Cooper pair box to achieve a high noise immunity with a small sacrifice of anharmonicity. The suppression of charge noise sensitivity removes the need of tuning the device to optimal working points by electrostatic gating.

For transmon qubits, the C value in equation (13.21) is selected to restrict the E_J/E_C ratio between the values 10–50. The quantum information is encoded by the charge states on the superconducting island. The qubit splitting is given by

$$E_{01} \approx \sqrt{8E_C E_J} \sim 5\,\text{GHz} \qquad (13.23)$$

For the chosen E_J/E_C ratio ~10–50, a robust qubit is obtained with adequate anharmonicity, the difference between the qubit splitting and any other splitting in the system. Owing to the small dimensions of the transmon qubits in the range of a few tens of microns, they can be made by microelectronic fabrication techniques.

13.5.3 Trapped ion qubit devices

Here the quantum information is encoded in the electronic energy levels of ions suspended in vacuum. The ions are confined in vacuum chambers. They are isolated for manipulation using lasers or microwave signals. To get the requisite ions, metals such as calcium or ytterbium are evaporated by resistive heating by passing an electric current through the metal body. The vapors of the metals are directed towards ion traps. The vaporized atoms are electrically neutral. Therefore, during loading them into the ion traps, they are photoionized.

Photoionization is the interaction of a photon with an atom or molecule resulting in the ejection of an electron from the atom/molecule and its transformation into a charged ionic species. In photoionization, the electron in the outermost valence shell is removed. Consequently, ions with single valence electrons are obtained. As the ions are charged particles, application of appropriate potentials through proximately located gate electrodes is done to confine them inside the ion traps.

13.5.3.1 Paul ion traps: construction and operation

Paul traps are prime candidates for processing quantum information (Siverns *et al* 2012). The commonly used RF Paul trap gate electrode configuration for ion trapping is shown in figure 13.11.

The ion trap consists of four electrodes in the form of four parallel rods with their centers located on the vertices of a square (Seidelin *et al* 2006). A common RF potential is applied to the two continuous electrodes called the RF electrodes. The middle parts of the other two segmented electrodes known as the control electrodes

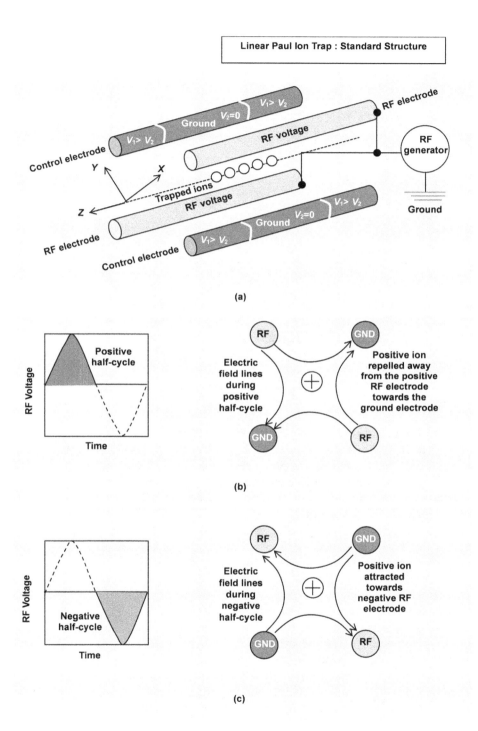

Figure 13.11. (a) A traditional Paul ion trap used for electron confinement in vacuum. (b) and (c): Radial (x–y) instantaneous electric field lines generated from the applied RF potential in the positive (b) and negative (c)

half-cycles of RF voltage. During the positive cycle of the RF voltage, the positively charged ion is pushed away from the RF electrodes towards the grounded control electrodes, whereas during the negative cycle of the RF voltage, the positively charged ion is pushed away from the grounded control electrodes toward the RF electrodes. (d) Creation of a linear trapping potential with the help of electrodes patterned on the planar surface of an insulating substrate. The electric field produces a quadrupole null resulting in the formation of a linear ion trap above the surface of the trap.

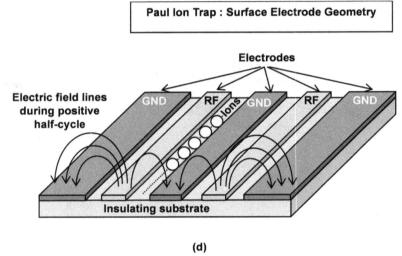

(d)

Figure 13.11. (Continued.)

are held at RF ground $V_2 = 0$ while their end parts are kept at a positive potential $V_1 > V_2$ (for positive ions) with respect to the middle parts. The RF signals produce a quadrupole RF field profile in the plane perpendicular to the axis of the rods. Hence, a nearly harmonic ponderomotive pseudopotential is created in the x–y plane. At the quadrupole 'null', the RF field vanishes. Hence the atomic ions are under the influence of a trapping potential.

This trapping potential typically takes the shape of a line serving as a one-dimensional linear potential where a chain of ions is readily trapped. For longitudinal confinement along the direction of z-axis of the trapping zone, appropriate static (DC) potentials V_1 are applied to the end parts of the segmented electrodes relative to $V_2 = 0$. Thus, the static end cap potentials V_1 and V_2 are necessary for axial trapping of the ions.

To put it in simple words: out of the four rods, two rods are subjected to oscillating potentials while the other two are grounded. As a result, an effective harmonic potential is produced in the x–y plane. If the RF of the signal is appreciably high, i.e., much larger than the natural frequency of motion of an ion in the chain known as its secular frequency, the ions are subjected to a confining potential. This is a situation in which a quadrupolar null is formed by the electric field. It is the region of zero electric field. Harmonic confinement in the z-plane is secured with the help of two extra DC electrodes.

The ions are laser-cooled to their vibrational ground states. This is imperative to make sure that thermal variations and electromagnetic field fluctuations do not induce any random excitation of these states, and hence cause motion of the ions. When the number of ions is small ~50, the ions are organized in the form of a chain along a straight line in the z-axis direction at gaps of 10 μm apart in such a manner that the sum-total forces from external fields counterbalance the Coulomb interaction forces between the ions.

The 3D design of four rods is simplified by placing the four rods in a common plane (figure 11.10(d)). The flat RF and control electrodes are alternately arranged in a planar structure. This structure is fabricated by photolithography and metal deposition on a low-loss polished fused quartz substrate (Seidelin *et al* 2006).

13.5.3.2 Eigenstates

A large number of eigenstates exist in the harmonic oscillator potential. These eigenstates pertain to the various modes of vibration of the trapped ions.

13.5.3.3 Types of trapped ion qubits

The energy levels of individual ions in the trap are used for defining the qubits and encoding the $|0\rangle$ and $|1\rangle$ states. According to the orbital energy levels used for encoding, the trapped ion qubits are subdivided into hyperfine and optical classes:

  (i) Hyperfine class: In the first class of qubits, the qubit sates are related to the hyperfine levels in the atomic s-orbital. For the ion $^{171}Yb^+$ which has a nuclear spin of 1/2, the qubit is encoded using the singlet $|S\rangle$ and $|T_0\rangle$ configurations of the electron and nuclear spins. The $|T_0\rangle$ state is separated from other triplet states $|T_-\rangle$ and $|T_+\rangle$ by the application of a small DC magnetic field. The qubit splitting is ~12.6 GHz for $^{171}Yb^+$. This splitting is determined by the hyperfine interaction between the electron and the nucleus. It is not disturbed by magnetic field changes up to the first order.

  (ii) Optical class: The second class of qubits depends on the s-orbital and d-orbital electronic energy levels. The energy splitting is ~411 THz for ^{40}Ca, which is equivalent to 729 nm. In absence of heterogenous magnetic and electric fields in the trap, the trapped ion qubits show high reproducibility. Presence of such non-uniform fields in the trap modifies the energy levels by Zeeman and Stark effects.

Zeeman effect is the splitting of atomic energy levels and the associated spectral lines of a molecule into several components when the molecule is subjected to a static magnetic field. Stark effect is the splitting and shifting of spectral lines of atoms, molecules or ions in an external electric field.

13.6 Summary and the way forward

13.6.1 Highlights of chapter 13 at a glance

  (i) Present-day computers are essentially non-quantum mechanical in operation although their structural elements, the transistors work on quantum-mechanical principles. The logic gates of a classical computer are irreversible in nature.

(ii) Quantum computers are based on quantum physics-based approaches to information processing. Quantum mechanics presents the enthralling concepts of wave–particle duality and the uncertainty, which bolster the distinctiveness of quantum computing from classical computing.

(iii) For easy understanding, elementary quantum mechanics was reviewed including Schrödinger's equation, the particle-in-a-box problem, state superposition and collapsing phenomena.

(iv) The basic information units of quantum computing are the bits compliant with the laws of quantum mechanics. These bits, called quantum bits or qubits, can exist in a superposition of 0 and 1 states concomitantly. They have different properties than the binary bits used in traditional digital computers. We know that existence in only one of the two states, 0 or 1 is permitted for binary bits.

(v) Bloch sphere representation of a quantum bit was explained.

(vi) Atomic and subatomic particles display interesting properties of quantum superposition, entanglement and interference. The meanings of these terms were elucidated.

(vii) Measurements carried out on qubits cause their wave functions to collapse forcing the qubit into a definite 0 or 1 state.

(viii) A major roadblock in quantum computing is the decoherence of the qubit on interaction with the environment which shortens the survival time of qubits.

(ix) The three types of quantum computing devices that were discussed in this chapter include the silicon, superconducting and trapped qubit devices.

(x) Silicon spin qubit is one of the primary modalities for representation and processing of information in quantum computing. In the silicon qubit modality, the coding of quantum information is done by exploiting the intrinsic spins of electrons.

(xi) Prominent silicon qubit devices that were considered herein are the Si quantum dot furnished with gate-control and phosphorous donor implanted Si substrates.

(xii) A comparison of the different approaches to build quantum computers reveals that superconducting qubits offer the most promising opportunities. In this approach, superconductive circuits are used to create qubits on a chip made of niobium or aluminum maintained at very low temperatures. Each qubit acts as an inductance–capacitance circuit resonator which is tuned to behave like an atom with two quantum energy levels by applying microwave pulses. These mechanisms allow manipulation and reading the state of the qubit.

(xiii) Superconductive quantum computers are the most prevalent type. This architecture is being developed by Google and IBM.

(xiv) In another method, the stable electronic states of trapped ions in an ultrahigh vacuum environment are controlled with precision lasers. Under the trapped ion qubit devices, construction and operation of Paul ion traps was discussed, and the types of trapped ion qubits were expounded.

(xv) Keywords for this chapter are: Qubit, Bloch sphere, silicon quantum dot, superconducting qubit, transmon, trapped ion.

13.6.2 Getting ready to begin chapter 14

After reviewing the basic properties of qubits and the principal methods of their realization and maneuvering, we move to logic gates used in quantum computing and the arithmetic circuits made by using them for addition and multiplication of qubits.

References

Celsi M R and Celsi L R 2024 Quantum computing as a game changer on the path towards a net-zero economy: a review of the main challenges in the energy domain *Energies* **17** 1–22

Dzurak A S *et al* 2003 Charge-based silicon quantum computer architectures using controlled single-ion implantation arXiv:cond-mat/0306265

Hollenberg L C L, Dzurak A S, Wellard C, Hamilton A R, Reilly D J, Milburn G J and Clark R G 2004 Charge-based quantum computing using single donors in semiconductors *Phys. Rev.* B **69** 113301-1–4

Khan T M and Robles-Kelly A 2020 Machine learning: quantum vs classical *IEEE Access* **8** 219275–94

Koch J, Yu T M, Gambetta J, Houck A A, Schuster D I, Majer J, Blais A, Devoret M H, Girvin S M and Schoelkopf R J 2007 Charge-insensitive qubit design derived from the Cooper pair box *Phys. Rev.* A **76** 1–21

Linke N M, Maslov D, Roetteler M, Debnath S, Figgatt C, Landsman K A, Wright K and Monroe C 2017 Experimental comparison of two quantum computing architectures *Proc. Natl Acad. Sci.* **114** 3305–10

Sahu H and Gupta H P 2023 Quantum computing toolkit: from nuts and bolts to sack of tools arXiv:2302.08884 [quant-ph]

Seidelin S *et al* 2006 Microfabricated surface-electrode ion trap for scalable quantum information processing *Phys. Rev. Lett.* **96** 1–5

Siverns J D, Simkins L R, Weidt S and Hensinger W K 2012 On the application of radio frequency voltages to ion traps via helical resonators *Appl. Phys.* B **107** 921–34

Vedral V and Plenio M B 1998 Basics of quantum computation *Prog. Quantum Electron.* **22** 1–39

Webb G 2024 The Schrödinger equation and the two-slit experiment of quantum mechanics *Discrete Cont. Dyn. Syst.* S **17** 1981–2008

Zajac D M, Sigillito A J, Russ M, Borjans F, Taylor J M, Burkard G and Petta J R 2017 Resonantly driven CNOT gate for electron spins *Science* **359** 439–42

IOP Publishing

AI-Processor Electronics
Basic technology of artificial intelligence
Vinod Kumar Khanna

Chapter 14

Quantum logic gates and circuits

A quantum gate is the quantum analog of the classical logic gate of digital computers. It is a basic structural unit of a quantum circuit which operates on a small number of qubits to deliver a predefined output. A quantum circuit is a graphical representation of the sequence of the quantum gates used and the measurements performed in a quantum computation. Various quantum logic gates, namely, the Hadamard, T phase, NOT, Pauli-X, controlled NOT (CNOT), controlled-Z, SWAP and controlled phase gates as well as higher-level Toffoli and Fredkin gates are described. Conventions of reading quantum circuit diagrams are explained. Operation of a quantum full adder consisting of registers and gates is discussed, elaborating the role of the NOT gate, the sum calculation with CNOT gates and the carry C-out calculation with Toffoli gates. A primary tool used for performing arithmetic operations on a quantum computer is the quantum Fourier transform (QFT). The QFT adder and multiplier circuits are presented.

14.1 Introduction

In this chapter, we study quantum gates which are classified into three main types:
- (i) single-qubit gates, operating on a single qubit and used for creating superpositions or performing rotations;
- (ii) two-qubit gates, operating on two qubits and used to produce entanglement between qubits; and
- (iii) measurement gates applied for extracting information from the quantum state of a qubit.

We also explore quantum circuits which are built by arranging quantum gates in specific orders, and are used to execute various quantum algorithms.

14.2 Quantum logic gates

In short called a quantum gate, this is an elementary quantum circuit. It works on a small number of qubits to perform defined quantum operations on them. A quantum

doi:10.1088/978-0-7503-6259-7ch14
14-1

gate is represented either by a schematic diagram illustrating its function or in the form of a unitary matrix (Muñoz-Coreas and Thapliyal 2022). A unitary matrix U is a square matrix of complex numbers whose inverse is equal to its complex conjugate transpose U^H. In symbols,

$$U^{-1} = U^H \qquad (14.1)$$

In other words, the multiplication of a unitary matrix U with its complex conjugate transpose U^H yields an identity matrix I of the same order as U, as expressed by the equation

$$UU^H = U^H U = I \qquad (14.2)$$

The unitary matrix is a unitary operator. The unitarity postulate in quantum mechanics states that the time evolution of a closed quantum system is represented by a unitary operator. Unitary operators are linear operators on a Hilbert space that preserve the inner product and norm of a vector. They ensure that the total probability of all possible quantum states remains constant and sums to 1, meaning that there is no loss of information during operations, thus leading to safeguarding and conservation of information, and hence reversibility of quantum gates.
Figure 14.1 presents the circuit diagram symbols of a variety of quantum gates.

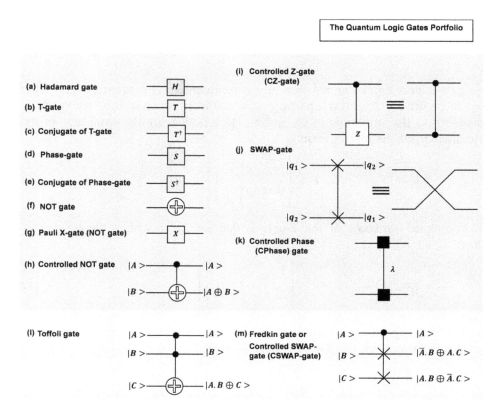

Figure 14.1. The quantum gate symbols.

14.2.1 The Clifford + T gate set

This is a commonly used set of gates because it comprises an approximate universal collection of gates in which fault tolerance can be provided with the codes for correction of quantum errors.

14.2.1.1 Hadamard gate

This is named after the French mathematician Jacques Hadamard. In the circuit diagram, it is represented by a square with the letter 'H' inscribed inside (figure 14.1 (a)). This gate is a single-qubit logic gate that turns a state of $|0\rangle$ or $|1\rangle$ into a superposition of $|0\rangle$ and $|1\rangle$. The input state $|0\rangle$ is turned by the Hadamard gate into the output state $(|0\rangle + |1\rangle)/\sqrt{2}$. Conversely, the input state $|1\rangle$ is turned by the Hadamard gate into the output state $(|0\rangle - |1\rangle)/\sqrt{2}$. In matrix form, the Hadamard gate is expressed as

$$H \equiv \frac{1}{\sqrt{2}}\begin{bmatrix} 1 & 1 \\ 1 & -1 \end{bmatrix} \tag{14.3}$$

The square of the Hadamard gate is the identity matrix

$$H^2 = I \tag{14.4}$$

Thus, application of the Hadamard gate two times to a qubit in a row produces no effect on it.

14.2.1.2 T gate

This gate is also known as $\pi/4$ gate and is represented by a square with a letter T inside it (figure 14.1(b)). It is a phase gate which working on a single qubit applies a phase shift to the amplitude of the qubit's $|1\rangle$ state leaving the amplitude of its $|0\rangle$ state unchanged. Its matrix form is

$$T = \begin{bmatrix} 1 & 0 \\ 0 & \exp\left(\dfrac{i\pi}{4}\right) \end{bmatrix} \tag{14.5}$$

The conjugate transpose of the T-gate is denoted by T with a dagger as shown below:

$$T^{\dagger} = \begin{bmatrix} 1 & 0 \\ 0 & \exp\left(-\dfrac{i\pi}{4}\right) \end{bmatrix} \tag{14.6}$$

On a circuit diagram, a square with $T^{\dagger}$ written inside is used for representing the conjugate transpose of the T-gate (figure 14.1(c)).

14.2.1.3 Phase gate

This is also called the S-gate or Z90 gate because it signifies a rotation by 90° about the z-axis. It defines a single qubit operation given by

$$S = \begin{bmatrix} 1 & 0 \\ 0 & i \end{bmatrix} \tag{14.7}$$

Its circuit symbol is a square inside which S is written (figure 14.1(d)).

The conjugate transpose of the S-gate is denoted by S with a dagger as

$$S^\dagger = \begin{bmatrix} 1 & 0 \\ 0 & -i \end{bmatrix} \tag{14.8}$$

A square with $S^\dagger$ inside is used to denote it on a circuit diagram (figure 14.1(e)).

14.2.1.4 NOT gate
This is a gate which turns a $|0\rangle$ state into a $|1\rangle$ state, and a $|1\rangle$ state into a $|0\rangle$ state. Its circuit diagram symbol is a circle with a cross inside or a square with the letter 'X' written inside it (figure 14.1(f)). In matrix notation, it is written as

$$X = \begin{bmatrix} 0 & 1 \\ 1 & 0 \end{bmatrix} \tag{14.9}$$

It is also called a Pauli-X gate (figure 14.1(g)) because its effect on a quantum state is equivalent to a 180° rotation about the X-axis of that state.

14.2.1.5 CNOT gate
The CNOT or controlled NOT gate is also known as the Feynman gate. This gate has two input pins and two output pins (figure 14.1(h)). It is a two-qubit gate in which a two-input operation takes place. One input is the control qubit $|A\rangle$. The second input is the target qubit $|B\rangle$. The result of the operation is determined by the control qubit $|A\rangle$. The result is observed on the target qubit $|B\rangle$. When the control qubit $|A\rangle$ is in $|0\rangle$ state, the target qubit $|B\rangle$ is left unchanged. However, when the control qubit $|A\rangle$ is in $|1\rangle$ state, the control qubit is left unchanged and a Pauli X-gate acts on the target qubit $|B\rangle$.

The Pauli-X gate is the quantum analog of the classical NOT gate. It flips a $|0\rangle$ state to $|1\rangle$ state and vice versa, i.e., $|1\rangle$ to $|0\rangle$.

$$|B\rangle = |B\rangle \text{ when } |A\rangle = 0 \tag{14.10}$$

$$|B\rangle = |1 \oplus B\rangle = |\bar{B}\rangle \text{ when } |A\rangle = 1 \tag{14.11}$$

where $\oplus$ is XOR operation.

The CNOT gate is a controlled gate. The reason is that its action, i.e., the NOT operation depends on the value of the qubit $|A\rangle$. It is a kind of NOT gate but its action is not exactly the same as a that of a NOT gate. It performs the NOT operation only subject to a particular condition. If this condition is not met, the NOT operation does not take place. The matrix version of this gate is

$$CNOT = \begin{bmatrix} 1 & 0 & 0 & 0 \\ 0 & 1 & 0 & 0 \\ 0 & 0 & 0 & 1 \\ 0 & 0 & 1 & 0 \end{bmatrix} \quad (14.12)$$

14.2.1.6 Controlled Z-gate

This is a Clifford and symmetric two-qubit gate in which one qubit acts as the control qubit and the other qubit as the target qubit (figure 14.1(i)). When the control qubit $= |0\rangle$, both qubits remain unaffected. But when the control qubit $= |1\rangle$, a phase flip is applied using a Pauli Z-gate to the target qubit. Its matrix representation is

$$CZ = \begin{bmatrix} 1 & 0 & 0 & 0 \\ 0 & 1 & 0 & 0 \\ 0 & 0 & 1 & 0 \\ 0 & 0 & 0 & -1 \end{bmatrix} \quad (14.13)$$

14.2.1.7 SWAP gate

This is a symmetric and Clifford gate which swaps the states of two qubits $|q_1\rangle$ and $|q_2\rangle$ (figure 14.1(j)). Its operation is mathematically described as

$$\text{SWAP}(|q_1 > \otimes |q_2>) = |q_2 > \otimes |q_1> \quad (14.14)$$

In matrix form, it is expressed as

$$SWAP = \begin{bmatrix} 1 & 0 & 0 & 0 \\ 0 & 0 & 1 & 0 \\ 0 & 1 & 0 & 0 \\ 0 & 0 & 0 & 1 \end{bmatrix} \quad (14.15)$$

14.2.1.8 Controlled phase gate

This is a diagonal and symmetric Clifford gate (figure 14.1(k)). It induces a phase on the state of the target qubit. This operation depends on the control state. Its matrix form is

$$CPhase = \begin{bmatrix} 1 & 0 & 0 & 0 \\ 0 & 1 & 0 & 0 \\ 0 & 0 & 1 & 0 \\ 0 & 0 & 0 & e^{i\lambda} \end{bmatrix} \quad (14.16)$$

14.2.2 Higher-level quantum gates

14.2.2.1 Toffoli gate

It is a controlled gate with three inputs $|A\rangle$, $|B\rangle$, $|C\rangle$ and three outputs $|A\rangle$, $|B\rangle$, $A . B \oplus C\rangle$ (figure 14.1(l)). The two qubits $|A\rangle$ and $|B\rangle$ are control qubits. The third qubit $|C\rangle$ is the target qubit. Because it resembles a CNOT gate with an extra control bit, it is often referred to as a CCNOT gate or doubly controlled CNOT gate. If both $|A\rangle = 1$ and $|B\rangle = 1$, then $|C\rangle$ is complemented. However, when $|A\rangle = 1$ and $|B\rangle = 0$, or $|A\rangle = 0$ and $|B\rangle = 1$, $|C\rangle$ is left unchanged. In matrix form, this gate is expressed as

$$CCNOT = \begin{bmatrix} 1 & 0 & 0 & 0 & 0 & 0 & 0 & 0 \\ 0 & 1 & 0 & 0 & 0 & 0 & 0 & 0 \\ 0 & 0 & 1 & 0 & 0 & 0 & 0 & 0 \\ 0 & 0 & 0 & 1 & 0 & 0 & 0 & 0 \\ 0 & 0 & 0 & 0 & 1 & 0 & 0 & 0 \\ 0 & 0 & 0 & 0 & 0 & 1 & 0 & 0 \\ 0 & 0 & 0 & 0 & 0 & 0 & 0 & 1 \\ 0 & 0 & 0 & 0 & 0 & 0 & 1 & 0 \end{bmatrix} \tag{14.17}$$

14.2.2.2 Fredkin gate

This is a three-qubit controlled gate. It takes three inputs $|A\rangle$, $|B\rangle$, $|C\rangle$ and returns three outputs $|A\rangle$, $|\overline{A} \cdot B + A \cdot C\rangle$, $|A \cdot B + \overline{A} \cdot C\rangle$ (figure 14.1(m)). The control qubit is $|A\rangle$ and the target qubits are $|B\rangle$ and $|C\rangle$. When the control qubit $|A\rangle = 0$, the target qubits $|B\rangle$ and $|C\rangle$ are left unchanged. But when the control qubit $|A\rangle = 1$, the target qubits $|B\rangle$ and $|C\rangle$ are interchanged or swapped. Hence, this gate swaps the two target qubits depending on the state of the control qubit $|A\rangle$. Bearing this property in mind, it is called a controlled swap gate or CSWAP gate. Another name for it is a conservative logic gate. Its matrix form is

$$CSWAP = \begin{bmatrix} 1 & 0 & 0 & 0 & 0 & 0 & 0 & 0 \\ 0 & 1 & 0 & 0 & 0 & 0 & 0 & 0 \\ 0 & 0 & 1 & 0 & 0 & 0 & 0 & 0 \\ 0 & 0 & 0 & 1 & 0 & 0 & 0 & 0 \\ 0 & 0 & 0 & 0 & 1 & 0 & 0 & 0 \\ 0 & 0 & 0 & 0 & 0 & 0 & 1 & 0 \\ 0 & 0 & 0 & 0 & 0 & 1 & 0 & 0 \\ 0 & 0 & 0 & 0 & 0 & 0 & 0 & 1 \end{bmatrix} \tag{14.18}$$

14.3 Quantum circuits

In a quantum computer, the qubits are stored in a quantum register. The quantum register is a sequence of distinct qubit variables that is identified with the set of qubit variables inside it. A quantum register is practically realized by one of the several technologies, e.g.,

(i) qubits represented by electron spins in a quantum dot structure fabricated by lithographic positioning,

(ii) qubits based on the nonlinear resonant superconducting circuits containing Josephson junctions, or

(iii) qubits stored in the stable electronic states of ions confined in free space using electromagnetic fields.

Quantum operations are performed on qubit registers by activation with laser or microwave pulses to manipulate the qubits.

14.3.1 Conventions of reading a quantum circuit diagram

We look at the quantum circuit from the viewpoint of its electrical analog in computing. An electrical circuit is a model of computation representing the transport of information through wires.

Placed across the wires are several gates performing the operations on the information for carrying out necessary transformations *en route*.

Similarly, a quantum circuit consists of a linear arrangement of quantum logic gates for implementing a quantum algorithm. On receiving one or more qubits at its input, it applies unitary transformations on them via quantum gates. Consequently, the state of the qubit(s) is altered. As already said in section 14.2, a unitary transformation is a mathematical operation which changes the state of a qubit without changing the length of its quantum state vector and the inner product between vectors, thus preserving the probability amplitude of the quantum state. It acts like a rotation in the Hilbert space. Also, it results in a unitary matrix. Finally, a measurement gate is used to obtain the output as a classical bit.

In a quantum circuit diagram (figure 14.2), the wires represent the qubits of information. The gates act as the manipulators of information by acting on the

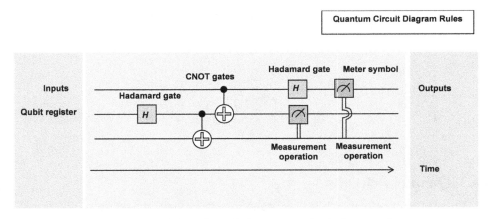

Figure 14.2. Familiarization with the quantum circuit diagram rules.

qubits. The information flows from left towards the right direction. The time increases in this direction. Earlier operations in time are therefore performed towards the left, and those occurring later are shown rightwards. It must be emphatically noted that:

(i) Horizontal lines in the diagram represent qubit registers. The direction of computation is from left to right.

(ii) An input value is written at the left end of the horizontal qubit register line.

(iii) The output value is written at the right end of the horizontal qubit register line.

(iv) Gates acting on one qubit register are displayed as squares or boxes over the horizontal line for the register. The letters for the names of the gates are inscribed inside the squares, e.g., the Hadamard gate is shown with an H symbol inside its square.

(v) Gates acting on more than one qubit register are shown in a different manner, e.g., an N-qubit gate is shown as a square spanning over the N-qubit registers represented by their horizontal lines.

A controlled gate, e.g., the CNOT gate is shown as a vertical line joining the horizontal lines for the two qubit registers. A dot is marked at one end of the vertical line connected to the control qubit register shown as a horizontal line. A cross is placed inside a circle at the other end of the vertical line linked to the target qubit register, shown by the other horizontal line.

(vi) A meter symbol placed on a horizonal line for a qubit register is an indicator of a measurement operation. This operation takes a qubit register as the input and supplies classical information as the output.

(vii) The number of inputs to a quantum gate must be equal to the number of outputs. This is anticipated because of the unitary and reversible nature of quantum operations. Any square or box on the circuit diagram must have as many outputs as there are inputs feeding it. If the number of outputs exceeds the number of inputs, the quantum operation will be irreversible. Irreversible operations are contradictory to the prescribed norms of quantum computing.

14.3.2 Quantum full adder

Full adder circuits are generally made by stringing together logic gates of the present-stage full adder with Carry-in from the previous-stage full adder (Coggins 2020). So, a full adder circuit has three inputs. Let us denote the three inputs by A, B and C-in (figure 14.3). The symbol 'C-in' stands for 'Carry-in'. The Carry-in' means that the carry is brought forward as an input from the previous full adder to the present full adder.

A full-adder circuit has two outputs: Sum and C-out. The symbol 'C-out' stands for 'Carry-out'. Hence, the carry is taken forward as an input from the present full adder to the next full adder in the same way that C-in was brought from the preceding full adder. This C-in was the C-out for the preceding full adder.

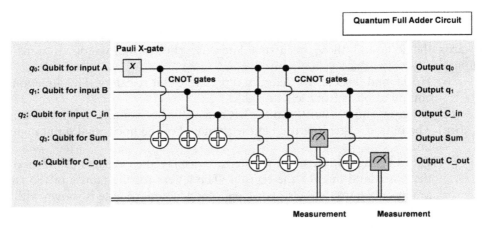

Figure 14.3. The quantum full adder circuit consisting of one Paul X-gate, two CNOT gates, three Toffoli (CCNOT) gates, along with inputs and outputs.

Five qubit registers are arranged in a proper configuration for implementation of a full adder operation on a quantum computer.

14.3.2.1 Registers
The registers used for building the full adder are named as follows in accordance with the inputs, A, B, C-in, and outputs, Sum and C-out:

q_0 qubit register for input A,
q_1 qubit register for input B,
q_2 qubit register for C-in,
q_3 qubit register for Sum, and
q_4 qubit register for C-out.

14.3.2.2 Gates
The quantum circuit of the full adder contains one NOT gate, three CNOT gates and three Toffoli (CCNOT) gates. The various steps executed by the NOT gate, CNOT gates and Toffoli gates for computation and measurement of sum and carry are given in the subsections below and described with shorthand notation for explanatory purpose.

14.3.2.3 Role of the NOT gate
A Not gate is applied to inputs q_2 for C-in, q_1 for B and q_0 for A because they are all initialized to $|0\rangle$. We denote these steps by writing the quantum circuit with symbol x placed before the corresponding input, e.g., Quantum circuit $x(q[?])$:

Quantum circuit $x(q[2])$,
Quantum circuit $x(q[1])$,
Quantum circuit $x(q[0])$

This means that $q_2 = 0$, $q_1 = 0$ and $q_0 = 0$ will be flipped to 1.

14.3.2.4 Sum calculation with CNOT gates
For sum calculation, the CNOT gate is applied to the target q_3 from all the inputs, i.e., input q_0 for A, q_1 for B, and q_2 for C-in as control inputs, as shown below:
  (i) A CNOT gate is applied to target q_3(Sum) from input q_0 for A as control bit. This means that if A = 1 then q_3 will be flipped to 1; we write this step as
        Quantum circuit $cx(q[0], q[3])$
  (ii) A CNOT gate is applied to target q_3 (Sum) from input q_1 for B as a control bit. This means that if B = 1 then q_3 will be flipped to 1; we write this step as
        Quantum circuit $cx(q[1], q[3])$
  (iii) A CNOT gate is applied to target q_3 (Sum) from input q_2 for C-in as control bit. This means that if C-in = 1 then q_3 will be flipped to 1; we write this step as
        Quantum circuit $cx(q[2], q[3])$

Note: If all the inputs A = B = C = 0, then q_3 will remain 0.

14.3.2.5 Carry C-out calculation with Toffoli gates
For carry C-out calculation, the Toffoli gate is applied to q_4 with input combinations q_0 for A, q_1 for B; q_0 for A, q_2 for C-in; and q_1 for B, q_2 for C-in as the control qubits, as under:
  (i) A Toffoli gate is applied with target q_4 (Sum) and input combination q_0 for A, q_1 for B as the control qubits, written as
        Quantum circuit $ccx(q[0], q[1], q[4])$
  (ii) A Toffoli gate is applied with target q_4 (Sum) and input combination q_0 for A, q_2 for C-in as the control qubits, written as
        Quantum circuit $ccx(q[0], q[2], q[4])$
  (iii) A Toffoli gate is applied with target q_4 (Sum) and input combination q_1 for B, q_2 for C-in as the control qubits, written as
        Quantum circuit $ccx(q[1], q[2], q[4])$

14.3.2.6 Measurements
Two measurements are done:
  (i) After the calculation for sum is completed, a measurement is performed, written as
        Quantum circuit measure $(q[3], c[0])$
  (ii) After the calculation for C-out is completed, a measurement is performed, written as
        Quantum circuit measure $(q[4], c[1])$

Thus, we obtain the results of the computation as the two measurements giving the outputs. q = quantum register, c = classical register.

14.4 Quantum Fourier transform (QFT) as a primary tool for performing arithmetic operations on a quantum computer

After the initially suggested quantum circuits for arithmetic operations that were derived by creating reversible versions of classical circuits (Vedral *et al* 1996), an attractive alternative route to quantum computing was opened by the quantum Fourier transform and the QFT adder circuit. In the realm of quantum computing, the Quantum Fourier transform (QFT) occupies a prestigious place because it acts as the pivotal support of many arithmetic operations performed on a quantum computer (Ruiz-Perez and Garcia-Escartin 2017).

As an example of QFT-based arithmetic circuit, we choose the quantum adder circuit using QFT. To facilitate understanding of the QFT adder, we peruse the QFT adder step-by-step in the order: discrete Fourier transform, quantum Fourier transform and inverse quantum Fourier transform, controlled phase gate, extension to the desired number of inputs, signed addition operation and subtraction, and non-modular additions. Then we discuss a variant on Draper's QFT adder (Draper 2000) followed by its extension to non-modular additions.

It is relevant to note that modular addition of two numbers is done by adding together the given numbers, dividing the sum by a modulus (absolute value), and writing the remainder as a result, e.g., taking the modulus 2

$$(5 + 3) \bmod 2 = 8 \bmod 2 = 0 \qquad (14.19)$$

The QFT adder is the precursor to build a QFT multiplier, which will be undertaken next.

Non-modular addition is that done without using modular arithmetic. By following this order, we will be able to easily understand the QFT adder construction and its operation as well as the QFT multiplier circuit, two main components of a quantum computer, which hold the key to performing various arithmetic operations on it.

14.4.1 The DFT

The impetus for quantum Fourier transform has its underpinnings in the adaptation of Fourier transform to its discrete version known as the discrete Fourier transform (DFT). It is recalled that the Fourier transform is an extension of the Fourier series. The Fourier series itself is an expression of a periodic function as the sum of sine and cosine functions.

The QFT representing the quantum analog of DFT acts as a quintessence part of many quantum algorithms. The DFT transforms a sequence consisting of N complex numbers

$$\{X_k\} = X_0, \; X_1, \; X_2, \; \dots, \; X_{N-1} \qquad (14.20)$$

into the ordered N-number list given by

$$\{x_j\} = x_0, \; x_1, \; x_2, \; \dots, \; x_{N-1} \qquad (14.21)$$

by the relational equation

$$X_k = \sum_{j=0}^{N-1} x_j \exp\left(-\frac{2i\pi}{N}kj\right) \tag{14.22}$$

14.4.2 The QFT and IQFT

14.4.2.1 The QFT

Identically to the DFT, the action of QFT on the state $|x\rangle$ in a computation basis

$$\{|0\rangle, \ |1\rangle, \ \dots, \ |d-1\rangle\} \tag{14.23}$$

of a system having d dimensions, is described by the equation

$$\text{QFT of } |x\rangle \to \text{QFT}|x\rangle = \frac{1}{\sqrt{d}}\sum_{k=0}^{d-1} \exp\left(i\frac{2\pi xk}{d} \ |k\rangle\right) \tag{14.24}$$

The advantage gained by using QFT is that it allows the encoding of a number x in the relative phases of the states of a uniform superposition. The main part of the superposition is the sum of all the states $|k\rangle$ in the computational basis. Each state has the amplitude $1/\sqrt{d}$.

Actually, the QFT is a change of basis. The state QFT $|x\rangle$ encoding $|x\rangle$ in this new transformed basis is represented by $|\theta(x)\rangle$.

14.4.2.2 The IQFT

The inverse quantum Fourier transform (IQFT) is defined as

$$\text{IQFT of } |k\rangle \to \text{IQFT}|k\rangle = \frac{1}{\sqrt{d}}\sum_{x=0}^{d-1} \exp\left(-i\frac{2\pi xk}{d} \ |x\rangle\right) \tag{14.25}$$

The direct and inverse quantum Fourier transforms serve as the conveyances for interconversion between the computational basis and the phase representations. Phase encoding is an essential common feature in all QFT-based calculations

14.4.3 Controlled phase gate

Recall the controlled phase gate discussed in section 14.2.1.8. It is the basic element in the quantum operators that work on the phases of quantum states. The Controlled-Z or CZ gate is a two-qubit gate which applies a phase flip to the target qubit using the Pauli-Z gate when the control qubit is $|1\rangle$. Otherwise, i.e., for the control qubit $=|0\rangle$, the target qubit remains unaffected. If the two input qubits are $|x\rangle$ and $|y\rangle$, we can write

$$\text{CZ of } |x\rangle y\rangle \to \text{CZ }|x\rangle y\rangle = \exp\left(i\pi xy\right)| \ x\rangle y\rangle \tag{14.26}$$

Generalization of the gate for a system of d dimensions, i.e., for qudits gives

$$\text{CZ of } |x\rangle y\rangle \text{ for } d \text{ dimensions} \to \text{CZ }|x\rangle y\rangle \text{ for } d \text{ dimensions} = \exp\left(i\frac{2\pi xy}{d}\right)| \ x\rangle y\rangle \tag{14.27}$$

These gates are the key structural units frequently used in quantum arithmetic formulations.

Two numbers encoded in the computational basis are added together by phase encoding one of the numbers, and thereafter applying a controlled phase shift on it. The sequence of three operations, namely *IQFT*, *CZ* ad *QFT* involved in addition are expressed as

$$IQFT_2.\ CZ.\ QFT_2\ |x\rangle y\rangle = |x\rangle x + y(\text{mod } d) \tag{14.28}$$

where the subscript '2' indicates that two numbers are being added.

14.4.3.1 The first operation (QFT)

This is encoding the number $|y\rangle$ from computational basis into the phase basis through the equation

$$\text{QFT}_2 \text{ of } |x\rangle y\rangle \rightarrow \text{QFT}_2 |x\rangle y\rangle = \frac{1}{\sqrt{d}} \sum_{k=0}^{d-1} \exp\left(i\frac{2\pi yk}{d}\ |x\rangle\ |k\rangle \right) \tag{14.29}$$

14.4.3.2 The second operation (phase shift)

This is the introduction of a phase shift using a phase gate. This operation is tantamount to a modulo-d addition in the phase basis. In mathematical parlance, we write

$$\text{CZ of} \frac{1}{\sqrt{d}} \sum_{k=0}^{d-1} \exp\left(i\frac{2\pi yk}{d} x\ |k\rangle \right)$$

$$\rightarrow \frac{1}{\sqrt{d}} \sum_{k=0}^{d-1} \exp\left(i\frac{2\pi yk}{d} \right) \exp\left(i\frac{2\pi xk}{d} \right) |x\rangle\ |k\rangle \tag{14.30}$$

$$= \frac{1}{\sqrt{d}} \sum_{k=0}^{d-1} \exp\left\{ i\frac{2\pi (x + y)k}{d} \right\} |x\rangle\ |k\rangle$$

14.4.3.3 The third and final operation (IQFT)

This is the inverse quantum Fourier transform for restoration of the result into the computational basis. The mathematical representation of this operation is:

$$\text{IQFT}_2 \text{ of } \frac{1}{\sqrt{d}} \sum_{k=0}^{d-1} \exp\left\{ i\frac{2\pi (x + y)k}{d} \right\} |x\rangle\ |k\rangle$$

$$\rightarrow \text{IQFT}_2 \frac{1}{\sqrt{d}} \sum_{k=0}^{d-1} \exp\left\{ i\frac{2\pi (x + y)k}{d} \right\} |x\rangle\ |k\rangle \tag{14.31}$$

$$= \frac{1}{d} \sum_{k,l=0}^{d-1} \exp\left\{ i\frac{2\pi (x + y)k}{d} \right\} \exp\left(-i\frac{2\pi kl}{d} \right) |x\rangle\ |l\rangle = |x\rangle x + y(\text{mod } d)$$

14.4.4 Extension to the desired number of inputs

The quantum adder is stretchable to the desired number of inputs. For N integers x_1, x_2, ..., x_N encoded into the state $|x_1\rangle$, $|x_2\rangle$, ..., $|x_N\rangle$, repetition of the summation process in equation (14.28) is done. On carrying out the repetition with the operation

$$\text{IQFT}_N.\ CZ_{1,\,N} \ ... \ CZ_{N-2,\,N}.\ CZ_{N-1,\,N}.\ QFT_N\ |x_1\rangle\ |x_2\rangle,\ ...\ ,\ |x_{N-1}\rangle\ |\ x_N\rangle \quad (14.32)$$

an output state

$$|x_1\rangle\ |x_2\rangle,\ ...\ ,\ |x_{N-1}\rangle\ |x_1 + x_2 + ... + x_N(\text{mod } d)\rangle \quad (14.33)$$

is produced.

During the application of a quantum gate on a subset of all the possible states, subindices are introduced to indicate the states that are being acted upon by the gate. To elaborate, the subindex c in $CZ_{c,t}$ is the index of the control state c. Similarly, the subindex t in $CZ_{c,t}$ is the index of the target state t.

14.4.5 Non-modular additions

In place of modulo d addition, non-modular additions are possible. This is done by encoding the data in a system of a larger dimension d'. The modulo d' addition is the same as regular arithmetic addition for the range of values under consideration. Taking the case of two integers x and y between 0 and $d - 1$, we note that the sum will always be confined between 0 and $(2d - 2)$. This remark is suggestive of the adequacy of a system of dimension $d' = 2d - 2 + 1 = 2d - 1$.

Likewise, for adding N numbers, a system with dimension $d' = Nd - N + 1$ is necessary. Therefore, the QFT and CZ circuits are revised for this dimension.

14.4.6 Signed addition operation and subtraction

A signed addition for numbers up to $d/2$ is realized using these circuits by encoding the positive numbers $x < d/2$ into states $|x\rangle$ and the negative numbers $- x$ into states $|d - x\rangle$. This is done in both cases in the computational basis. After evaluation of QFT, positive numbers are correlated with the phases $\exp(i2\pi x/d) < \pi$, corresponding to a phase $\exp(i2\pi xk/d)$ supplementing each state $|k\rangle$. The negative numbers are correlated with negative phases, which are equivalent to phases above π for $k = 1$. Signed addition will then be achieved with the QFT adder, or we can say a subtraction has been done.

14.5 QFT adder for numbers encoded in n qubits

14.5.1 A variant on Draper's QFT adder

Draper (2000) presented a method for performing addition on a quantum computer using the quantum Fourier transform. Draper *et al* (2006) gave an efficient addition circuit, adopting techniques from the standard carry-lookahead reckoning circuit. Ruiz-Perez and Garcia-Escartin (2017) describe a modified form of Draper's technique for computing full arithmetic additions instead of modular additions. We shall explain their formulation here.

Let the numbers to be added be α, β, which are integers from 0 to $2^n - 1$. In binary representation, the equations for α and β are written in the form

$$\alpha = \alpha_1 \alpha_2 \alpha_3 \cdots \alpha_n \tag{14.34}$$

and

$$\beta = \beta_1 \beta_2 \beta_3 \cdots \beta_n \tag{14.35}$$

where

$$\alpha = \alpha_1 2^{n-1} + \alpha_2 2^{n-2} + \ldots + \alpha_n 2^0 \tag{14.36}$$

and

$$\beta = \beta_1 2^{n-1} + \beta_2 2^{n-2} + \ldots + \beta_n 2^0 \tag{14.37}$$

Then

$$\alpha = |\alpha_1\rangle \otimes |\alpha_2\rangle \otimes \cdots \otimes |\alpha_n\rangle \tag{14.38}$$

and

$$\beta = |\beta_1\rangle \otimes |\beta_2\rangle \otimes \cdots \otimes |\beta_n\rangle \tag{14.39}$$

The symbol $\otimes$ stands for the product measure. Considering two measurable spaces along with measures on them, one can obtain a product measurable space and a product measure on that space analogous to the definition of the Cartesian product of sets.

The QFT of α is determined to evolve $|\alpha\rangle$ into $|\theta(\alpha)\rangle$

$$|\theta(\alpha)\rangle = \text{QFT}|\alpha\rangle = \frac{1}{\sqrt{N}} \sum_{k=0}^{N-1} \exp\left(i\frac{2\pi\alpha k}{N}|k\rangle\right) \tag{14.40}$$

where

$$N = 2^n \tag{14.41}$$

The sum is calculated using the n qubits representing the number β. Thus $|\theta(\alpha)\rangle$ is taken into $|\theta(\alpha + \beta)\rangle$.

The addition is done by fragmentation of the CZ gates into conditional rotation phase gates given by the equation

$$P_l = \begin{bmatrix} 1 & 0 \\ 0 & \exp\left(\dfrac{2\pi i}{2^l}\right) \end{bmatrix} \tag{14.42}$$

which are controlled by the n qubits representing the number β. All the gates together introduce a total phase $\exp(2\pi i\beta k/N)$ for each state $|k\rangle$. Hence, the qubits containing β keep the same value. At the same instant, the qubit register containing the QFT of α now stores $|\theta(\alpha + \beta)\rangle$.

14.5.2 Extension of the plan to non-modular additions

For this purpose, the number α is encoded into a larger register. It is represented with $n + 1$ qubits; so

$$| \alpha \rangle = | 0 \rangle | \alpha_1 \rangle | \alpha_2 \rangle \ldots | \alpha_n \rangle \tag{14.43}$$

The QFT of $|\alpha\rangle$ is found using the circuit in figure 14.4.

$$\text{QFT of } |\alpha\rangle \rightarrow QFT |\alpha\rangle = \frac{1}{\sqrt{2^{n+1}}} \sum_{k=0}^{2^{n+1}-1} \exp\left(i\frac{2\pi\alpha k}{2^{n+1}} | k\rangle\right) \tag{14.44}$$

The states $|\theta_j(\alpha)\rangle$ symbolize the jth qubit of the phase state $|\theta(\alpha)\rangle$ encoding α.

Now the number β is added by means of controlled phase rotation gates as in Draper's arrangement. The numbers α and β are added by application of the conditional phase rotation

$$\exp 2\pi i \frac{(\alpha_j + \beta_j)2^{n-j}k_s 2^{n+1-s}}{2^{n+1}} = \exp 2\pi i (\alpha_j + \beta_j)k_s 2^{-j+n-s} = \exp 2\pi i \frac{(\alpha_j + \beta_j)k_s}{2^{j+s-n}} \tag{14.45}$$

that depends on the jth qubits of the representation of the numbers to be added. It is applied on the sth qubit in the transformed register enclosing the superpositions of states

$$k = | k_1 \rangle \otimes | k_2 \rangle \otimes \cdots \bigotimes | k_{n+1} \rangle \tag{14.46}$$

The jth qubit of $|\beta\rangle$ controls the gate. A change is produced only when $\beta_j = 1$. The conditional phase rotation gates

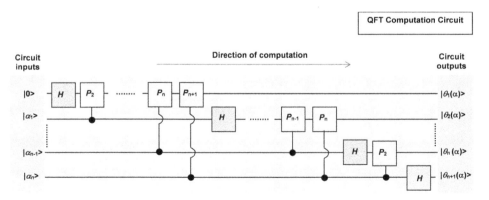

Figure 14.4. The quantum Fourier transform circuit showing the circuit inputs/outputs, Hadamard and conditional rotation phase gates. The direction of computation is indicated by an arrow.

$$P_l = P_{j+s-n} \tag{14.47}$$

are selected when

$$j + s - n > 0 \tag{14.48}$$

But if

$$j + s - n \leqslant 0 \tag{14.49}$$

the phase

$$\exp\left(2\pi 2^{n-j-s}i\right) = 1 \tag{14.50}$$

is applied, and the state is not changed. Figure 14.5 shows the resulting circuit.

The state $|\theta(\alpha+\beta)\rangle$ is now stored in the register containing the QFT of α. Subsequently performing inverse QFT and returning to the computational basis, the correct result is obtained in the qubit register with the sum.

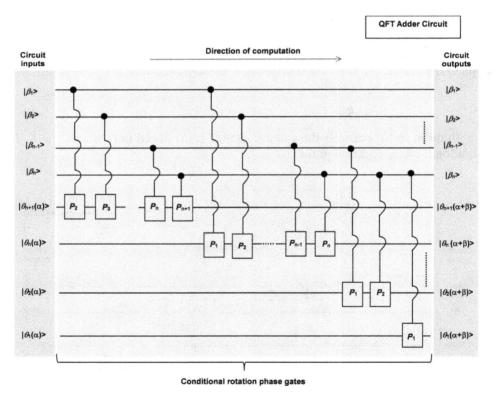

Figure 14.5. The quantum adder circuit consisting of the circuit inputs/outputs and conditional rotation phase gates. Computation takes place in the direction of the arrow.

14.6 QFT multiplier

Multiplication of two n-bit numbers is done by carrying out n consecutive controlled QFT additions, yielding the result as a $2n$-qubit register encoding the number $\alpha \cdot \beta$, as shown in figure 14.6.

In this calculation, the inputs to the first adder block $2^0\Sigma$, are the n qubits representative of a number α and $2n$ qubits for the number 0.

14.6.1 The ancillary state

An ancillary state is prepared taking the QFT of number 0, i.e., $|\theta(0)\rangle$ to support the process. A series of conditional phase rotation gates are applied by the $2^0\Sigma$ block, thereby evolving the state into $|\theta(0 + \alpha)\rangle$. This block, controlled by the least significant qubit of $|\beta\rangle$, produces the output state

$$| \alpha \rangle \, | \, \theta(0 + \beta_n \, 2^0\alpha)\rangle \qquad (14.51)$$

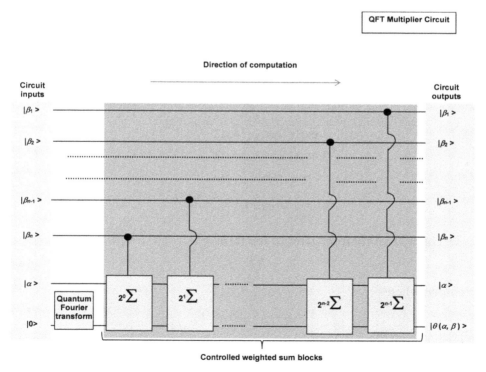

Figure 14.6. The quantum multiplier circuit consisting of the QFT block and a chain of controlled weighted sum blocks. Circuit inputs/outputs are given and the direction of computation is shown by an arrowhead.

14.6.2 The output state

The second QFT adder is controlled by β_{n-1}. At this point, the phase addition is scaled by a factor 2^1 leading to the output state

$$| \theta(0 + \beta_n\, 2^0\alpha + \beta_{n-1}\, 2^1\alpha)\rangle \tag{14.52}$$

A similar procedure is followed for the further blocks. On application of the last QFT adder, the output state is

$$| \theta(0 + \beta_n\, 2^0\alpha + \beta_{n-1}\, 2^1\alpha + \ldots + \beta_2\, 2^{n-2}\alpha + \beta_1\, 2^{n-1}\alpha)\rangle = | \theta(0 + \alpha\beta)\rangle = | \theta(\alpha\beta)\rangle \tag{14.53}$$

Computation of the product $\alpha\cdot\beta$ relies on the judicious selection of the conditional phase rotation gates for running each QFT adder block.

The output state obtained after computing the QFT of $|0\rangle$, is

$$QFT \text{ of } |0\rangle \rightarrow QFT\,|0\rangle = \frac{1}{\sqrt{2^{2n}}} \sum_{k=0}^{2^{2n}-1} \exp\left(j\frac{2\pi 0 k}{2^{2n}}\,|k\rangle\right) = |\theta(0)\rangle \tag{14.54}$$

$$k = k_1\, 2^{2n-1} + k_2\, 2^{2n-2} + \ldots + k_{2n}\, 2^0 = \sum_{s=1}^{2n} k_s 2^{2n-s} \tag{14.55}$$

The state $|\theta(0)\rangle$ is taken to

$$| \theta(0 + \beta_j 2^{n-j}\alpha)\rangle \tag{14.56}$$

by using the phase rotation gates controlled by β_j and by picking up each α_i in order to apply a phase rotation.

$$\begin{aligned}
\exp\, i\frac{2\pi\,(\alpha_i 2^{n-i}\beta_j 2^{n-j})k_s 2^{2n-s}}{2^{2n}} \\
= \exp\, i\, 2\pi\,(\alpha_i\beta_j k_s)2^{n-i+n-j+2n-s-2n} \\
= \exp\, i\, 2\pi\,(\alpha_i\beta_j k_s)2^{n-i+n-j+2n-s-2n} \\
= \exp\, i\, 2\pi\,(\alpha_i\beta_j k_s)2^{2n-i-j-s} \\
= \exp\, i\, \frac{2\pi\,(\alpha_i\beta_j k_s)}{2^{i+j+s-2n}}
\end{aligned} \tag{14.57}$$

Hence, conditional rotation gates of the form

$$P_l = P_{i+j+s-2n} \tag{14.58}$$

are chosen where

$$i + j + s - 2n > 0 \tag{14.59}$$

for execution of the QFT adder block controlled by β_j. The size of the ancillary register is taken such that the exact value of $\alpha\cdot\beta$ is obtained, rather than a modular multiplication. Note that the size of the ancillary register can be changed at one's pleasure. The P_l gates are altered accordingly to accomplish any desired modular multiplication in moduli which are powers of two of the ancillary register size (Ruiz-Perez and Garcia-Escartin 2017).

14.7 Summary and the way forward

14.7.1 Highlights of chapter 14 at a glance

(i) Quantum logic gates act as the counterparts of logic gates used in classical computing. They are quantum-mechanical equivalents of the classical logic gates.

(ii) Working as the quantum substitute of the classical logic gate of digital computers, the quantum gate is a basic structural unit of a quantum circuit which operates on a small number of qubits to deliver a predefined output. It is used to manipulate the quantum states of qubits to perform computational operations.

(iii) Various quantum logic gates, namely, the Hadamard gate, T Gate, phase gate, NOT and CNOT gates as well as higher-level Toffoli and Fredkin gates were described.

(iv) Quantum circuits are made by connecting quantum gates, in different topologies to perform specific computational operations. A quantum circuit is a graphical depiction of a sequence of quantum gates used for executing a quantum computation, inclusive of the measurements performed.

(v) Conventions of reading quantum circuit diagrams were explained.

(vi) Operation of quantum full adder consisting of registers and gates was discussed elaborating the role of the NOT gate, sum calculation with CNOT gates and carry C-out calculation with Toffoli gates.

(vii) A frequently used component in many quantum algorithms is the QFT, which is a linear transformation on quantum bits. Therefore, the QFT has become a primary tool used for performing arithmetic operations on a quantum computer. It is the quantum replica of the DFT.

(viii) Quantum adders and multiplier circuits are constructed using QFTs which drastically diminishes the number of gates used in comparison to classical adders and multipliers (Yao and Liu 2024). The QFT adder and multiplier circuits were presented.

(ix) Keywords for this chapter are: Hadamard gate, T Gate, phase gate, NOT and CNOT gates, Toffoli and Fredkin gates, QFT, QFT adder, multiplier.

14.7.2 Getting ready to begin chapter 15

(i) Equipped with the knowledge of performing arithmetic operations with quantum logic gates, it is time now to enquire into the working of a quantum processing unit (QPU). The QPU is the brain of a quantum computer where the qubits reside and take part in various computations. This comparison is similar to the role played by the central processing unit (CPU) on the bits of a binary digital computer because here the bits assist in calculations. The CPU is the matching part of a classical computer corresponding to the QPU.

(ii) But unlike a CPU, the QPU needs complex infrastructural support for its operation because the qubits need to be maintained in an isolated system at extremely low temperatures and are controlled using lasers or microwave pulses.

(iii) The succeeding chapter will cast a look into the internal construction and details of working of the QPU.

References

Coggins M 2020 *Quantum Programming 101: How to Perform Addition on Quantum Computers* (Resonance Alliance Inc.) https://thequantuminsider.com/2020/01/10/quantum-programming-101-how-to-perform-addition-on-quantum-computers/#:~:text=To%20implement%20a%20Full%20Adder,output%20of%20the%20Full%20Adder).&text=For%20calculating%20the%20Sum%20we,will%20be%20flipped%20to%201

Draper T G 2000 Addition on a quantum computer arXiv:quant-ph/0008033v1

Draper T G, Kutin S A, Rains E M and Svore K M 2006 A logarithmic-depth quantum carry-lookahead adder *Quantum Inf. Comput.* **6** 351–69

Muñoz-Coreas E and Thapliyal H 2022 Everything you always wanted to know about quantum circuits *Wiley Encyclopedia of Electrical and Electronics Engineering* ed J G Webster (New York: Wiley) pp 1–18

Ruiz-Perez L and Garcia-Escartin J C 2017 Quantum arithmetic with the quantum Fourier transform *Quantum Inf. Process.* **16** 152

Vedral V, Barenco A and Ekert A 1996 Quantum networks for elementary arithmetic operations *Phys. Rev.* A **54** 147–53

Yao J and Liu D 2024 Matrix multiplication on quantum computer arXiv:2408.03085 [quant-ph]

IOP Publishing

AI-Processor Electronics
Basic technology of artificial intelligence
Vinod Kumar Khanna

Chapter 15

Quantum processing unit

Incompatibility of classical memory for quantum computing is explained. Efficient storage/modification of quantum or classical data is accomplished by using a quantum random-access memory (QRAM). A QRAM applies the principle of quantum superposition for simultaneously accessing all the data stored in the memory cell to assure the superior performance of quantum algorithms. Data reading and writing in a bucket-brigade QRAM are described and the feasibility of its optical implementation is discussed. Quantum computers are understood through layered architectures, viz., 3-, 4-, and 5-layer architectures. Organization of a superconducting quantum computer is illustrated by considering a generic architecture. The superconducting transmon qubits produced inside a dilution refrigerator are manipulated by sequences of microwave pulses. Electronic systems used for producing the control signals and reading the quantum information are described. Quantum computers have been built by several industry giants. Improved quantum processors are being assembled with continuous advancements and refinements. Future computing is likely to follow a hybrid paradigm in which the quantum computers will work in tandem with classical computers in a seamless, integrated environment. Each type of computing will complement the strengths and weaknesses of its counterpart.

15.1 Introduction

This final chapter reconnoiters the functioning of QRAM, and investigates the distinctly different structural organization of a quantum computer from a normal central processing unit (CPU), with the quantum computer completely isolated from external thermal, electrical and magnetic disturbances. Appearance wise, a quantum computer looks vastly different from an ordinary diminutive computer. Its shape is contemplated as a large-size lighting fixture with branches, much like a giant chandelier.

doi:10.1088/978-0-7503-6259-7ch15　　　15-1　　　© IOP Publishing Ltd 2024. All rights,

15.2 Quantum random-access memory (QRAM)

15.2.1 Incongruity of the classical memory for quantum computing

Conventional memory used by a classical computer working on binary digits 0 or 1 is incompatible with a quantum computer whose operation is based on quantum bits or qubits. The qubits can store a tremendously large amount of information compared to binary bits. A mere 500 qubits can epitomize the amount of information unrepresentable with $>2^{500}$ classical bits. Determination of prime factors of a 2048-bit number would require millions of years with a standard computer in common use today (Andika 2023).

15.2.1.1 The classical RAM
In a classical computer, the RAM chip is mounted in the vicinity of the CPU on the motherboard to provide a quickly accessible temporary memory space where data for immediate retrieval and use are saved. The traditional RAM stores data as long as the computer is not rebooted.

15.2.1.2 Problems to be addressed by a QRAM
In analogy to the RAM chip in a common computer, a quantum RAM or QRAM is necessary as an equivalent replacement for a quantum computer. But in a quantum computer, the memory issues are different. The main cause is that the quantum memory storage is impersistent, typically lasting for a short duration of time ~ 100 ms. Furthermore, a quantum state cannot be read without altering it. Therefore, the classical RAM becomes unusable for the quantum processing unit (QPU). It cannot be straightway applied to the QPU because the reading of the data causes the quantum bit to emancipate all the states stored by it at that point. A value of either zero or one is obtained because the measurement operation done by a classical memory device collapses the wavefunction.

Keeping the aforementioned aspect in view, the QRAM is specially designed for a quantum computer as an arrangement to allow storage and recovery of quantum states without collapsing their superposition. Unlike the RAM, the QRAM applies quantum mechanics for encoding information in a decoherence-resistant fashion.

15.2.2 Perambulation from RAM to bucket-brigade QRAM

The traditional RAM consists of an input register and an output register. These registers are connected to an array of memory cells. The memory cells are organized in rows and columns. Each memory cell holds one bit of data. Data accessing and manipulation are done with the help of address lines, data lines, and control lines. These lines are energized by read and write enable signals. The CPU sends the memory address via the address lines to the memory cells. For a read signal, the data at that address is read from the relevant memory cell. This operation is called the read operation. In the case of a write signal, data is stored in the particular memory cell of a given address. This is the write operation.

Unlike the usual RAM, a QRAM does not have a lattice of memory cell arrays. Instead, it has a bifurcation graph structure based on the binary tree. The reader

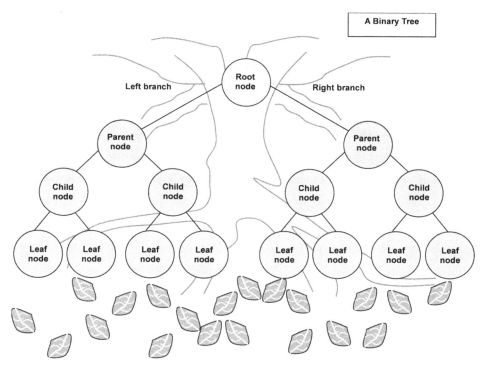

Figure 15.1. A binary tree. This is a nonlinear data structure with a maximum of two children for each parent; hence called a 'binary tree'. The node of a tree represents a termination point. Its root is the topmost node while its parent node is any node, besides the root, having at least one sub-node of its own. The child node of a tree is a node that directly originated from a parent node when moving away from the root, whereas its leaf node is a node with no child.

may call to mind that a binary tree is a tree-shaped data structure consisting of several nodes or intersection points (figure 15.1). In this QRAM called a bucket-brigade QRAM, the root of the tree is its topmost node. Each node has a maximum of two children. These children are referred to as the left child and the right child of the node. Nodes with no children are known as leaves. Obviously, the non-leaf nodes are the internal nodes. The leaf nodes are the memory cells. The remaining nodes act as switches. These switches are actually logic gates that can route the address state to the correct memory cell.

The name 'bucket brigade' is given to this QRAM because it works like a human chain in which a water bucket is transferred from one human participant to the next. In this transference of the bucket, each participant stands at his/her own position and does not move from the place occupied by the participant. Nonetheless, the water bucket is sent from one end of the human chain to the other end. In the case of QRAM, the address state is routed through the tree structure in place of a water bucket.

Like the RAM, a QRAM has three components: an input register, an output register and the QRAM itself. The QRAM consists of memory cells located at the leaf nodes. The rest of the nodes serve as switches that are used for routing the address state to the correct cell. In dissimilarity to a qubit, which is a two-level system with states $|0\rangle$ and $|1\rangle$, each switch is a three-level system. It has the states $|\cdot\rangle$, $|0\rangle$, and $|1\rangle$, and is called a qutrit. The state with a dot is earmarked as the wait state. The necessity of the wait state $|\cdot\rangle$ arises because starting from all switches in this state, a switch changes from $|\cdot\rangle$ state to the received state. The next time the same switch receives another qubit state, then this switch will route it to one of its children node switches. Hence, the memory cells that were not accessed previously will not be disturbed (Phalak *et al* 2023).

15.2.2.1 Data reading and data writing
The data is read/written from the memory by moving along the following steps (figure 15.2):

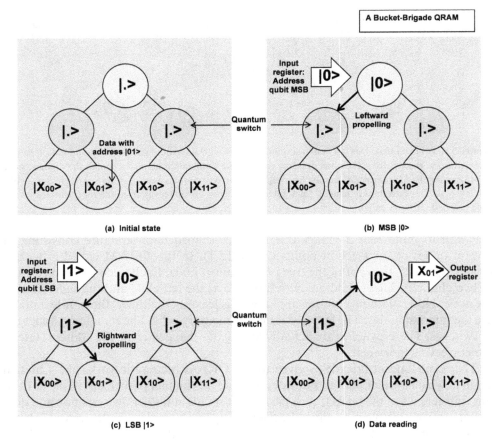

Figure 15.2. Working of a QRAM with two address lines, and four memory cells, and single address in the input register: (a) initial state, (b) and (c) activation of switches by the input register and (d) readout of data by the output register. The address qubits are sequentially loaded in the input register in a top-down style, beginning from the MSB and moving to the LSB. The MSB qubit $|0\rangle$ is loaded first. It changes the state of the root quantum switch. The LSB qubit $|1\rangle$ follows the MSB bit. It routes the switch in the direction of the memory cell $|X_{01}\rangle$.

Step (i): Incipiently, all the qutrit states are put in the wait state (figure 15.2(a)).

Step (ii): The root node launches the address qubits as a quantum bus signal. The launching is done in a sequential top-down fashion starting with the most significant bit (MSB) (figure 15.2(b)), and ending with the least significant bit (LSB) (figure 15.2(c)). The switches are triggered in accordance with the address bits. On reception of a qubit state by the switch, it changes from the $|\cdot\rangle$ state to either $|0\rangle$, or $|1\rangle$ state. To which of these two states this changeover takes place depends on the input qubit.

Step (iii): A switch is propelled to left for a $|0\rangle$ state (figure 15.2(b)). It is pushed towards the right for a $|1\rangle$ state (figure 15.2(c)). The connected superposition of the address is interweaved with a set of switches of the tree network. Hence it portrays a superposition of routes through the graph to sketch the total route followed along the switches in the direction of the target memory cell.

Step (iv): After examining the activated quantum switches in the sketched route, the output register reads the data. The data reading is performed via the superposition of the address contour linking all the quantum switches to acquire the superposition of all the data. It is enabled by the change in the state of the quantum bus according to the quantum information of the memory cells. For doing so, an n qubit operation of controlled NOT (CNOT) gate is carried out. A possible circuit description for an $N = 2^3$ qubit bucket-brigade QRAM is presented by Arunachalam *et al* (2015).

Step (v): For data reading, the bus returns following the same route (figure 15.2 (d)). As requested by the RAM, it brings back with it the superposition of quantum information as output.

Step (vi): For data writing, the bus writes new data by reversal of the above-created route.

15.2.2.2 A conceivable optical implementation of QRAM
The address qubits are incorporated in the input register of QRAM as photons that can be transmitted in a sequential order (Giovannetti *et al* 2008). The encoding of photons is embedded in their polarization into one of the two states, either a polarization state $|0\rangle$ or a polarization state $|1\rangle$. The polarization state $|0\rangle$ of a photon is able to mobilize the $|\text{wait}\rangle \rightarrow |\text{left}\rangle$ atomic transition of the qutrit. In dissimilarity, the polarization state $|1\rangle$ of a photon has the ability to organize the $|\text{wait}\rangle \rightarrow |\text{right}\rangle$ atomic transition of the qutrit.

The qutrits are placed as trapped atoms inside cavities. At the outset, all the qutrits are initialized to the lowest energy state $|\cdot\rangle$. The qubits encoded in the photons traverse the cavity by encountering the trapped atom-based qutrits on their path. In this encounter, the three states of the qutrits are realized as three different energy levels. The $|\cdot\rangle$ state is at the lowest energy. The two higher energy levels are: $|0\rangle$ and $|1\rangle$. The level $|0\rangle$ is coupled along the left spatial path, i.e., associated with the left qutrit along the bifurcation graph whereas the level $|1\rangle$ is teamed along the right spatial path with the right qutrit.

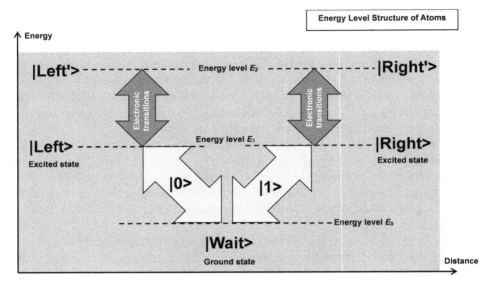

Figure 15.3. Energy level structure of the atoms in a possible optical bucket-brigade QRAM comprising a ground state |wait⟩ and two excited states |left⟩ and |right⟩.

Still higher energy states representing the coupling are: |left⟩ (for left spatial path) and |right⟩ (for right spatial path). Figure 15.3 depicts the energy level diagram of these qutrit switches.

Following are the steps adopted for data reading/writing in the QRAM:

Step (i): Suppose the first photon propagates through the cavity and reaches the root node switch. It is absorbed into the higher energy state of the qutrit, either a |0⟩ or |1⟩ state. Consequently, the state of the qutrit changes depending on the quantum state encoded in the photon. This process is called a Raman transition.

The Raman effect or Raman scattering is an optical phenomenon. It takes place when photons interact with the molecules of a substance. As a repercussion, there is a change in their energy and direction of motion. The scattered photons have different energies than the incident photons. The molecules either gain or lose energy. So, their vibrational states are altered. Use of a strong laser source helps in changing the state of the qutrit from |·⟩ to |0⟩ if the photon state is |0⟩. The other possibility is that the qutrit state changes from |·⟩ to |1⟩ if the photon state is |1⟩.

Step (ii): Upon arrival of the second photon at the root node switch, it undergoes absorption and experiences a Raman transition. At this time, the changeover is either from |0⟩ to |left⟩ or from |1⟩ to |right⟩. The photon is re-emitted to the qutrit along the correct spatial path based on the state of the qutrit. If it is |0⟩, the left path is followed. If it is |1⟩, the right path is pursued.

Step (iii): Following a similar procedure, the third and all subsequent photons of the input register set the qutrit switches one by one. Ultimately, a path is created beginning from the root node switch and terminating at the desired memory cell. A photon striking an atom which is in a |left state⟩ causes the excitation of a cyclic transition using the level |left'⟩. It is re-emitted by the atom. The |left⟩ → |left'⟩ transition is unresponsive to the polarization of the photon. It is linked to an outward-bound spatial mode leaving the trapped atom in the left direction. This indicates that a photon in any polarization state that impacts an atom in the |left state⟩ is deviated along the graph towards the left. Similarly, a photon in any state incident on an atom in the |right state⟩ undergoes rightward divergence.

Step (iv): There are two likely possibilities. Either the contents from the memory cell (read operation) are loaded into the output register or new data are stored in the memory cell (write operation) via the created path of qutrit switches. When the load/store (read/write) operation is complete, all the qutrits sequentially undergo a final Raman transition. This commences from the last node and finishes at the root node for restoration of the QRAM to the |·⟩ state.

15.3 Layered architecture of a quantum computer

The quantum computer is a complex machine with a sophisticated structural organization. The behavior of this machine becomes easily comprehensible by decomposition of the system behavior into manageable groups of operations (Jones *et al* 2012). Abstraction is a keystone of computer science. A layered architecture simplifies the system through abstraction of layers. Each layer is assigned a set of correlated functions. As one climbs from the bottom layer to the topmost layer, one reaches closer to the quantum computing environment. The layered architecture furnishes a systematic framework which can be used as a scaffolding to solve problems in computer development. It also aids in preparation of a cohesive computer design.

15.3.1 Three-layer architecture

The constituent layers of this model (figure 15.4) are (Rietsche *et al* 2022):

(i) The hardware layer: This embodies the physical apparatus. The apparatus of interest here consists of the qubits, the controller, the readout and the QPU. Because the information is stored as qubits, the advantages of their quantum-mechanical properties are leveraged. Notably, the effects of superposition and entanglement are exploited.

(ii) The software layer: The system software layer sitting over the hardware layer contains the main program. It also contains the quantum error correction and noise suppression software. It coordinates the processes to utilize the superposition and entanglement abilities of the qubits. Undoubtedly, the problems of the thermodynamically unstable quantum states must be solved prudently. This is taken care of by decreasing the thermal noise beleaguering the quantum system to the minimum level possible by cooling it to around 0.02 K.

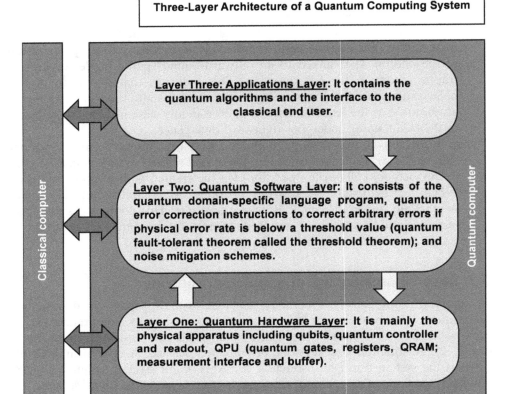

Three-Layer Architecture of a Quantum Computing System

Layer Three: Applications Layer: It contains the quantum algorithms and the interface to the classical end user.

Layer Two: Quantum Software Layer: It consists of the quantum domain-specific language program, quantum error correction instructions to correct arbitrary errors if physical error rate is below a threshold value (quantum fault-tolerant theorem called the threshold theorem); and noise mitigation schemes.

Layer One: Quantum Hardware Layer: It is mainly the physical apparatus including qubits, quantum controller and readout, QPU (quantum gates, registers, QRAM; measurement interface and buffer).

Classical computer

Quantum computer

Figure 15.4. Three-layered architecture of a quantum computer from the applications perspective. From bottom upwards, the three layers in this model are the hardware, software and application layers. Each layer is engaged in a two-way communication with the succeeding layer. All the three layers communicate bidirectionally with the classical computer.

Nevertheless, environmental noise from temperature fluctuations, vibrations, etc cannot be completely eliminated, and the remnant noise causes qubit decoherence leading to chaotic variations in quantum states. So, quantum error correction is done by representing logical qubits by a group of physical qubits. Physical qubits are the actual atoms or superconducting circuits used for storage of quantum information while the logical qubits are those created by grouping several physical qubits together using error correction codes to improve the reliability of quantum computing. Typically, 5–9 physical qubits are used to obtain one error-free logical qubit. The clear-sighted idea is that a group of physical qubits are less likely to introduce errors than one single physical qubit.

(iii) The application layer: In this topmost layer of the model, provisions and arrangements for user interaction are made. The quantum states of qubits cannot be stored in a memory for a long time for use in future calculations.

So, the data is transferred from a classical computer to a quantum computer. The calculations are done by the quantum computer. Then the results are read or measured by the classical computer before the information in qubits is lost.

Important quantum algorithms include:

(i) Grover's algorithm: This is the quantum search algorithm used for a thorough quest to track down a particular record which satisfies a given property in an unstructured database or an unordered list of N records. This algorithm can complete the search in $\sqrt{N}$ steps. Obversely, on treading the classical path, one has to go through $N/2$ records to find the correct record. The Grover's algorithm is very quick for large values of N, enabling a quadratic speedup over classical methods (Grover 1996).

(ii) Shor's algorithm. This is an integer factorization algorithm. Its use is rewarding because of its ability to factorize integers almost exponentially faster than the best-known classical algorithm (Shor 1994).

(iii) HHL (Harrow–Hassidim–Lloyd): This is the quantum algorithm for linear systems of equations. This algorithm can find a vector $\vec{x}$ such that $A\,\vec{x} = \vec{b}$ given a matrix A and a vector $\vec{b}$ (Harrow *et al* 2009).

15.3.2 Four-layer architecture

According to the Layered Architecture for Quantum Computing (2024), there are four layers as follows (figure 15.5):

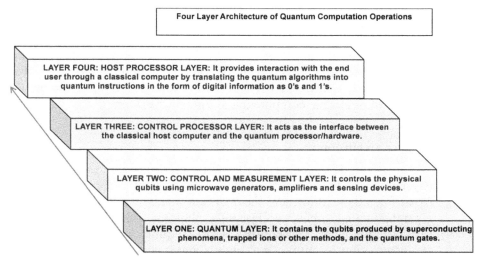

Figure 15.5. Four-layered architecture of a quantum computer from the viewpoint of computation operations. Each layer is assigned different responsibilities and functionalities. From bottom upwards, the layers in the model are the quantum layer, the control and measurement layer, the control processor layer and the host processor layer.

(i) Layer one: Quantum layer: The main components of this layer are the qubits produced by one of several technological options. It also includes the quantum gates used for manipulating the qubits. Cryogenic control of the quantum layer is crucial. The space surrounding a quantum computer is efficiently cooled to prevent losss of information because the qubits are vulnerable to thermal and environmental noises. Stability and reliability of operations is ensured by preservation of the quantum coherence of the qubits. The chances of quantum decoherence are forestalled by maintaining temperatures close to absolute zero.

(ii) Layer two: Control and measurement layer: The physical qubits are controlled by this layer. To achieve this aim, microwave generators, amplifiers and sensors are used. Each qubit is connected to this layer via wires.

Analog signals from the control processor layer are used in this layer to produce microwave/IR signals of specified frequency. These are delivered to the qubits for executing the quantum algorithms. Posterior to the execution of quantum operations, detection of the resultant responses of qubits is scheduled and completed by performance of measurements via application of microwave pulses to the qubits. The altered outputs of the states that have been read out after carrying out quantum operations on the qubits are conveyed as the measurement results to the classical computer.

(iii) Layer three: QPU layer (control processor layer): The task of this layer is to provide interfacing between the classical host computer and the quantum processor/hardware. It manipulates the physical qubits and gates in the QPU. It does so by identification and activation of the sequence of quantum gate operations and measurements to execute the instructions given in the program. The instructions are converted into analog signals that can be understood by the quantum hardware for implementation of the quantum algorithm.

(iv) Layer four: CPU layer (host processor layer): This is the layer of interaction with the end user through a classical computer. It also collects the results received after execution of the quantum algorithm. Room temperature is controlled between 20 °C and 25 °C for environmental comfort of electronic computers and human operators.

15.3.3 Five-layer architecture

This architecture (figure 15.6) consists of the layers described below (Jones *et al* 2012):

(i) Physical layer: This consists of physical qubits, and their cool enclosures. Apart from these, it includes the quantum gates, microwave irradiation for superconducting qubits, and lasers for photonic technology.

(ii) Virtualization layer: This consists of virtual qubits and virtual gates. Here, the information primitives, e.g., qubits and quantum registers are derived from the quantum states.

(iii) Quantum error correction layer: In this layer, the noise-induced errors are eliminated. The entropy is driven out from the system as an error syndrome to produce logical qubits and logical gates.

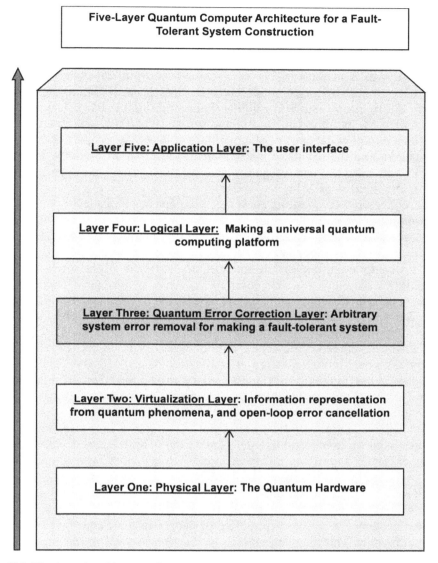

Figure 15.6. Five-layered architecture of a quantum computer for creating a foundation for fault-tolerant system development. Proceeding from bottom to top, the layers in this model are: the physical layer, the virtualization layer, the quantum error correction layer, the logical layer and the application layer.

(iv) Logical layer: The fault-tolerant structures of logical qubits and logical gates are processed to form the gates to be used by the application layer.

(v) Application layer: Quantum encoded data are fed into the system followed by the execution of quantum algorithms.

15.4 Organization of a superconducting quantum computer

15.4.1 A quick look into the general architecture of a quantum computer

The quantum processing unit works as a performance accelerating coprocessor in association with a classical CPU as a close friendship colleague. Therefore, a quantum computer operates under classical control with digital inputs/outputs although information is stored in qubits (Sahu and Gupta 2023). The information is represented by the physical state of the system. The QPU translates the digital instructions into physical control of a quantum system. The properties of the system are measured, and the results of measurements are reported digitally. Preparation, evolution and measurement of qubits are the three principal stages of quantum computing (Schuld *et al* 2014).

The architecture of a quantum computer encompasses a vast technological field at one extremity of which are positioned the complicated quantum algorithms and at the opposite end sit the physically functioning qubits (figure 15.7). Let us describe the architecture with reference to a superconducting quantum computer (Das *et al* 2023). Focusing attention on this computer, our first observation is that the participating technologies in this computing system require stringent temperature regulation in four different temperature zones, viz., <20 mK, 4.2 K, 77 K and 300 K. The strict temperature maintenance demands the installation of the required cooling equipment and accurate control instrumentation. Since the QPU is isolated in dedicated supercooled containers, communication channels between the CPU and QPU are extremely limited (Brassel *et al* 2019).

A high-level programming language is used for writing the quantum algorithms. The high-level language uses easily understandable syntax and command words identical to natural human language. The syntax and command words chosen make computer programming much simpler and easier than if a low-level language was used. The algorithms work on qubits and quantum logic gates. This mission is accomplished by translation of the algorithms into sequences of instructions with a compiler. The instructions give directions to the physical qubits for execution of the planned operations. The tasks of the circuits are scheduled for optimal performance taking into consideration the restrictions imposed by the connectivity of qubits. To meet this goal, the sequences of pulses and the flow of instructions produced by the compiler are conveyed to the control electronics module. This module generates analog signals corresponding to the instructions. The analog signals are delivered to the qubits at correctly timed intervals for initiating the desired actions. Computations can be performed in the presence of errors by employing error mitigation and correction methods.

15.4.2 Superconducting qubits production inside a dilution refrigerator

The qubit production system is mounted inside a dilution refrigerator. This refrigerator uses helium and its isotopes in a closed-loop system to lower temperatures down to <20 mK without any moving parts. The refrigerator is equipped with additional components such as pulse tube coolers for precooling, and a mixing chamber where the electronic assembly for making qubits is placed. The chamber is

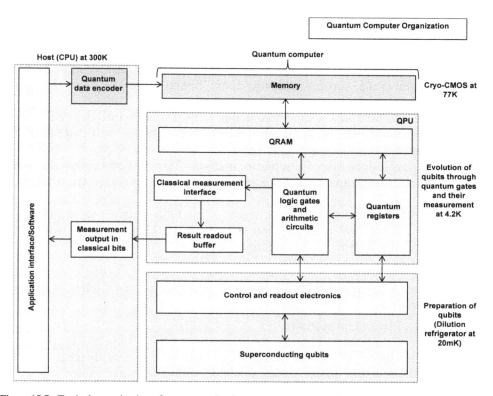

Figure 15.7. Typical organization of a superconducting quantum computer in which the classical von Neuman architecture classical computer encodes input data and receives output measurement results from a quantum computing system. Various components of the computer require different controlled temperatures for operation, as mentioned in the diagram. The computer has four operating temperature regions at 20 mK, 4.2 K, 77 K and 300 K. In the quantum computing system, a QPU (comprising the quantum logic gates, quantum registers, QRAM unit, classical measurement interface and the output buffer) operates at 4.2 K to supply instructions to the qubits. Components preparing qubits for the computer viz., superconducting qubits and control and readout electronics, are the most critical parts of this organization. They are housed in a dilution refrigerator maintained at 20 mK. Memory locations at 77 K and 300 K are also shown.

cooled by the continuous circulation of helum-3 and helium-4 (He-3 and He-4). Its temperature is measured with ruthenium oxide temperature sensors. The system also contains the circuitry including amplifiers, attenuators and filters for signal conditioning. Among other facilities, mention may be made of:

    (i) the temperature controllers to counteract thermal deviations,
    (ii) state-of-the-art integrated superconducting magnets with associated electronics for regulation of intensity and direction of magnetic fields,
    (iii) mechanical provisions for ensuring a vibration-free platform, and
    (iv) the software and telecommunication interfaces for remote monitoring.

The circuits for making qubits are manufactured as chips on special substrates. Advanced thin film deposition and lithographic technologies are applied for their fabrication. The circuits mainly comprise Josephson junctions made with aluminum or niobium metals and aluminum oxide tunneling barrier. The three frequently used types of qubits are the charge, flux and phase qubits.

The charge qubits function by encoding information in the charge state of Cooper pairs on a miniscule superconducting island linked to a reservoir through a Josephson junction.

The flux qubits comprise a superconducting loop. The loop includes one or more Josephson junctions. The magnetic flux passing through the loop determines the quantum states of the Josephson junctions.

The nonlinear inductance of Josephson junctions forms the basis of phase qubits. The junctions are shunted by a large capacitor, which renders their frequencies noise-insensitive.

At the working temperature range close to few milli-Kelvins, these qubits can survive and preserve their quantum superposition states. They can be maneuvered in a coherent fashion with the aid of microwave pulses.

15.4.3 Electronic apparatus for producing the control signals and readout of quantum information

This consists of two main sections:

(i) Control electronics: This orchestrates the working of the QPU by producing and sending signals for actuating qubit operations. As already mentioned, the superconducting qubits are manipulated with microwave pulses whose frequencies match with those of the qubits. The microwave pulses are produced by waveform generators having sampling rates >1 GHz. The pulses are mixed with microwave carrier waves. The amplitude, frequency and phase of the microwave carrier waves are controlled by in-phase/quadrature modulation techniques. After mixing, the frequencies of the pulses are either increased or decreased to precisely equalize them to the qubit frequencies. The power levels of the pulses are fine-tuned using amplifiers/attenuators. These amplifiers/attenuators can be programmed for automatically carrying out the necessary sequences of steps. The pulses are routed and directed to particular qubits with the help of switches. By gating the pulses in time, the switches confirm that the operations are performed on the designated target qubits only.

(ii) Readout electronics: The readout process of the received weak qubit information-bearing microwave signals is triggered by accurate timing circuits for coordination of signal capturing. The readout commences by uplifting the magnitudes of captured signals using cryogenic amplifiers. Superconducting parametric amplifiers are deployed for this purpose. These amplifiers are quantum-limited amplifiers. They add negligible noise to the circuit restricting it below the limit permitted by quantum mechanics, which is around half a photon per unit bandwidth. Hence, they are able to raise the signal magnitudes above the thermal noise levels without introducing additional noise in the signals. After transmission of the strengthened signals outside the cryogenic assembly, the signals are subjected to further amplification. The noise-filtered signals are fed to analog-to-digital converters for digitization. Digital signal processing methods are applied. High-fidelity readout is achieved by adopting careful calibration procedures and utilizing error correction schemes.

15.5 Summary and the way forward

15.5.1 Highlights of chapter 15 at a glance

(i) Inappropriateness of classical memory for quantum computing was explained and the reasons of incongruity were indicated.

(ii) Efficient storage/modification of quantum or classical data is accomplished by using a QRAM, which applies the principle of quantum superposition for simultaneously accessing all the data stored in the memory cell to assure the superior performance of quantum algorithms.

(iii) Data reading and writing in a bucket-brigade QRAM were described and the feasibility of its optical implementation was discussed.

(iv) The working of quantum computers is readily understood through layered architectures, viz., three-layer, four-layer and five-layer architectures. This approach was followed for easy comprehension.

(v) Organization of a superconducting quantum computer was illustrated by considering a general architecture. The superconducting transmon qubits produced inside a dilution refrigerator are manipulated by sequences of microwave pulses. Electronic systems used for producing the control signals and reading the quantum information were described.

(vi) Quantum computers have been built by several industry giants, and improved quantum processors are being built with continuous advancements and refinements.

(vii) Generally, a classical digital computer is considered to be an efficient universal computing device. It can simulate any physical computing device with an increase in computation time by at the most a polynomial factor. Shor (1997) showed that this may be untrue when quantum mechanics is taken into consideration. A quantum computer is not limited to polynomial-time algorithms. It is 'better than polynomial' in the context of computational complexity. Efficient randomized algorithms were developed for solving two problems on a hypothetical quantum computer. These problems are generally believed to be difficult on a classical computer. The problems are factoring integers and finding discrete logarithms.

(viii) Keywords for this chapter are: QRAM, bucket-brigade memory, layered architectures, superconducting quantum computer.

15.5.2 Future perspectives

(i) To envision the future computing possibilities, let us recall the achievements of classical computing. The CPUs perform efficiently in certain tasks such as data preparation and entry, visualization, error correction; and in

AI-Processor Electronics

organizing and accessing memory. GPUs excel in rendering sophisticated 3D graphics. Over the years, they have become more flexible. Reinforcements with programming capabilities led to their increasing utilization in high-performance computing. The tensor processing unit is an AI accelerator application-specific integrated circuit (ASIC) developed by Google for neural network machine learning, using their own TensorFlow software. Convoluted neural network (CNN) processors and neural processing units (NPUs) can handle workloads primarily involving calculations of neural network layers embracing scalar, vector, and tensor mathematics followed by a nonlinear activation function. Graph analytics processors execute graph algorithms efficiently. APUs work similar to associative memories but revolutionize the concept of computing from serial data processing, i.e., shuttling the data back and forth between the processor and memory, to massive parallel data processing, computing, and searching in-place directly in the memory array.

All these classical computing platforms have their merits and short-comings but do not rival quantum computing because they perform their tasks earnestly with full proficiency within their domains meeting their focused objectives and goals. So, the question of their being obsolete is *prima facie* ruled out.

(ii) After reviewing the pros and cons of classical and quantum scenarios, it appears that future computing systems are likely to adopt a hybrid strategy leveraging the capabilities of classical and quantum computers (Keesling 2023). This hybrid paradigm means that classical and quantum computers will work side-by-side, hand-in-hand, shoulder-to-shoulder, in close collaboration, with either member of the alliance complementing and supplementing the functioning of its counterpart in tackling the hurdles on the rocky road of computing. The quantum computers will work in tandem with classical computers in a seamless, integrated environment with each type of computing offering solutions for problems where its partner fails and supporting its partner in an auxiliary role in an area where the latter is superior to boost the overall capabilities of the combined system. A heterogeneous all-inclusive computing architecture employing diverse processors gives the opportunity to tap each processor's strengths and lessen weaknesses.

(iii) Hybrid algorithms, classical plus quantum, offer glimpses into the future of integrated computing. In these algorithms, classical and quantum resources will work harmoniously and complementarily in close partnership to maximize the potencies and minimize the frailties of respective type of computing. This will create a symbiotic relationship between them through a systematically planned division of labor. The united approach promises to advance all the fields to deal with some of the most challenging problems confronting the world.

15-16

Futuristic hybrid computing model
The future of computing hovers around classical plus quantum integration,
With mutual cooperation and collaboration,
Leveraging the strengths of both modalities,
To mitigate their weaknesses and inabilities,
While classical computers will handle data preparation and visualization,
The quantum computers will perform complex calculations,
Classical computing will never retire,
But will work in a new role and attire,
It is not a single choice between two technologies,
But a blended and unified ideology!
Every ending is the beginning of a new story,
Welcome to the quantum age of grandeur and glory!

References

Andika M 2023 What is a qubit? https://medium.com/@tamanpulogebang/what-is-a-qubit-dfef18b7080d

Arunachalam S, Gheorghiu V, Jochym-O'Connor T, Mosca M and Srinivasan P V 2015 On the robustness of bucket brigade quantum RAM *New J. Phys.* **17** 123010 1–16

Brassel A, Kuo J and Young J 2019 A survey of quantum control architecture, pp. 1–11. https://cs.umd.edu/class/fall2019/cmsc657/projects/group_4.pdf

Das S, Chatterjee A and Ghosh S 2023 SoK: a first order survey of quantum supply dynamics and threat landscapes *HASP'23: Proc. of the 12th Int. Workshop on Hardware and Architectural Support for Security and Privacy held in Conjunction with the 56th Int. Symp. on Microarchitecture, MICRO 2023 (October 29, 2023) (Toronto, Canada)* ACM Int. Conf. Proc. Series (New York: Association for Computing Machinery) pp 82–90

Giovannetti V, Lloyd S and Maccone L 2008 Quantum random access memory *Phys. Rev. Lett.* **100** 160501 1–4

Grover L K 1996 A fast quantum mechanical algorithm for database search *Miller G. L. (Chairman), Proc. of the 28th Annual ACM Symp. on Theory of Computing: STOC96, Philadelphia, Pennsylvania (May 22–24)* (New York: Association for Computing Machinery) pp 212–9

Harrow A W, Hassidim A and Lloyd S 2009 Quantum algorithm for linear systems of equations *Phys. Rev. Lett.* **103** 150502

Jones N C, Meter R V, Fowler A G, McMahon P L, Kim J, Ladd T D and Yamamoto Y 2012 Layered architecture for quantum computing *Phys. Rev.* **2** 031007

Keesling A 2023 The future of computing is hybrid: why quantum computers will work alongside classical systems *Forbes Technology Council* https://forbes.com/sites/forbestechcouncil/2023/11/10/the-future-of-computing-is-hybrid-why-quantum-computers-will-work-alongside-classical-systems/#:~:text=The%20future%20of%20computing%20is%20not%20an%20either%2Dor%20proposition,and%20compensates%20for%20its%20weaknesses

Layered Architecture for Quantum Computing 2024 https://geeksforgeeks.org/five-layered-architecture-of-quantum-computing/

Phalak K, Chatterjee A and Ghosh S 2023 Quantum random access memory for dummies *Sensors* **23** 7462

Rietsche R, Dremel C, Bosch S, Steinacker L, Meckel M and Leimeister J-M 2022 Quantum computing *Electron. Mark.* **32** 2525–36

Sahu H and Gupta H P 2023 Quantum computing toolkit from nuts and bolts to sack of tools *arXiv preprint* arXiv:2302.08884

Schuld M, Sinayskiy I and Petruccione F 2014 The quest for a quantum neural network *Quantum Inf. Process.* **13** 2567–86

Shor P W 1994 Algorithms for quantum computation: discrete logarithms and factoring *Proc. of the 35th Annual Symp. on Foundations of Computer Science. IEEE Computer Society (20–22 November) (Santa Fe, NM, USA)* pp 124–34

Shor P W 1997 Polynomial-time algorithms for prime factorization and discrete logarithms on a quantum computer *SIAM J. Comput.* **26** 1484–509